R 自然语言处理导论

Introduction to Natural Language Processing with R

鲍 贵　莫俊华　陆俊花　**编著**

东南大学出版社
SOUTHEAST UNIVERSITY PRESS
·南京·

图书在版编目(CIP)数据

R 自然语言处理导论 / 鲍贵, 莫俊华, 陆俊花编著. --南京：东南大学出版社, 2023.11
ISBN 978-7-5766-0941-7

Ⅰ. ①R… Ⅱ. ①鲍… ②莫… ③陆… Ⅲ. ①程序语言-自然语言处理 Ⅳ. ①TP312.8

中国国家版本馆 CIP 数据核字(2023)第 209670 号

R 自然语言处理导论
R Ziran Yuyan Chuli Daolun

编　　著：	鲍　贵　莫俊华　陆俊花
出版发行：	东南大学出版社
社　　址：	南京市四牌楼 2 号　邮编：210096　电话：025-83793330
网　　址：	http://www.seupress.com
电子邮箱：	press@seupress.com
出 版 人：	白云飞
经　　销：	全国各地新华书店
印　　刷：	苏州市古得堡数码印刷有限公司
开　　本：	787mm×1092mm　1/16
印　　张：	20
字　　数：	512 千字
版　　次：	2023 年 11 月第 1 版
印　　次：	2023 年 11 月第 1 次印刷
书　　号：	ISBN 978-7-5766-0941-7
定　　价：	78.00 元

本社图书若有印装质量问题，请直接与营销部联系。电话(传真)：025-83791830
责任编辑：刘庆楚　责任校对：子雪莲　封面设计：王玥　责任印制：周荣虎

前　言

随着信息科学技术的快速发展，人类社会已经大踏步迈进大数据时代。浩如烟海的数据不仅是我们生活中的"家常便饭"，而且也是我们开展科学研究的基础。自然语言处理（Natural Language Processing，NLP）综合利用计算机科学、人工智能（Artificial Intelligence，AI）和计算语言学等学科的知识和技术，旨在开发一整套程序使计算机处理人类自然语言，实现人机互动，推动知识创新，而这种知识创新是传统的细读和人工综合等定性方法难以企及的。在大数据时代，自然语言处理技术已经渗透到我们生活的方方面面。我们多数人都使用过谷歌翻译，都有与聊天机器人交流的经历，都使用过邮件过滤系统删除过垃圾邮件，都使用过 Word 的语言拼写自动校对功能。这些都是自然语言处理技术的实际应用。毋庸置疑，在新文科建设和数字人文的大背景下，自然语言处理技术的应用领域已经不再囿于自然科学，而是快速蔓延到人文和社会科学各个领域，已经成为我国高等学校人文与社会科学人才培养的优选项。

"工欲善其事，必先利其器。"使用什么样的科研"利器"来高效地处理科学研究中的海量数据成为科研工作者们孜孜以求的目标。虽然人文和社会科学研究也越来越多地利用这些技术推动学科发展，但是自然语言处理课程在大多数高校的大多数专业课程建设中被束之高阁，课程改革与创新往往是"雷声大，雨点小"，很难落到实处。目前，市面上已有一些相关著作和教材，如基于 Python 编程语言介绍自然语言处理方法。但是，这些书籍专业性过强、缺乏可读性和可操作性，导致学习门槛过高，使得许多人文和社会科学专业的本科生和研究生望而却步。

Python 和 R 都是编程语言。Python 是程序员的语言，而 R 是统计学家的语言。根据我们的经验，和 Python 语言相比，R 语言相对简单，学习曲线不那么陡峭。马修·乔克斯在其著作《文学研究中的 R 文本分析》（Text Analysis with R for Students of Literature）的"致谢"中回忆自己从 Python 和 Perl 等编程语言转向 R 语言的难忘经历（Jockers，2014）。乔克斯教授曾经很长一段时期都用 Java 和 Perl 等教授文本分析课程。大约在 2005 年，他开始接触和钻研 R 语言。此后的几年里，乔克斯教授一直用 Java 和 Perl 等编写文本分析代码，再把结果输入 R 开展统计分析。把数据从一种编程环境转向另一种编程环境耗时耗力，于是在 2008 年乔克斯教授决定只用 R 开展文本分析，自此之后，几乎所有的问题都在 R 语言中解决。乔克斯教授在"致谢"中特别提到 R 出色的在线帮助文档、非常有用的 R 数据包、强大的 R 社区、R 论坛上的答疑解惑、斯蒂芬·格赖斯（Stefan Gries）（详见本书第七章）和哈德利·威克姆（Hadley Wickham）（详见本书第九章）等学者对其著作做出的贡献。

长期以来，统计学家利用 R 从事统计分析和制图新技术的开发利用，因而在自然语言处理方面 R 数据包的开发起步较晚，但是发展势头迅猛，近期开发出 koRpus、quanteda、tidytext 和 udpipe 等著名的自然语言处理数据包。

当前，国内出版的基于 R 的自然语言处理类著作和教材非常稀缺。这不仅制约了不同学科领域专业研究者对 R 编程语言的开发与利用，而且也不利于拓宽本科生和研究生的学术视野，难以适应大数据时代对通用型和专业型人才的培养要求。

为了紧跟时代发展的潮流和学科建设的步伐，更为了回应广大研究者和师生的诉求，我们通过协作攻关、反复讨论和实践，历时五年，几易其稿，编写出这部名为《R 自然语言处理导论》的著作。本著作以主题为导向，以问题为抓手，用十二章的篇幅系统介绍与讨论了 R 在自然语言处理中的主要应用，帮助读者在较短的时间内了解 R、熟悉 R 和使用 R。

本著作具有三大特点：

第一，内容全面。本著作不但介绍 R 软件的安装要领和各种功能，而且介绍常见的文本统计与测量方法，特别是基于整洁数据的自然语言处理技术。

第二，实践性强。本著作充分践行"发现式学习"和"做中学"的教学理念，通过设计和实施一个个具体的研究任务，帮助读者在真实情景中掌握 R 自然语言处理技术。

第三，使用便捷。本著作对用户友好，为读者提供了多种可下载的数据包和详细的代码与说明，凸显各个自然语言处理方法的重点和难点。

本著作至少在以下三个层面有重要的价值：

第一，在教材开发层面，本著作能够填补国内在基于 R 编写自然语言处理教材方面的不足，全面展现 R 在自然语言处理方面取得的重要成就。

第二，在人才培养层面，本著作能够拓宽学生的思维空间，提升他们的信息素养、实践能力和创新能力。

第三，在学科建设层面，本著作能够推进数据驱动的科学研究，促进不同学科之间的交叉研究，使学术研究牢固地建立在数据基础之上。

"纸上得来终觉浅，绝知此事要躬行。"驾驭自然语言处理技术离不开实践。《R 自然语言处理导论》用深入浅出的语言、通俗易懂的案例和清晰明了的程序，把理论与实践有机地结合在一起，使读者在操作实践中通晓自然语言处理方法，学会融会贯通，为学术创新蓄力赋能。我们期盼读者能够从本著作的学习中得到丰厚的回报。这种回报不仅表现在对一门新程序语言的掌握，而且还在于以本著作提供的新技术、新方法为原动力创建新项目、催生新成果。本著作不仅可以作为人文和社会科学各个学科不同专业高年级本科生和研究生学习 R 自然语言处理的应用性教材，还可以作为研究者开展 R 自然语言处理的参考书。

本著作在出版过程中承蒙东南大学出版社和刘庆楚分社长鼎力支持，在此一并致谢。

鲍　贵（南京工业大学）
莫俊华（苏州大学）
陆俊花（南京工业大学）
2023 年 10 月

导　读

　　自然语言处理（Natural Language Processing，NLP）是涵盖计算机科学、人工智能和计算语言学等多个学科的交叉研究领域，旨在开发一整套程序使计算机能够处理人类自然语言，实现人机互动。自然语言处理技术广泛用于语言学、文学、经济学、管理学、政治学、社会学和医学等各个学科。近年来 R 语言在自然语言处理技术的开发和利用方面取得了令人瞩目的成就。但是，国内外有关 R 语言的著作多面向传统的统计学，主要以数值型数据（numeric data）为研究对象，鲜有著作系统介绍如何利用自然语言处理技术对文本数据（text data）开展整理、加工和分析，从中获取有价值的信息。《R 自然语言处理导论》由此诞生。

　　这部著作包括十二章内容，遵循由易到难、循序渐进的原则，各章内容简介如下：

　　第一章是 R 语言入门，主要内容包括 R 简史、R 软件和数据包安装、基本 R 对象（如向量、数据框和矩阵）和 R 循环。

　　第二章介绍字符串处理，主要内容包括不同文档（如纯文本文档和 word 文档）的读取、正则表达式和 R 基础包中的正则表达式函数。本章还介绍了字符串操作数据包 stringr，讲授如何编写 R 代码开展语境中的关键词检索（Keyword in Context，KWIC）。

　　第三章介绍常见的文本统计方法。本章依据 R 数据包 koRpus 先介绍如何计算常见的文本统计量，包括文本词数、句子数和平均句长等，随后讨论如何利用停用词表计算文本中的词汇密度。本章最后介绍如何编写 R 命令制作不同类型的词频表（如利用词目和停用词的词频表）和绘制词频分布图（如词频分布条形图和词云图）。

　　第四章依据 R 数据包 koRpus 介绍与讨论文本词汇多样性测量方法，包括传统的类符-形符比、平均分段类符-形符比、移动平均类符-形符比、Herdan's C、Guiraud's R、尤伯指数（U）、萨默指数（S）、Yule's K、Maas 指数、HD-D、文本词汇多样性测量（MTLD）和文本词汇多样性移动平均测量（MTLD-MA）。

　　第五章同第四章一样利用 R 数据包 koRpus，重点介绍文本可读性多种测量方法和一种听力文本难易度的测量方法，包括自动化可读性指数（ARI）、科尔曼-廖（Coleman-Liau）指数、戴尔-乔尔（Dale-Chall）可读性新公式、弗莱什（Flesch）阅读难易度、法尔-詹金斯-帕特森（Farr-Jenkins-Paterson）指数、弗莱什－金凯德（Flesch-Kincaid）年级水平、安德森（Anderson）可读性指数、复杂词词频（FOG）修正指数、复杂词简单测量（SMOG）和听力难易度公式。

　　第六章聚焦于 N 元组（N grams）和关键词提取方法。本章首先简要介绍了文本数据计量分析数据包 quanteda、quanteda.textstats 和 quanteda.textplots，然后以二元组和三元组为例介绍 n 元组在文本挖掘中的应用。本章最后讨论了关键词提取方法，内容包括文本词语关键性检验方法(如卡方检验和似然比检验)、比较学生故事复述文本与原文本中使用的关键词以及比较美国总统就职演说文本中的关键词,同时编写 R 命令绘制关键词比较条形图。

第七章介绍两种特殊形式的 N 元组,即搭配(collocations)和搭配构式(collostructions)。搭配构式是对搭配的拓展。本章继续使用 R 数据包 quanteda 和 quanteda.textstats 开展搭配分析,并利用 R 数据包 Coll.analysis 开展搭配构式分析。本章详细介绍了搭配和搭配构式的统计分析方法,包括点互信息(Pointwise Mutual Information,PMI)、z 值和 t 值、G^2 和 λ 值。本章随后以第二次世界大战结束后的美国总统就职演说文本为例介绍了搭配分析方法,最后详细介绍如何调用 Coll.analysis 数据包开展搭配构式分析。

第八章介绍文档或文本相似度(similarity)和距离(distance)测量。文本相似度测量包括文本词汇相似度(text lexical similarity)测量和文本语义相似度(text semantic similarity)测量。文本词汇相似度测量包括皮尔逊相关系数、余弦相似度和 Jaccard 相似度等,文本距离测量包括欧式(Euclidean)距离、闵可夫斯基(Minkowski)距离和曼哈顿(Manhattan)距离等。文本语义相似度测量利用潜在语义分析(Latent Semantic Analysis,LSA),重点介绍词项频率-逆文档频率(Term Frequency-Inverse Document Frequency,TF-IDF)。本章在介绍各种相似度和距离测量之后,利用学生故事复述文本和原文本开展文本相似度案例分析。

第九章着重介绍基于整洁文本的自然语言处理,是本著作内容最为丰富的一章。本章包括七节内容。第一节介绍新型数据框 tibble 格式,第二节介绍整洁数据的特征和整洁数据包 tidyr 的使用。第三节介绍管道操作和数据包 dplyr 的使用。第四节介绍如何利用数据包 tidytext 对整洁数据开展形符化处理。第五节简要介绍数据包 ggplot2 的主要特点,通过案例详细说明如何利用这个数据包绘制文本词频分析条形图和情感分析(sentiment analysis)条形图。第六节介绍几个重要的情感词库,举例说明如何对短篇故事开展情感分析。本章的最后一节是第七节,综合利用基于整洁数据的自然语言处理技术对简·奥斯汀的小说《傲慢与偏见》开展文本特征分析,内容包括:把小说读入 R、对小说开展基础描述性统计以及对小说开展情感分析。

第十章介绍中文自然语言处理技术,包括六节内容。第一节简要介绍中文分词的两个重要数据包 quanteda 和 jiebaR,重点介绍调用 jiebaR 包时如何通过自定义词典进行分词。第二节以一则故事为例,调用 jiebaR 包开展基础统计分析,计算文本长度、句子数和平均句长,绘制高频词分布条形图。第三节介绍如何利用 quanteda.textplots 包、wordcloud 包和 wordcloud2 包绘制词云图。第四节和第五节介绍如何对中文文本开展关键词和短语检索与提取。最后一节即第六节,以短篇故事为例,介绍如何调用整洁数据包中的函数开展中文情感分析。

第十一章介绍如何对文本词汇开展词类标注和句法成分依存分析,分五节,内容包括:数据包 udpipe 的安装与初试、文本中的短语提取、句法分析、词语共现和快速自动关键词提取。

第十二章是本著作的最后一章,介绍如何利用 R 数据包 reticulate 实现 R 与 Python 的无缝对接。主要内容包括安装 Python 和 Python 库、R 数据包 reticulate 的安装和基础操作、利用 NLTK 和 spaCy 库开展自然语言处理。

  本著作把握学术前沿，融理论与实践于一体，以问题为导向，深入浅出，通俗易懂，通过"发现式学习"和"做中学"提升学习者的理论知识和实践能力。本著作可作为研究者开展 R 自然语言处理的参考书，也可作为各个学科不同专业高年级本科生和研究生的应用性教材。

目 录

第一章　R 语言基础 … 1
- 1.1　R 简史 … 1
- 1.2　R 软件和数据包安装 … 2
- 1.3　基本 R 对象 … 5
- 1.4　R 循环 … 14

第二章　字符串处理 … 19
- 2.1　文档读取 … 19
- 2.2　字符串分割 … 21
- 2.3　正则表达式 … 22
- 2.4　R 基础包中的正则表达式函数 … 33
- 2.5　字符串操作数据包 stringr … 37
- 2.6　语境中的关键词检索 … 41

第三章　文本基础统计 … 44
- 3.1　数据包 koRpus 的安装与调试 … 44
- 3.2　描述性统计 … 47
- 3.3　词汇密度 … 49
- 3.4　词频表 … 51
- 3.5　词频分布图 … 56

第四章　文本词汇多样性测量 … 59
- 4.1　传统的类符-形符比 … 59
- 4.2　平均分段类符-形符比 … 61
- 4.3　移动平均类符-形符比 … 63
- 4.4　Herdan's C … 65
- 4.5　Guiraud's R … 66
- 4.6　尤伯指数（U） … 68
- 4.7　萨默指数（S） … 70
- 4.8　Yule's K … 71
- 4.9　Maas 指数 … 73
- 4.10　HD-D … 77
- 4.11　文本词汇多样性测量（MTLD） … 78

4.12 文本词汇多样性移动平均测量（MTLD-MA） ………… 83

第五章 文本可读性测量 ………… 86
5.1 自动化可读性指数（ARI） ………… 86
5.2 科尔曼-廖指数 ………… 87
5.3 戴尔-乔尔可读性新公式 ………… 89
5.4 弗莱什阅读难易度 ………… 90
5.5 法尔-詹金斯-帕特森指数 ………… 92
5.6 弗莱什-金凯德年级水平 ………… 94
5.7 安德森可读性指数 ………… 95
5.8 复杂词词频修正指数 ………… 96
5.9 复杂词简单测量（SMOG） ………… 98
5.10 听力难易度公式 ………… 99

第六章 N元组和关键词 ………… 101
6.1 文本数据计量分析数据包安装与初试 ………… 101
6.2 两个文本的二元组比较 ………… 105
6.3 两个文本的三元组比较 ………… 108
6.4 文本比较关键词提取 ………… 108

第七章 搭配和搭配构式 ………… 124
7.1 搭配和搭配构式分析方法 ………… 124
7.2 文本中的搭配分析案例 ………… 128
7.3 文本中的搭配构式分析案例 ………… 131

第八章 文本相似度和距离测量 ………… 135
8.1 词汇相似度测量 ………… 135
8.2 语义相似度测量 ………… 143
8.3 文本相似度测量举例 ………… 147

第九章 基于整洁文本的自然语言处理 ………… 152
9.1 作为新型数据框的tibble ………… 152
9.2 整洁数据和数据包tidyr ………… 157
9.3 管道操作和数据包dplyr ………… 160
9.4 基于整洁数据的数据包tidytext ………… 169
9.5 精美制图数据包ggplot2 ………… 172
9.6 利用整洁数据的文本情感分析 ………… 179
9.7 对小说《傲慢与偏见》的文本特征分析 ………… 189

第十章 中文自然语言处理 ………… 210
10.1 中文分词 ………… 210
10.2 文档基本描述性统计量 ………… 214

10.3	文本词频分布词云图	221
10.4	关键词检索	225
10.5	关键词提取	227
10.6	中文情感分析	230

第十一章 词性与句法分析 237
11.1	数据包 udpipe 的安装与初试	237
11.2	文本中的短语提取	247
11.3	句法分析	251
11.4	词语共现	260
11.5	快速自动关键词提取	263

第十二章 在 R 中调用 Python 开展自然语言处理 271
12.1	安装 Python 和 Python 库	271
12.2	R 数据包 reticulate 的安装和基础操作	272
12.3	利用 NLTK 库的自然语言处理	281
12.4	利用 spaCy 库的自然语言处理	292

参考文献 304

第一章　R 语言基础

本章简要回顾 R 的发展历程，重点介绍 R 软件和数据包的安装、向量（vector）、数据框（data frame）、矩阵（matrix）、列表（list）和 R 循环（R loops）。

1.1　R 简史

约翰·钱伯斯（John Chambers）于 1976 年在位于美国新泽西州的贝尔实验室（Bell Labs）领衔开发了统计编程 S 语言。随后，约翰·钱伯斯、约翰·图基（John Tukey）和威廉·克利夫兰（William Cleveland）等统计学家出版了一批有关数据分析和图形绘制的杰作，如图基的著作《探索性数据分析》（*Exploratory Data Analysis*）（Tukey，1977）和钱伯斯等人的著作《数据分析图示方法》（*Graphical Methods for Data Analysis*）（Chambers et al.，1983）。这些著作为统计学家，特别是 R 社区，如何以清晰、开放与包容的方式开展技术交流播下了种子。这也是 R 社区有深厚渊源的鲜明标志（Scavetta et al.，2021）[4]。S 语言的最初发行版是开源的，在 Unix 系统上运行，但是后来 S 语言被授权在 S-PLUS 软件下运行。这促使奥克兰大学统计学系高级讲师罗斯·伊哈卡（Ross Ihaka）和罗伯特·杰特曼（Robert Gentleman）在 1991 年开发了另一种开源语言（Ihaka et al.，1996）。这种新开发的语言称作 R，一方面是为了对 S 语言的贡献予以认可，另一方面是为了庆祝开发者的成果（两位开发者名字的首字母都是 R）。R 语言使用了 S 语言的句法结构，因为这种结构为统计学家能够轻松自在地表达统计学思想提供了强有力的手段（Ihaka et al.，1996）[300]。但是，R 语言在语义方面继承了 Scheme 语言（Lisp 语言家族的一员），与 S 语言截然不同。因此，R 语言可以被视作 S 语言的一种方言（dialect）。

发端于 S 语言，经过约半个世纪的发展，R 语言现已成为一个融合统计计算、制图和自然语言处理等为一体的编程语言。在 R 语言的发展过程中，至少有十件大事值得一提（Scavetta et al.，2021）。一是在 1997 年组建了 R 核心团队（R Core Team），现有 20 名成员，包括约翰·钱伯斯、罗斯·伊哈卡和罗伯特·杰特曼。二是在组建核心团队的同一年搭建了 R 综合典藏网 CRAN（https://cran.r-project.org/mirrors.html）。三是在 1999 年建立了 R 网站（https://cran.r-project.org/）。四是在 2000 年 R 1.0.0 版正式发布，在 2022 年 R 4.2.0 版正式发布。五是在 2002 年 BioConductor（生物导体）被确立为处理多样化生物学数据的新型 R 数据包仓库（https://www.bioconductor.org/），对生物信息学的发展产生重大的影响。六是在 2007 年哈德利·威克姆（Hadley Wickham）发表的博士论文包括两个 R 数据包 reshape 和 ggplot2，为后来 R 集成数据包 tidyverse 的创建奠定了基础。数据包

reshape 首次让我们领略了数据结构何以影响我们对数据的思考方式和处理方法；ggplot2 包提供了直观的高层绘图方法，大大简化了前期已有的制图工具(Scavetta et al., 2021)[6]。七是 2009 年专业期刊 *The R Journal* 第一期正式面世。八是 2011 年 RStudio 由同名公司公开发布，大大提升了 R 语言使用的便捷性。九是在 2015 年成立了 R 联盟(the R Consortium)，致力于用户推广和其他服务于用户和开发者社区的项目。十是在 2017 年 CRAN 数据包超过 1 万个，到 2022 年 CRAN 数据包超过 1.8 万个，充分体现了 R 语言的知名度和 R 社区的活力。

R 语言既利于用户利用现有的数据包开展数据分析，又利于用户开发新的数据包，实现从用户的角色向开发者角色的转变。瑞奇·斯卡维他和博洋·安吉洛夫(Rick Scavetta & Boyan Angelov)称 R 是 FUBU 编程语言(Scavetta et al., 2021)。FUBU 原本是 20 世纪 90 年代成立的一家街头服饰公司名，是"For Us, By Us"(有我由我)的首字母缩写。FUBU 意味着社区合作，意味着"人人为我，我为人人"。这也正是创建 R 语言的目的。R 是最受欢迎的统计编程语言之一，既有数据包和函数说明网，如 RDocumentation（http://www.rdocumentation.org/），又有强大的社区提供技术支持和学术交流，如 R-bloggers（http://www.r-bloggers.com/）。

1.2 R 软件和数据包安装

R 软件是国际通用软件，下载和安装都很简便。R 软件适用的操作系统包括 Linux、macOS 和 Windows。我们使用的操作系统是 Windows 11。进入 R 软件官网(https://cran.r-project.org/)，点击 Download R for Windows，进入下一页界面，再点击 install R for the first time，进入另一个界面。在该界面上，点击 Download R 4.2.2 for Windows，便可下载 R 软件。R 软件更新很快，本书所用的版本是 R 4.2.2(R Core Team 2022)。要使 UTF-8(8-bit Unicode Transformation Format，8 位统一码转换格式)成为本地编码格式，至少需要安装 R 4.2.0 和 Windows 10 版本。UTF-8 使用 8 位字符单位(8-bit character units)来编码字符。8 位是 1 个字节(byte)，UTF-8 最多使用 4 个字节的空间来编码某个字符(Hart-Davis et al., 2022)[183]。

如果在 macOS 系统安装 R，为了能够处理 UTF-8 格式，首次使用时需要打开 R 工作界面，输入：system("defaults write org.R-project.R force.LANG en_US.UTF-8")，按回车后关闭，再重新打开 R 工作界面。建议用户下载或更新最新 R 版本。将下载的 R 软件保存在本地硬盘文件夹里(如 D:\)，点击 R-4.2.2-win 图标，按照安装提示即可完成操作。安装过程中，提示你使用何种语言，你可以选择英语或中文。对于 Message translations(信息翻译)选项，你可以根据自身需要勾选或不勾选。系统默认安装在 C:\目录下。建议采用默认安装路径，不过你也可以修改路径和文件夹，如 D:\R。R 安装的路径可以在 R 操作界面输入:R.home()，再按回车键查看。

安装完成后，我们可以双击桌面上的 R 图标，进入 R 操作界面。R 操作界面提供 R 的基本信息,包括运行的 R 版本、许可信息、帮助信息和命令提示符(>)等。要了解 R 软件安装的文件夹信息,在 R 界面输入:shell.exec(R.home()),再按回车键即可。R 默认的当前工作簿可以利用 R 命令 getwd() 查询。例如,在我们使用的电脑上,R 当前工作簿是 C:/Users/DELL/Documents。

> getwd()
[1] "C:/Users/DELL/Documents"

注意,以上 R 命令前的提示符(>)是 R 自带的,表示 R 已经准备就绪,用户无须输入,否则会报错。如果我们在提示符后输入代码(代码是执行某个任务所写的命令),如 3*8,按回车键,就会得到以下结果:

> 3*8
[1] 24

其中,[1]是 R 对输出做的索引,告诉用户这个结果是执行前面的表达式得到的第一个值。

R 基础系统默认安装了基础数据包(base)和其他一些推荐的数据包(如 stats 和 lattice),安装位置可以利用以下方式查看:find.package("数据包名称"),如 find.package("base")。由于统计方法的更新,有一些外置的数据包需要加载到 R 软件中,以便调用。例如,我们需要安装 readtext 数据包,则在 R 操作界面输入 install.packages("readtext"),然后选择一个就近的镜像点,即可完成自动安装。我们也可以指定数据包的安装位置。以 Windows 11 为例,在 R 界面输入命令:install.packages("readtext", lib = "C:/Program Files/R/R-4.2.2/library/"),把安装的数据包放在系统 C:/盘安装路径文件夹 library 中。看到 R 界面提示"Would you like to use a personal library instead?",我们点击"是"。接下来,R 数据包安装提示我们选择一个站点。我们选择"China(Lanzhou)[https]",然后点击"ok"。如果电脑正常联网,数据包的安装通常会自动完成。需要注意的是,在自定义安装数据包的路径前,需要鼠标右击 R 图标,在弹出的菜单中选择"以管理员身份运行(A)"的方式打开 R 操作界面。要查找数据包安装的路径,可以在 R 操作界面输入:system.file(package="")(在引号内输入数据包的名称),再按回车键即可。例如,针对我们的操作系统,在 R 操作界面输入:system.file(package="readtext"),按回车键后得到以下路径:[1] "C:/Users/DELL/AppData/Local/R/win-library/4.2/readtext"。要了解已经安装的数据包,我们可以在 R 工作界面输入命令:.packages(TRUE),或者输入命令:installed.packages(),然后按回车键查看。

安装好的数据包仍然处于非工作状态。如果要调用 readtext 数据包,在 R 操作界面直接输入 library(readtext)或者 require(readtext)即可,也可以通过 R 界面上的 Packages 菜单来完成(Packages→Load package…→Select one)。要查看在 R 工作空间已经加载的数据包,可以输入命令:search(),按回车键即可。如果要快速了解数据包的信息,那么可以在

R 操作界面输入以下命令：library(help="readtext")，或者输入以下命令：help(package="readtext")，再按回车键即可查看。注意，这里的 library()是函数，而 package 仅指数据包名称。例如，如果我们执行以下 R 命令：library(help="readtext")，可以得到如图 1.1 所示的数据包信息。

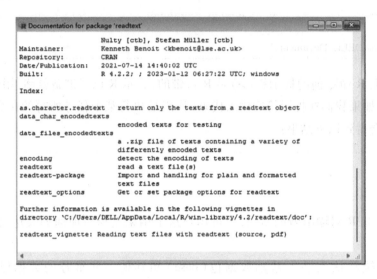

图 1.1　数据包 readtext 说明

如果要卸载数据包，可以在 R 工作界面输入：remove.packages("")（在引号内输入要卸载的数据包名称），按回车键后即可完成自动卸载。如果要更新数据包，可以在 R 工作界面输入：update.packages("")（在引号内输入要更新的数据包名称），按回车键后选择就近的镜像点即可完成自动更新。要关闭或退出 R，在工作界面输入：q()，再按回车键，或者点击右上方的图标" "。R 会提示是否保存工作空间镜像(workspace image)，通常选择"否(N)"。

鉴于 R 软件更新很快，你可能希望过一段时间后更新软件版本。安装新版 R 软件的最简单方法是前往 R 软件官网下载新版本，直接安装即可。旧版本可以保留，也可以删除。你可能在旧版本上安装了很多数据包，如"readtext"，且希望在新版中不用重新安装这些包。一个简单可行的方法是把在旧版中存放下载数据包的文件夹复制到新版文件夹中。例如，你可以把旧版文件夹"library"（存放数据包的路径："C:\Program Files\R\R-4.2.0\library"）复制到新版文件夹"R-4.2.2"中（路径："C:\Program Files\R\R-4.2.2\"），替换现有的文件。如果需要更新复制过来的数据包，那么可以进入 R 新版界面，输入：update.packages(checkBuilt=TRUE, ask=FALSE)，按回车键选择就近镜像点更新数据包即可自动完成更新。

另外一种更新方法是下载 R 数据包 installr，然后调用函数 library(installr)，最后输入 updateR()，按回车键即可完成更新。如果当前使用的版本已是最新版，那么 R 会在工作界面显示 FALSE。

1.3 基本 R 对象

基本 R 对象包括向量、数据框、矩阵和列表。

1.3.1 向量

向量是最基本的 R 对象,由单维数据元素组成。它是矩阵的特例。只有一排的矩阵称作排向量(row vector),只有一列的矩阵则称作列向量(column vector)。向量分为三个类别:数值向量(numeric vector)、字符向量(character vector)和逻辑向量(logical vector)。

1.3.1.1 数值向量

数值向量的基本存储方式利用 R 函数 c( ),其中 c(小写字母)表示组合(concatenate),括号中包括用逗号隔开的元素。例如:

```
> x<-c(1,2,4,5,7,9,10);x
[1]  1  2  4  5  7  9 10
```

在以上 R 命令中,我们创建了一个向量,利用分配符(<-)为向量指定一个 R 存储对象 x。另外一个不太常用的分配符是等于号(=)。要查看 x,只需在 R 界面输入 x,再按回车键执行即可。由于我们把两段代码放在了同一行,因此用分号(;)把它们隔开了。不过,我们通常每行输入一段代码,特别是在代码较长时。例如:

```
> x<-c(1,2,4,5,7,9,10)
> x
[1]  1  2  4  5  7  9 10
```

向量长度的计算利用 R 函数 length( ),在括号里输入 R 存储对象或变量名。例如,在 R 操作界面输入命令 length(x),按回车键后得到 7。我们可以对数值向量进行简单的数学运算。例如:

```
> x+10 # add 10 to each element in a vector
[1] 11 12 14 15 17 19 20
> x-1 # each element in a vector minus 1
[1] 0 1 3 4 6 8 9
> x*5 # multiply each element in a vector by 5
[1]  5 10 20 25 35 45 50
> x/2 # divide each element in a vector by 2
[1] 0.5 1.0 2.0 2.5 3.5 4.5 5.0
> x^2 # square each element in a vector
[1]   1   4  16  25  49  81 100
```

在上面的计算中,符号"#"表示注解,R 忽略注解内容。R 是解释性语言(interpreted language)。当我们把代码或可执行的命令输入 R 工作界面,R 解释器(interpreter)就会分析与执行代码,输出结果。从以上例子可以发现,R 按照元素顺序依次进行计算。如果向量长度不等,那么 R 使较短向量中的数值循环出现,直至与较长向量的长度匹配。例如:

```
> c(1:5) * c(5,10)
[1]  5 20 15 40 25
Warning message:
In c(1:5) * c(5, 10):
  longer object length is not a multiple of shorter object length
```

这里,第一个向量 c(1:5)等同于 c(1,2,3,4,5),长度为 5;第二个向量中的元素 5 和 10 循环得到长度相等的向量 c(5,10,5,10,5),按元素顺序依次相乘。以上结果中出现了警告信息:长对象的长度不是短对象长度的整倍数。R 有自动循环的功能。如果尝试执行一些 R 通常希望对象具有相同长度的操作,R 会自动扩展较短的对象,使两个对象具有相同的长度。如果较长对象的长度是较短对象长度的整倍数,这就相当于简单地重复较短对象的倍数。但是,如果它不是整倍数,R 就会提醒这是否是有意为之还是由错误造成的。

除了 c(1:5)之外,还有其他一些函数能够很方便地创建向量,如 seq()和 rep()。函数 seq(from,to,by)中的 from 和 to 设置数值序列的起点和终点,by 设置间隔增值,默认的间隔增值为 1。只要按照变元顺序设置数值,变元名称可以省略。例如:

```
> seq(from=1,to=20,by=2)
[1]  1  3  5  7  9 11 13 15 17 19
> seq(1,20,2)
[1]  1  3  5  7  9 11 13 15 17 19
```

函数 rep(x,times)中的第一个变元 x 是向量,第二个变元 times 表示重复的次数。例如:

```
> rep(x=c(1,2,5),times=2)
[1] 1 2 5 1 2 5
> rep(c(1,2,5),2) # same as above
[1] 1 2 5 1 2 5
```

1.3.1.2 字符向量

我们可以利用函数 rep()创建字符向量。例如:

```
> rep("R is an interpreted language.",2)
[1] "R is an interpreted language." "R is an interpreted language."
```

注意,在 R 操作界面输入字符时,需要添加双引号(" ")或单引号('),其中以双引号更常见。

当我们用函数 length( )计算文本长度时,却得到 1:

```
> text1<-"Once there was a little boy who was hungry for success in sports."
> length(text1)
[1] 1
```

我们期待 R 输出的长度为 13(13 个词),但是 R 却把向量视作一个字符串(character string)。我们将在第二章介绍字符串的切分问题。我们可以利用 R 函数 nchar( )计算文本字符数,包括空格数(因为 R 不能区分空格和字符)。例如,计算 text1 字符数的 R 命令和输出结果如下:

```
> nchar(text1)
[1] 65
```

以上结果显示,这个字符串由 65 个字符组成,包括 52 个字母、1 个句号和 12 个空格。

当文本词数较多时,我们通常不直接在 R 操作界面输入文本,而是将之放在本地硬盘中,使用 R 命令调入。如果要了解 R 默认的工作簿,可以在 R 操作界面输入命令:getwd( ),再按回车键即可。如果我们按照 Windows 11 系统(DELL 电脑)默认的 R 软件安装方式,键入 R 命令 getwd( )后就会得到:[1] "C:/Users/DELL/Documents"。如果我们把待处理文件或 R 脚本放在非 R 默认的路径下,那么需要利用函数 setwd( )重新设定当前工作簿。例如,要把待处理文件或 R 脚本放在本地 D:\Rpackage 文件夹中,则在 setwd( )的括号里添加引号,将路径放在引号里,如 setwd("D:/Rpackage")。如果我们要把保存在文件夹 Rpackage 中的文本文件 text1 读入 R 操作界面,那么可以输入 R 命令:

```
text1<-scan(choose.files( ),what="char",sep="\n",quiet=TRUE); text1
```

得到以下结果:

```
[1] "Once there was a little boy who was hungry for success in sports."
```

注意,执行第一个命令弹出一个窗口,要求选择一个文本文件(扩展名为.txt),并赋予变量名 text1;第二个命令输入变量名,R 输出结果。两个命令之间用分号(;)隔开。在第一个命令中,左向箭头(<-)为对象或变量分配符,由小于号(<)和连字号(-)组成,符号之间没有间隔。变元(argument)what 设定为"char",表示提取字符;sep 设为"\n",表示换行(next line);quiet 设为 TRUE,表示不显示执行结果。

利用因素函数 factor( )也可以存储字符向量。因素与名义或类别变量同义,由两个或两个以上水平(levels)构成。函数 factor( )的主要结构是 factor(x, levels, labels),其中变元 x 是数值向量,通常包括几个不同的值。变元 levels 是由向量 x 中唯一值构成的一组值,如 levels=c(1,2)指向量包括 1 和 2 两个值。凡是接受函数默认设置方式的变元，都可以在 R 命令中省略。变元 labels 指用字符设定因素水平的标签。例如，因素"性别"有男性（Male）和女性（Female）两个水平，每个水平有 10 人，则以下 R 命令可以创建"性别"向量：

```
> Gender<-c(rep(1,10),rep(2,10))
> Gender<-factor(Gender,levels = c(1,2),labels = c("Male","Female"));Gender
 [1] Male   Male   Male   Male   Male   Male   Male   Male   Male   Male
[11] Female Female Female Female Female Female Female Female Female Female
Levels: Male Female
# 或者省略 levels=c(1,2)
> Gender<-factor(Gender,labels=c("Male","Female"));Gender
 [1] Male   Male   Male   Male   Male   Male   Male   Male   Male   Male
[11] Female Female Female Female Female Female Female Female Female Female
Levels: Male Female
```

执行以上第一个命令得到 10 个 1 和 10 个 2，存储的变量名为 Gender。在第二个命令中，利用函数 factor()，将上一个 Gender 向量中 1 和 2 分别替换为"Male"和"Female"，并再次分配给同一个变量名 Gender 作为替换。我们也可以利用函数 gl(n, k, labels)，其中变元 n 是水平数，k 是重复数，labels 利用字符设定因素水平名称。例如：

```
> gl(n=2, k=10, labels=c("Male","Female"))
 [1] Male   Male   Male   Male   Male   Male   Male   Male   Male   Male
[11] Female Female Female Female Female Female Female Female Female Female
Levels: Male Female
```

1.3.1.3 逻辑向量与操作符

逻辑向量包括 TRUE 和 FALSE 两个元素（用于真假陈述判断），或者包括 NA（表示缺失值，not available）。利用函数 as.logical()很容易把字符"TRUE""FALSE""T""F""true"和"false"等转换为逻辑符号。例如：

```
> x<-c("TRUE","FALSE","T","F","true","false","na")
> x<-as.logical(x);x
[1] TRUE FALSE TRUE FALSE TRUE FALSE NA
```

我们可以利用逻辑操作符对向量元素进行筛选。逻辑向量常用操作符包括：<（小于号）、<=（小于或等于号）、>（大于号）、>=（大于或等于号）、==（等于号）、!=（不等于号）、x|y（或者符）、x||y（或者符，只评估两个对象的第一个元素）、x & y（与符）、&&（与符，只评估两个对象的第一个元素）、x%in%y（包含符）。例如：

```
> W<-c(20,30,50,15,60,50)
> W>30 # which elements are more than 30
[1] FALSE FALSE TRUE FALSE TRUE TRUE
> W<-c(10,20,30,45,60,80)
> W[W>30]# which elements satisfy a given condition
[1] 45 60 80
> W[W==60] # which elements are equal to 60
[1] 60
> W[W!=30] # which elements are not equal to 30
[1] 10 20 45 60 80
> W[W<30|W>50] # which elements are less than 30 or larger than 50
[1] 10 20 60 80
> W[which(W<30)] # which() outputs the elements that satisfy a given condition
[1] 10 20
> W[W>30&W<60] # which elements are larger than 30 and less than 60
[1] 45
> (c(11,2,5,6,9)>3)|(c(12,4,11,8,10)>7)
[1] TRUE FALSE TRUE TRUE TRUE
> (c(11,2,5,6,9)>3)||(c(12,4,11,8,10)>7)
[1] TRUE
> (c(1,2,5,6,9)>3)&(c(1,4,11,8,10)>7)
[1] FALSE FALSE TRUE TRUE TRUE
> (c(1,2,5,6,9)>3)&&(c(1,4,11,8,10)>7)
[1] FALSE
> c(11,2,5,6,9)%in%c(12,4,11,8,10)
[1] TRUE FALSE FALSE FALSE FALSE
> text<-c("Once","there","was","a","little","boy","who","was",
  "hungry","for","success","in","sports")
> text[text!="was"] # which elements are not "was"
[1] "Once"  "there"  "a"  "little"  "boy"  "who"  "hungry"
[8] "for"  "success"  "in"  "sports"
```

如果要筛选哪些词(如功能词)不在文本中,那么可以将符号!和%in%并用。例如:

```
> stopwords<-c("was","a","who","for","in")
> text[!text%in%stopwords]
[1] "Once"  "there"  "little"  "boy"  "hungry"  "success"  "sports"
```

关于|等逻辑操作符的更多说明,用户可以在 R 工作界面输入：?Logic,按回车键后在弹出窗口点击 R 基础包 base 中的 Logical Operators 进行查阅。关于!=等逻辑操作符的更多说明,用户可以在 R 工作界面输入：?Comparison,按回车键查询。

1.3.2 数据框

数据框是 R 常见的数据集存储方法,由多列数据构成,每列代表一个变量的测量。数据框可以被看作只有列名称、没有排名称的矩阵。数据框的构建使用 R 函数 data.frame(),把不同向量组合在一起。例如：

```
> x<-data.frame(A=c(1,2,3,4,5),B=c("D","E","F","G","H"),C=c(TRUE,TRUE,FALSE,FALSE,NA));x
    A   B   C
1   1   D   TRUE
2   2   E   TRUE
3   3   F   FALSE
4   4   G   FALSE
5   5   H   NA
```

在这个例子中,A 是数值向量,B 是字符向量,C 是逻辑向量,它们构成 5 排 3 列数据,说明数据框可以包括不同性质的数据。我们可以用索引符"[,]"(逗号前面的变元为排,后面的变元为列)提取数据框中的元素。例如,[1:3,]表示提取所有列的前三排数据。[1:3,2]表示提取第二列的前三排数据。且看下面的例子：

```
> x[1:3,]
    A   B   C
1   1   D   TRUE
2   2   E   TRUE
3   3   F   FALSE
x[1:3, 2]
[1] "D"  "E"  "F"
```

提取数据框中某个变量的数据还可以使用美元操作符 $。譬如,提取 x 数据框中变量 B 的所有数值,输入命令 x $ B 可以得到：

```
> x $ B
[1] "D"  "E"  "F"  "G"  "H"
```

要了解数据框中数据的排数和列数,可调用函数 nrow() 和 ncol(),或者调用 dim() 显示数据框的维度(dimensions)。例如：

```
> nrow(x);ncol(x);dim(x)
[1] 5
[1] 3
[1] 5  3
```

如果我们要把数据框 x 保存在本地硬盘文件夹中（如 D:\Rpackage\），那么利用函数 write.table()，输入以下 R 命令即可：

```
write.table(x,file="D:/Rpackage/df.x.txt",quote=FALSE,sep="\t")
```

在这个命令中，file="D:/Rpackage/df.x.txt" 表示保存数据框 x 的路径和文件名（文件名包括扩展名.txt），quote=FALSE 表示字符（或因素）不用引号，sep="\t" 表示分隔符为制表符。如果要把数据框读入 R 操作界面，可以使用函数 read.table()。例如：

```
> read.table("D:/Rpackage/df.x.txt",header=TRUE,sep="\t")
   A  B  C
1  1  D  TRUE
2  2  E  TRUE
3  3  F  FALSE
4  4  G  FALSE
5  5  H  NA
```

在这个命令中，"D:/Rpackage/df.x.txt" 指要读取的数据框名称和路径，header=TRUE 表示输出变量名作为第一行。

1.3.3 矩阵

矩阵是 n 排和 p 列数值构成的长方形排列，是二维数组。假如有 4 名学生的口语听后故事复述文本（text1、text2、text3 和 text4），文本中重复原文名词的数量依次为：25、47、51、31，重复原文动词的数量依次为：11、16、27、11，重复原文形容词的数量依次为：23、21、26、20。R 输入各个向量的形式如下：

```
> text<-c("text1","text2","text3","text4")
> POS<-c("noun","verb","adjective")
> values<-c(25,11,23,47,16,21,51,27,26,31,11,20)
```

在以上 R 命令中，text 是文本分类，POS 是三个词性或词类（part of speech），它们都是字符向量。变量 values 是按照每个文本三个词性计数顺序输入的向量，为数值向量。R 创建矩阵利用函数 matrix(data=NA,nrow=1,ncol=1,byrow=FALSE,dimnames=NA)，其中 data 是数值向量，nrow 和 ncol 分别指排数和列数，byrow=FALSE 表示矩阵中数值输入的方式，默认按列输入，dimnames 表示由排名称和列名称向量构成的列表。本例 R 命令和输出结果如下：

```
> POS.m<-matrix(values,nrow = 3,ncol = 4,byrow = FALSE,dimnames = list(POS,text))
> POS.m
          text1  text2  text3  text4
noun       25     47     51     31
verb       11     16     27     11
adjective  23     21     26     20
```

要确定矩阵包括的排数和列数，可以调用函数 nrow( ) 和 ncol( )，或者调用函数 dim( )。例如：

```
> nrow(POS.m);ncol(POS.m);dim(POS.m)
[1] 3
[1] 4
[1] 3 4
```

对矩阵的排和列可以进行算术运算。譬如，利用 mean(POS.m[,1]) 计算矩阵 POS.m 第一列的平均数，得到 19.666 67；利用 mean(POS.m[1,]) 计算矩阵 POS.m 所有列第一排的平均数，得到 38.5。如果要计算所有排和列的平均数或其他统计量，一个简便的方法是调用 R 函数 apply( )。函数 apply( ) 既适用于矩阵又适用于数据框中排向量和列向量的计算。譬如，执行 R 命令 apply(POS.m,1,mean) 得到每排向量的平均数，执行 R 命令 apply(POS.m,2,mean) 得到每列向量的平均数：

```
> apply(POS.m,1,mean)
   noun       verb    adjective
   38.50      16.25    22.50
> apply(POS.m,2,mean)
   text1      text2    text3     text4
19.66667   28.00000   34.66667   20.66667
```

矩阵保存和读取方式与数据框相同。例如：

```
> write.table(POS.m,file="D:/Rpackage/POS.m.txt",quote=FALSE,sep="\t")
> read.table("D:/Rpackage/POS.m.txt",header=TRUE,sep="\t")
          text1  text2  text3  text4
noun       25     47     51     31
verb       11     16     27     11
adjective  23     21     26     20
```

1.3.4 列表

列表同数据框一样可以存储不同性质的数据,但与数据框不同的是,列表可以存储不同长度的 R 对象。列表数据存储的 R 函数为 list(),可以将多个对象合并在一个对象之下。假如一项实验研究设计有两个被试间因素,即材料呈现方式(format)和作文指令(instructions)。材料呈现方式包括文本格式(text format,简称 Text)和网页格式(web format,简称 Web)。作文指令包括记叙文(narrative,简称 N)、概述(summary,简称 S)和议论文(argument,简称 A)。测量结果是内容推理得分(Scores)。该研究使用 6 个实验组,即 6 个单元格,如文本格式+记叙文组(简称 TextN)。各个单元格的 R 输入结果如下:

```
> TextN<-c(60,70,80,80,70,65,50,85,90)
> TextS<-c(55,75,65,80,75,70,90,80)
> TextA<-c(80,70,65,50,70,90,75,80)
> WebN<-c(95,70,80,60,65,75,80,85,70,60)
> WebS<-c(70,60,85,60,80,50,100,70)
> WebA<-c(95,90,90,85,85,80,80,70)
```

如果我们用 Type 表示格式与指令类别,Size 表示各个单元格样本量,Score 表示得分,将上面的数值向量转化为列表格式的 R 命令如下:

```
> Type<-c("TextN","TextS","TextA","WebN","WebS","WebA")
> Size<-c(9, 8, 8, 10, 8, 8)
> Score<-c(TextN, TextS, TextA, WebN, WebS, WebA)
> Format_Instruct<-list(Type=Type,Size=Size,Score=Score)
> Format_Instruct
 $ Type
 [1] "TextN"   "TextS"   "TextA"   "WebN"   "WebS"   "WebA"
 $ Size
 [1] 9  8  8  10  8  8
 $ Score
 [1]  60  70  80  80  70  65  50  85  90  55  75  65  80  75  70  90
 [17] 80  80  70  65  50  70  90  75  80  95  70  80  60  65  75  80
 [33] 85  70  60  70  60  85  60  80  50  100  70  95  90  90  85  85
 [49] 80  80  70
```

提取列表成分可以利用提取符 $ 或者利用双中括号[[ ]]。例如:

```
> Format_Instruct $ Type
 [1] "TextN"   "TextS"   "TextA"   "WebN"   "WebS"   "WebA"
> Format_Instruct[[2]]
```

```
[1] 9 8 8 10 8 8
> Format_Instruct $ Score[1:9] # Scores for TextN
[1] 60 70 80 80 70 65 50 85 90
```

1.4 R 循环

循环指在某个对象上多次迭代或重复。R 循环包括重复循环、while 循环和 for 循环。

1.4.1 重复循环

在重复(repeat)循环中,代码被反复执行,直至出现停止的条件才退出循环。重复循环的基本结构如下:

```
repeat {
commands
if( condition) {
break
}
}
```

在该结构中,要循环的代码或命令(commands)、条件(condition)和中止陈述(break)置于大括号({})之内,中止陈述也放在大括号内,嵌套在外围的大括号中。例如,我们想要得到 5 个"Hello" "R!",则 R 命令和执行结果如下:

```
> Hello<-c("Hello","R!")
> count<-1
> repeat {
+ print(Hello)
+ count=count+1
+ if(count>5) {
+ break
+ }
+ }
[1] "Hello"  "R!"
[1] "Hello"  "R!"
[1] "Hello"  "R!"
[1] "Hello"  "R!"
[1] "Hello"  "R!"
```

在以上命令中,执行符>和换行符+是 R 自带的符号,用户无须输入,有 Hello 和 count

两个变量。第一个变量是字符串,第二个变量设置初始值1(即重复一次)。在循环中,我们首先要R返还Hello中的字符,然后计数从1开始,if条件句限定重复数不得超过5次。这意味着"Hello" "R!"将被打印5次。

我们再来看另外一个例子:

```
> y<-2
> repeat {
+ print(y)
+ y=y+2
+ if(y==10){
+ break
+ }
+ }
[1] 2
[1] 4
[1] 6
[1] 8
```

在这个例子中,我们先设置y的初始值为2,if条件句限定y值为10时,循环结束,因此该循环只打印2到8之间的偶数。

1.4.2 while 循环

如果某个条件为真(TRUE),while循环会一直执行大括号中的命令或代码。如果条件评估为误(FALSE),那么循环结束。while循环的基本结构如下:

```
while (condition) {
commands
}
```

试看下面的一个例子:

```
> Hello<-c("Hello","this is the R world")
> count<-1
> while (count<5) {
+ print(Hello)
+ count = count + 1
+ }
[1] "Hello"            "this is the R world"
[1] "Hello"            "this is the R world"
```

```
[1] "Hello"                    " this is the R world"
[1] "Hello"                    " this is the R world"
```

在以上命令中,执行符>和换行符+是 R 自带的符号,用户无须输入,有 Hello 和 count 两个变量。第一个变量是字符串,第二个变量设置初始值 1(即重复一次)。在循环中,我们首先要 R 返还 Hello 中的字符,然后计数从 1 开始,while 条件句限定重复数小于 5 次。只要循环的 count + 1 值小于 5,Hello 中的字符串就会一直被打印出来,所以我们看到字符串重复 4 次。

1.4.3 for 循环

从上一节可知,while 循环是非限定性迭代(indefinite iteration),只要条件成立,R 代码就会重复运行。for 循环则是限定性循环,按照限定的次数重复执行 R 代码。基本结构如下:

```
for (i in sequence) {
command 1
command 2
...
}
```

我们知道,标准差和标准误差都表示离散性,但是标准差反映样本原始数值离散趋势,标准误差则是样本统计量(如平均数)抽样分布的标准差,反映对总体参数(如总体平均数)估计的精确度(鲍贵 等,2020b)[50]。我们利用 for 循环从总体平均数为 10、标准差为 2 的正态分布总体中模拟 10 000 个样本量为 25 的样本,利用函数 mean() 计算每个样本的平均数,再利用函数 sd() 计算 10 000 个样本的标准差,看看结果是否等于理论上的标准误差 0.4(总体标准差 2 与样本量平方根 5 的商)。本例 R 命令和统计分析结果如下:

```
> set.seed(2)
> N=10000
> means<-NULL
> for (i in 1: N){
+ x<-rnorm(25, 10, 2)
+ means[i]=mean(x)
+ }
> round(sd(means),2)
[1] 0.4
```

这一结果与理论上的标准误差完全相同。在这个例子中,set.seed(2) 设置随机种子为 2(其他任何数值均可),是为了使研究结果可以重复。N = 10 000 设定循环量,means<-

NULL 用于存放每次循环执行大括号中的命令得到的平均数,for(i in 1:N)表示循环序列值(从第一次、第二次直到第一万次)。正态分布随机抽样函数 rnorm()中的第一个变元是样本量,第二个变元是总体平均数,第三个变元是总体标准差。四舍五入函数 round()中的第一个变元是数值向量,第二个变元是保留的小数位数。

我们也可以利用 for 循环编写函数。假如我们要计算以下一个向量的最小值:

```
y<-c(21.2,19.2,21.2,20.2,18.6,21.2,21.3,19.5,17.9,19.0,21.1,21.6,20.1,20.2,19.8,
14.4,21.9,20.0,21.2,23.0,20.0,22.8,21.4,20.5,22.5,22.5,19.0,18.9,19.7,21.3)
```

我们可以编写以下 R 函数 Min 计算向量 y 的最小值:

```
> Min<-function(x){
  low<-Inf
  for (i in 1:length(x)){
  if(x[i]<low){
  low<-x[i]
  }
  }
  print(low)
  }
```

在这个函数中,x 是向量,low<-Inf 设定初始值为无穷大值,for(i in 1:length(x))表示循环序列值,从 1 开始一直递增到样本量值。如果样本中的第一个值小于无穷大值(这是必然的),那么保留样本第一个值。如果样本第二个值小于样本第一个值,就保留第二个值,否则保留第一个值,依次迭代下去,直至样本的最后一个值。print(low)要求返还计算结果。执行 R 命令 Min(y)得到 y 向量的最小值 14.4。读者也可以利用 R 自带函数 min()得到向量最小值。

如果我们希望保存自定义函数,以备后期调用,那么可以将函数保存为 R 数据文件格式(扩展名为.rds)。例如,下面的命令将函数 Min()保存在 D:\Rpackage 文件夹中:

```
saveRDS(Min,file="D:/Rpackage/Min.rds")
```

如果我们要调用保存的函数 Min,重新计算向量 y 的值,R 命令和统计分析结果如下:

```
> Min<-readRDS("D:/Rpackage/Min.rds")
> y<-c(21.2,19.2,21.2,20.2,18.6,21.2,21.3,19.5,17.9,19.0,21.1,21.6,20.1,20.2,19.8,
14.4,21.9,20.0,21.2,23.0,20.0,22.8,21.4,20.5,22.5,22.5,19.0,18.9,19.7,21.3)
> Min(y)
[1] 14.4
```

在前面的 if 条件陈述中,如果条件为真,那么 R 执行循环代码;如果条件为误,那么 R

结束工作。我们可以使用 if...else 条件陈述执行在条件为误时的命令。下面看一个例子：

```
> Number<-c(-1,2,-3,0,7,10,15,-9,-10,6,4,0)
> for (i in 1:length(Number)){
+ if (Number[i]<0) print(paste(Number[i]," -> negative"))
+ else if (Number[i]==0) print (paste(Number[i], " -> zero"))
+ else print (paste(Number[i], " -> positive"))
+ }
[1] "-1 -> negative"
[1] "2 -> positive"
[1] "-3 -> negative"
[1] "0 -> zero"
[1] "7 -> positive"
[1] "10 -> positive"
[1] "15 -> positive"
[1] "-9 -> negative"
[1] "-10 -> negative"
[1] "6 -> positive"
[1] "4 -> positive"
[1] "0 -> zero"
```

在这个例子中，我们利用 for 循环把向量 Number 中的每个元素按照正数、零和负数进行分类。在 Number[i]<0 为真时，R 执行命令 print(paste(Number[i]," -> negative"))，把符合条件的向量元素和" -> negative"放在一起打印。由于这个代码只有一行，因此略去了可以包括多个条件陈述的大括号。条件陈述 else if 对其他条件进行描述。在向量元素为 0 时，print (paste(Number[i], " -> zero"))要求 R 把符合条件的向量元素和" -> zero"放在一起打印。最后一行代码 else print (paste(Number[i], " -> positive"))表示在其他条件下要求 R 打印符合条件的向量元素和" -> positive"。

如果我们想要清理前期的操作，转向新的操作，可以利用函数 rm (list = ls (all = TRUE))，其中函数 rm()表示"清除"(remove)，函数 ls()指列出工作空间的所有对象。如果清理成功，那么在 R 操作界面输入 ls()，按回车键后会显示没有字符，即 character(0)。如果只想清理部分数据结构，可以键入 ls()查看待清理的数据结构，将选择结果作为变元放在函数 rm()的括号内按回车键即可，如执行命令 rm(Number)会删除已有向量 Number。

第二章 字符串处理

将文档读入 R 是开展自然语言处理(Natural Language Processing,NLP)的基本要求。由于文档格式不同,读取方式也不同。本章主要介绍纯文本文档、word 文档和 PDF 文档的读取方法。在此基础上,本章简要介绍正则表达式、R 正则表达式函数和语境中的关键词检索方法,调用的 R 数据包包括 readtext 和 stringr。

2.1 文档读取

2.1.1 纯文本文档读入 R

在第一章1.3.1.2节介绍字符向量时,我们简要介绍了利用函数 scan(choose.files(),what="char",sep="\n",quiet=TRUE)直接把保存在本地 D:\Rpackage 文件夹中的纯文本文档 text1 读入 R。除此之外,我们可以利用 R 函数 readLines()将纯文本文档 text1 读入 R。在该函数的括号中直接输入纯文本文档的路径,纯文本文档的扩展名为.txt。例如,利用以下命令读入文本文档 text1:

```
> text1<-readLines("D:/Rpackage/text1.txt");text1
Warning message:
In readLines("D:/Rpackage/text1.txt"):
  incomplete final line found on 'D:/Rpackage/text1.txt'
[1] "Once there was a little boy who was hungry for success in sports."
```

在上面的输出结果中,文档路径中使用了顺向斜杠(/),但是本地电脑使用的路径符为反斜杠(\)。如果我们要在 R 中使用反斜杠加载文档,那么需要使用双反斜杠(\\)。例如:

```
> text1<-readLines("D:\\Rpackage\\text1.txt");text1
Warning message:
In readLines("D:\\Rpackage\\text1.txt"):
  incomplete final line found on 'D:\Rpackage\text1.txt'
[1] "Once there was a little boy who was hungry for success in sports."
```

在以上结果中,我们看到了一条警告(warning message),提示文档 text1.txt 的最后一行不完整。这仅仅是一个提醒而已。如果不想看到这条警告,可以在函数 readLines()中设置变元 warn=FALSE。例如：

> text1<-readLines("D:/Rpackage/text1.txt",warn=FALSE);text1
[1] "Once there was a little boy who was hungry for success in sports."

我们也可以利用函数 setwd() 设定 R 当前文件夹,在 readLines() 的括号里只添加文档名,并且把它放在引号里。例如,读入文本文档 text1 的 R 命令如下：

> setwd("D:/Rpackage")
> text1<-readLines("text1.txt",warn=FALSE);text1
[1] "Once there was a little boy who was hungry for success in sports."

把纯文本文档读入 R 的另外一种方法是调用 R 数据包 readtext。这个数据包由肯尼斯·贝诺特等(Benoit et al.,2021)开发,是专门为纯文本和格式化文本文档读入 R 而设计的,当前版本号为 0.81,查看版本号的 R 命令为：packageVersion("readtext")。关于 R 数据包的安装,见第一章。数据包的调用利用命令 require(readtext) 或 library(readtext)。调用文本数据的函数是 readtext(),直接在函数的括号中输入文档的路径和文档名。在使用当前工作簿时,如果文档在工作簿里,那么只需输入文档名。例如,读入文本文档 text1 的 R 命令如下：

> library(readtext)
> text1<-readtext("D:/Rpackage/text1.txt");text1
readtext object consisting of 1 document and 0 docvars.
Description: df [1 x 2]
  doc_id   text
  〈chr〉   〈chr〉
1 text1.txt "\"Once there\"..."

我们看到,readtext() 输出的结果是一个数据框(df),包括两个列变量：doc_id(文档编号)和 text(文本)。如果我们只需要文本,那么可以使用提取符 $ 或使用索引符[ ]：

> text1 $ text
[1] "Once there was a little boy who was hungry for success in sports."
> text1[,'text']
[1] "Once there was a little boy who was hungry for success in sports."

2.1.2 word 文档读入 R

数据包 readtext 中的函数 readtext() 不仅能够读取纯文本文档,还能够读取 word 文档。读取 word 文档的方式与读取文本文档的方式相同,只是需要使用后缀.doc 或.docx。文档扩展名是.doc 还是.docx，可以用鼠标右击文档名查看属性。我们将 text1 中的文字以 word 的形式（扩展名为.docx）保存在 D:\Rpackage 文档夹中，文档名不变。读入 word 文档的 R 命令如下：

```
> library(readtext)
> text1<-readtext("D:/Rpackage/text1.docx")
> text1<-text1 $ text;text1
[1] "Once there was a little boy who was hungry for success in sports."
```

2.1.3 PDF 文档读入 R

数据包 readtext 中的函数 readtext()不仅能够读取纯文本文档和 word 文档，还能够读取 PDF 文档，读取方式相同，只是需要修改扩展名。读入保存在 D:\Rpackage 文档夹中的 PDF 文档 text1 的 R 命令如下：

```
> library(readtext)
> text1<-readtext("D:/Rpackage/text1.pdf")
> text1<-text1 $ text;text1
[1] "Once there was a little boy who was hungry for success in sports.\n"
```

我们看到，输出结果中多出了换行符"\n"(n 指 next line)。如果要除去换行符，那么 R 命令和处理结果如下：

```
> text1<-gsub("\n","",text1);text1
[1] "Once there was a little boy who was hungry for success in sports."
```

在以上命令中，我们使用了正则表达式（regular expression）函数 gsub()把换行符"\n" 替换为空格，并消除其痕迹（""）。关于正则表达式，2.3 节和 2.4 节再作介绍。

2.2 字符串分割

在上节中，我们把文档调入了 R 控制台或操作界面，但是显示的结果是一个字符串，没有把字符分开。我们通常需要把字符分开，以便进行后续的自然语言处理。最常用的字符串分割函数是 R 基础函数 strsplit(x," ")，其中变元 x 是字符串，" "（引号之间空一格）设置分割方式为空格。例如：

```
> library(readtext)
> text1<-readtext("D:/Rpackage/text1.docx") $ text
> strsplit(text1," ")
[[1]]
```

```
[1] "Once"    "there"    "was"    "a"    "little"    "boy"    "who"
[8] "was"    "hungry"    "for"    "success"    "in"    "sports."
```

从输出结果给出的双中括号（[[ ]]）可以看出,分割结果是一个列表。如果要得到字符向量,那么使用 R 基础函数 unlist()。例如:

```
> unlist(strsplit(text1," "))
[1] "Once"    "there"    "was"    "a"    "little"    "boy"    "who"
[8] "was"    "hungry"    "for"    "success"    "in"    "sports."
```

如果我们要计算文本词数,可以使用 R 基础函数 length()。例如:

```
> text1<- unlist(strsplit(text1," "))
> length(text1)
[1] 13
```

计算字符长度的 R 基础函数是 nchar()。例如:

```
> nchar(text1)
[1] 4 5 3 1 6 3 3 3 6 3 7 2 7
```

细心的读者会发现,在计算字符长度时,nchar()把" sports. "中的词" sports"和标点"."视作一个字符进行计算,而实际上我们希望计算" sports"的字符数（6 个字母）。这就牵涉到正则表达式的问题了。

2.3 正则表达式

2.3.1 字面符和元字符

正则表达式（regular expression,又称 regex）是对文本模式的标注,是一种特殊的字符或字符串。利用正则表达式的目的就是进行模式匹配或替换。在 R 中输入 help(regex),按回车键后可以查阅关于正则表达式的说明。正则表达式包括字面意义上的字符（literal character,字面符）和元字符（metacharacter）。字面意义上的正则表达式是按照字面意义理解的正则表达式,而元字符则有特殊含义。如果我们想在 R 中把元字符转化为其字面意思,那么需要在该字符前添加两个反斜杠（\\,退出符,表示转义）。例如,点或句号（.）作为元符号匹配任何字符,但是如果将其用作字面上的正则表达式,那么在 R 中需要输入"\\."。试比较：

```
> text<- "We paid a lot of money to buy the house. They gave us a large sum of money."
> gsub(".","!",text)
[1]
```

```
"!!!!!!!!!!!!!!!!!!!!!!!!!!!!!!!!!!!!!!!!!!!!!!!!!!!!!!!!!!"
> gsub("\\.","!",text)
[1] "We paid a lot of money to buy the house! They gave us a large sum of money!"
```

在上面的第一个命令 gsub(".","!",text) 中,gsub(g 代表 global)是 R 替换函数,"."是元字符,"!"是替换符,执行该命令把文本 text 中的所有字符替换为感叹号。在第二个命令 gsub("\\.","!",text)中,双反斜杠"\\."把作为元符号的点号还原为字面意义上的点号或句号,因而执行该命令只把文本 text 中的句号替换为感叹号。再如:

```
> gsub(pattern = "$","!",c("You love R $","Me $ too."))
[1] "You love R $!" "Me $ too.!"
> gsub(pattern = "\\$","!",c("You love R $","Me $ too."))
[1] "You love R!" "Me! too."
```

作为有特殊意义的元字符,"$"表示字符串的结尾。在上面的第一个例子中,正则表达式为特殊元字符"$",替换符为感叹号"!"。这意味着在每个字符串的结尾增加一个感叹号。在第二个例子中,正则表达式"\\$"表示通常意义上的美元符号,因而字符串中的两个美元符号"$"都被替换成了感叹号"!"。

元字符是非字母型或非数字型的符号。除了.、\\和$之外,其他一些常用元字符包括:|、+、?、*、()、{ }、^、[ ]。|是析取符,表示"或"的意思,即提供选项,把一个表达式分割成若干字符,然后匹配。例如:

```
> Text<-c("As you know, blood is the red liquid that flows inside your body.","You sometimes bleed.","When you bleed, you lose blood from your body as a result of injury or illness.","There are some idioms.","Let us take an example.","If someone draws first blood, they have had a success at the beginning of a competition.")
> grep("bl(oo|ee)d",Text,value=TRUE)
[1] "As you know, blood is the red liquid that flows inside your body."
[2] "You sometimes bleed."
[3] "When you bleed, you lose blood from your body as a result of injury or illness."
[4] "If someone draws first blood, they have had a success at the beginning of a competition."
> grep("blood|bleed",Text,value=TRUE)
[1] "As you know, blood is the red liquid that flows inside your body."
[2] "You sometimes bleed."
[3] "When you bleed, you lose blood from your body as a result of injury or illness."
[4] "If someone draws first blood, they have had a success at the beginning of a competition."
```

以上两个命令的结果相同。在第一个析取命令 grep("bl(oo|ee)d", Text, value =

TRUE)中,"g"代表 global,"rep"表示重复,匹配模式"bl(oo|ee)d"把选择项"oo|ee"放在括号(())里,目的是匹配 blood 和 bleed。即是说,匹配词以"bl"开头,中间部分可以是"oo",也可以是"ee",最后以"d"结尾。在第二个命令 grep("blood|bleed",Text,value=TRUE)中,匹配词 blood 和 bleed 之间用元符号"|"隔开。value=TRUE 表示返还包括匹配元素的向量;value=FALSE 则表示返还包括匹配的索引向量。

加号"+"表示前面的字符或元素出现一次或多次。例如,在下面的命令中,"b+"表示提取的词符必须至少包括一个字母"b","bb+"表示提取的词符中字母"b"至少连续出现两次:

```
> grep("b+",x=c("b","abc","bcd","bbbc","c"),value=TRUE)
[1] "b"    "abc"   "bcd"   "bbbc"
> grep("bb+",x=c("bc","bb","abcb","bcd","bbbc"),value=TRUE)
[1] "bb"   "bbbc"
```

问号"?"表示前面的字符出现零次或一次。由于这个元符号表示前面字符可以不出现,因此下例中向量 x 包括的各个元素(字符)均被提取:

```
> grep("b?",x=c("a","bc","bb","abc","bcd","f"),value=TRUE)
[1] "a"    "bc"   "bb"   "abc"   "bcd"   "f"
```

星号"*"表示前面的字符出现零次或多次。由于"*"同"?"一样表示前面的词符可以不出现,因而下面的结果与执行 R 命令 grep("b?",x=c("a","bc","bb","abc","bcd","f"),value=TRUE)的结果相同:

```
> grep("b*",x=c("a","bc","bb","abc","bcd","f"),value=TRUE)
[1] "a"    "bc"   "bb"   "abc"   "bcd"   "f"
```

再如:

```
> grep("you.*blood",Text,value=TRUE)
[1] "As you know, blood is the red liquid that flows inside your body."
[2] "When you bleed, you lose blood from your body as a result of injury or illness."
> grep("b*d",x=c("bc","bb","bbd","abc","bcd"),value=TRUE)
[1] "bbd"   "bcd"
```

在第一个命令中,查找模式首先包括 you,后面可以跟任何长度的字符,最后以 blood 结束。同样,在第二个命令中,查找模式首先包括 b,后面可以跟任何长度的字符,最后以 d 结束。

大括号{}表示字符出现的频次范围。{n}表示前面的字符出现 n 次,{n,}表示前面的

字符至少出现 n 次，{,n} 表示前面的字符最多出现 n 次，{n,m} 表示前面的字符出现的次数在 n 和 m 之间。例如：

```
> character<-c("bb","b","abm","bdmn","bc","bbbb","bbbf","bcg")
> grep("b{2}",character,value=TRUE)
[1] "bb"    "bbbb"   "bbbf"
> grep("b{2,}",character,value=TRUE)
[1] "bb"    "bbbb"   "bbbf"
> grep("b{,4}",character,value=TRUE)
[1] "bb"   "b"   "abm"   "bdmn"   "bc"   "bbbb"   "bbbf"   "bcg"
> grep("b{2,4}",character,value=TRUE)
[1] "bb"    "bbbb"   "bbbf"
```

尖角符^用于匹配行首、句首或字符串开头的字符。例如：

```
> grep("^When", Text,value=TRUE)
[1] "When you bleed, you lose blood from your body as a result of injury or illness."
```

以上命令中，"^When"表示在文本 Text 中查找以 When 开头的句子。尖角符^可以和表示选择范围的方括号[]连用。尖角符放在方括号外面表示匹配行(或字符串)必须以紧接在尖角符后面的模式开始。尖角符放在方括号里面表示不使用方括号内的任何字符。试比较：

```
> grep("^[W|I]",Text,value=TRUE)
[1] "When you bleed, you lose blood from your body as a result of injury or illness."
[2] "If someone draws first blood, they have had a success at the beginning of a competition."
> grep("^[WI]",Text,value=TRUE)
[1] "When you bleed, you lose blood from your body as a result of injury or illness."
[2] "If someone draws first blood, they have had a success at the beginning of a competition."
> grep("^[^W|I]",Text,value=TRUE)
[1] "As you know, blood is the red liquid that flows inside your body."
[2] "You sometimes bleed."
[3] "There are some idioms."
[4] "Let us take an example."
```

第一个命令 grep("^[W|I]",Text,value=TRUE) 和第二个命令 grep("^[WI]",Text,value=TRUE) 均从 Text 中查找出以 W 或 I 字母开头的两个句子。第三个命令中的"^[^W|I]"则排除了以 W 和 I 字母开头的句子，查找出满足条件的其他 4 个句子。

我们已经知道，元字符^和 $ 分别用于匹配字符串开头和结尾的位置，而不是匹配特定字符,因而我们又把它们称作锚点(anchors)。就像用船锚固定船只一样,我们用锚点为正则表达式提供匹配的起点或终点。试看下面的几个例子:

```
> grep("bleed.$",Text,value=TRUE)
[1] "You sometimes bleed."
> x<-c("cat","catfish","catastrophe","my favorite cat","wild cat","I love my cat")
> grep("^cat",x,value=TRUE)
[1] "cat"    "catfish"    "catastrophe"
> grep("cat$",x,value=TRUE)
[1] "cat"    "my favorite cat"    "wild cat"    "I love my cat"
> grep("^cat$",x,value=TRUE)
[1] "cat"
```

实际研究中,我们可能需要混合使用元字符。例如,我们对一个文本标注了名词(NOUN)和动词(VERB),将之放在尖括号(〈〉)中,在后续文本统计时希望去除标注内容。我们可以利用 R 基础函数 gsub(),其中 sub 为 substitute 的缩写,表示替换。我们想到使用表达式"〈.*〉"。该表达式的意思是,先查找左括号,接着查找出现零次或多次的任意字符,最后查找右括号。请看下面的例子:

```
> text<-"As you know 〈VERB〉, blood 〈NOUN〉 is the red 〈〉 liquid 〈NOUN〉 that flows 〈VERB〉 inside your body 〈NOUN〉."
> gsub("〈.*〉","",text)
[1] "As you know ."
```

命令 gsub("〈.*〉","",text) 找到第一个符合条件的表达式后就停止了。这种查找最小量的字符匹配,称作懒惰匹配(lazy matching)。就本例而言,上面的结果不是我们希望得到的。我们可以在表达式"〈.*?〉"后添加一个加号(+),告诉 R 查找所有符合条件的表达式,直至匹配完为止。这种查找最大量的字符匹配,称作贪婪匹配(greedy matching)。针对上例,R 命令如下:

```
> gsub("〈.*?〉+","",text)
[1] "As you know, blood is the red liquid that flows inside your body ."
```

需要注意的是,R 对字母大小写敏感。如果我们要在匹配中忽略字母大小写问题,那么可以在函数 grep()或 gsub()中添加变元设置 ignore.case=TRUE。例如:

```
> grep("b+",x=c("A","BC","BB","ABC","bcD"),value=TRUE,ignore.case=TRUE)
[1] "BC"    "BB"    "ABC"    "bcD"
> gsub("b+","*",x=c("A","BC","BB","ABC","bcD"),ignore.case=TRUE)
[1] "A"    "*C"    "*"    "A*C"    "*cD"
```

表 2.1 概括了本节所学的元符号。

表 2.1　常用元字符

元符号	说明
.	匹配任何字符
\\	退出符
|	表示"或",把一个表达式分割成若干字符
+	匹配一次或多次
?	匹配零次或一次
*	匹配零次或多次
( )	匹配整个字符串
{n}	匹配 n 次
{n,}	匹配至少 n 次
{n,m}	匹配至少 n 次,至多 m 次
[ ]	匹配括号内任意字符,如[ab]表示匹配 a 或 b
^	匹配字符串开头
$	匹配字符串结尾

根据表 2.1,我们发现正则表达式 a*匹配 a 零次或多次,a+匹配 a 一次或多次,a? 匹配 a 零次或一次,a{m}匹配 m 个 a。在表 2.1 的元字符中,+、?、*、{n}、{n,}和{n,m}用于限定模式匹配的次数,因而又称作量化符(quantifiers)。

R 自带数据集"state.name"包括美国 50 个州。下面举例说明一些元符号的用法:

```
# 匹配名称包括字母"z"的美国州
> grep(pattern = "z+", state.name,value=TRUE)
[1] "Arizona"
# 匹配名称以字母"A"开头的美国州
> grep(pattern = "^A+", state.name,value=TRUE)
[1] "Alabama"   "Alaska"   "Arizona"   "Arkansas"
# 匹配名称中字母"s"连续出现两次的美国州
> grep(pattern = "s{2}", state.name, value = TRUE)
[1] "Massachusetts"   "Mississippi"   "Missouri"   "Tennessee"
# 匹配名称中至少出现一次"ss"的美国州
> grep(pattern = "(ss)+", state.name,value=TRUE)
[1] "Massachusetts"   "Mississippi"   "Missouri"   "Tennessee"
# 匹配名称中至少包括一次字母组合"as"的美国州
```

```
> grep(pattern = "(as){1,}", state.name, value = TRUE)
[1] "Alaska"    "Arkansas"    "Kansas"    "Massachusetts"    "Nebraska"
[6] "Texas"    "Washington"
# 匹配名称中包括或不包括字母"A"并且以"a"结尾的美国州
> grep(pattern = "A?a$", state.name, value = TRUE)
 [1] "Alabama"    "Alaska"    "Arizona"    "California"    "Florida"    "Georgia"    "Indiana"
 [8] "Iowa"    "Louisiana"    "Minnesota"    "Montana"    "Nebraska"    "Nevada"    "North Carolina"
[15] "North Dakota"    "Oklahoma"    "Pennsylvania"    "South Carolina"    "South Dakota"    "Virginia"
    "West Virginia"
```

2.3.2 字符序列

除了基本的元字符之外,另外一种正则表达式是字符序列(sequences of characters)。R 使用一些速写的字符序列。例如,"\\W"表示匹配词以外的任意字符,在本质上等同于"[^A-Za-z0-9]",其中置于中括号中的尖角符"^"表示匹配[A-Za-z0-9]之外的任意字符。我们在下一节会介绍字符类。"[A-Za-z0-9]"和"[a-z]"这两个字符类分别表示匹配任意大小写字母,"[0-9]"则表示匹配任意数字。例如:

```
> text<-"Text mining is one of the automated techniques used in natural language processing. It converts unstructured text to structured text that a computer can understand."
> strsplit(text,"\\W")
[[1]]
 [1] "Text"          "mining"        "is"          "one"
 [5] "of"            "the"           "automated"   "techniques"
 [9] "used"          "in"            "natural"     "language"
[13] "processing"    ""              "It"          "converts"
[17] "unstructured"  "text"          "to"          "structured"
[21] "text"          "that"          "a"           "computer"
[25] "can"           "understand"

> strsplit(text,"[^A-Za-z0-9]")
[[1]]
 [1] "Text"          "mining"        "is"          "one"
 [5] "of"            "the"           "automated"   "techniques"
 [9] "used"          "in"            "natural"     "language"
[13] "processing"    ""              "It"          "converts"
[17] "unstructured"  "text"          "to"          "structured"
[21] "text"          "that"          "a"           "computer"
[25] "can"           "understand"
```

表 2.2 列出了 R 操作中基本的字符序列。

表 2.2　基本字符序列

字符序列	说明
\\w	匹配任何词
\\W	匹配词以外的任何字符
\\d	匹配任何数字
\\D	匹配数字以外的任何字符
\\b	匹配词边界
\\B	匹配词边界以外的任何字符
\\s	匹配空格符
\\S	匹配空格符以外的任何字符

我们来看一个例子：

```
> text<-c("The war broke out in 1938.","Many people were killed.","It was a disaster!!!")
> gsub("\\b"," * ",text,perl=TRUE)
[1] " * The *  * war *  * broke *  * out *  * in *  * 1938 * ."
[2] " * Many *  * people *  * were *  * killed * ."
[3] " * It *  * was *  * a *  * disaster * !!!"
```

在这个 R 命令中,我们调用替换函数 gsub( ),利用" \\b"把所有的词边界替换为星号(*),perl=TRUE 表示使用与 Perl(一种脚本语言)兼容的正则表达式。但是,我们的目标是得到各个词语,以便开展统计分析。我们想到前面学过的函数 unlist(strsplit(x," ")),但是该函数没有把词语和标点符号分开：

```
> unlist(strsplit(text," "))
[1] "The"     "war"     "broke"   "out"     "in"     "1938."   "Many"    "people"
[9] "were"    "killed." "It"      "was"     "a"      "disaster!!!"
```

如果使用"\\s"来分割本例文本 text 中的词汇,那么得到以下结果：

```
>unlist(strsplit(text,"\\s"))
[1] "The"     "war"     "broke"   "out"     "in"     "1938."   "Many"    "people"
[9] "were"    "killed." "It"      "was"     "a"      "disaster!!!"
```

以上两个字符串分割方法都没有把词语和标点符号分隔开。我们现在尝试利用"\\W"来分割文本中的词语：

```
> words<-unlist(strsplit(text,"\\W"))
> words
[1] "The"     "war"     "broke"   "out"     "in"     "1938"    "Many"    "people"   "were"    "killed"
[11] "It"     "was"     "a"      "disaster" ""      ""
```

以上结果中,有两个空格符" "是分割两个感叹号(!)留下的位置。在文本分析中,如果我们不需要这些空格位置,可以利用贪婪的匹配模式"\\W+",即在"\\W"后用加号"+"排除空格。R 命令和结果如下:

```
> text<-unlist(strsplit(text,"\\W+"))
> text
 [1] "The"    "war"    "broke"    "out"    "in"    "1938"    "Many"    "people"    "were"    "killed"
[11] "It"    "was"    "a"    "disaster"
```

在得到的词语中,"1938"是数字,在文本分析时有时需要排除。我们可以利用"\\d",即:

```
> text1<-gsub("\\d","",text);text1
 [1] "The"    "war"    "broke"    "out"    "in"    ""    "Many"    "people"    "were"    "killed"
[11] "It"    "was"    "a"    "disaster"
```

以上结果中,"1938"被替换成了空格符" ",但是在计算词数时空格符却包含在内。除去空格符可以利用函数 which()和逻辑符"!"(表示否)。在本例中,which(text1!="")返还所有词语的位置,再利用索引符"[]"得到想要的词语:

```
> words<-text1[which(text1!="")];words
 [1] "The"    "war"    "broke"    "out"    "in"    "Many"    "people"    "were"    "killed"    "It"
[11] "was"    "a"    "disaster"
```

我们利用 R 函数 length()便可得到正确的文本词数:

```
> length(words)
[1] 13
```

2.3.3 字符类

上一节提到,R 使用字符序列匹配多个不同字符,第三种正则表达式是字符类或字符集(character class or set)。字符类是包含在中括号([],表示匹配范围)里的字符或字符串,表示匹配其中的任何字符。例如,"[0-9]"表示匹配任何数字,等同于"[0123456789]","[^0-9]"则表示匹配数字以外的任意字符。再如,"[a-z]"表示匹配任何小写字母,"[^a-z]"则表示匹配小写字母以外的任何字符。请看下面的例子:

```
> numbers = c("12345","17-October","I-II-III-V","R 4.2.2")
> grep(pattern = "[0-9]", numbers, value = TRUE)
[1] "12345"    "17-October"    "R 4.2.2"
```

```
> grep(pattern = "[^0-9]", numbers, value = TRUE)
[1] "17-October" "I-II-III-V"  "R 4.2.2"
> transport<-c("car","bike","plane","boat","motor","jeep","ship")
> # look for 'b' or 'p'
> grep(pattern = "[bp]", transport, value = TRUE)
[1] "bike"  "plane"  "boat"  "jeep"  "ship"
> grep(pattern = "[b|p]", transport, value = TRUE)
[1] "bike"  "plane"  "boat"  "jeep"  "ship"
```

另有一类字符类,如[[:alpha:]],使用双中括号,其中的内括号表示类别名,外括号表示范围。这种以双中括号([[ ]])表示的字符类称作 POSIX(Portable Operating System Interface,便携式操作系统界面)字符类。字符类和字符序列是有联系的。例如,"\\W"(匹配词以外的任何字符)等同于"[^[:alnum:]]"或"[^a-zA-Z0-9]",其中的尖角符(^)表示匹配不在列表中的任何字符。试看下面的例子:

```
> text<-c("The war broke out in 1938.","Many people were killed.","It was a disaster!")
> unlist(strsplit(text,"\\W"))
[1] "The"    "war"    "broke"   "out"    "in"    "1938"   "Many"   "people"
[9] "were"   "killed"  "It"     "was"    "a"     "disaster"
> unlist(strsplit(text,"[^[:alnum:]]"))
[1] "The"    "war"    "broke"   "out"    "in"    "1938"   "Many"   "people"
[9] "were"   "killed"  "It"     "was"    "a"     "disaster"
> unlist(strsplit(text,"[^a-zA-Z0-9]"))
[1] "The"    "war"    "broke"   "out"    "in"    "1938"   "Many"   "people"
[9] "were"   "killed"  "It"     "was"    "a"     "disaster"
> x<-c("He loves R programming.","Me too.")
> strsplit(x,"[[:space:]]")
[[1]]
[1] "He"  "loves"  "R"  "programming."
[[2]]
[1] "Me"  "too."
> strsplit(x,"[^[:alnum:]]")
[[1]]
[1] "He"  "loves"  "R"  "programming"
[[2]]
[1] "Me"  "too"
> strsplit(x,"[[:space:]]|\\.")
[[1]]
[1] "He"  "loves"  "R"  "programming"
```

```
[[2]]
[1] "Me"    "too"
> x <- c("3.1. Research questions","3.2. Research design","4. Results and discussion","5. Conclusion")
> gsub("[[:digit:]]+\\.","",x)
[1] "Research questions"   "Research design"   "Results and discussion"   "Conclusion"
```

表2.3列出了一些常见的字符类正则表达式。

表2.3 常用字符类

字符类	说明
[[:alpha:]]	匹配任何字母,即[:lower:]和[:upper:]
[[:alnum:]]	匹配任何字母和数字,即[:alpha:]和[:digit:]
[[:upper:]]	匹配任何大写字母
[[:lower:]]	匹配任何小写字母
[[:digit:]]	匹配任何数字,即0、1、2、3、4、5、6、7、8、9
[[:punct:]]	匹配任何标点符号
[[:graph:]]	匹配任何字符,即[:alnum:]和[:punct:]
[[:space:]]	匹配任何空格
[0-9]	匹配任何数字,同[[:digit:]]
[a-z]	匹配任何小写字母
[A-Z]	匹配任何大写字母
[a-zA-Z0-9]	匹配任何字母和数字,同[[:alnum:]]

在上一节关于文本text的例子中,我们利用字符序列对文本开展形符化(tokenization),即把字符串转化为形符或词。利用字符类对文本开展形符化的R命令和统计分析结果如下:

```
> text<-c("The war broke out in 1938.","Many people were killed.","It was a disaster!!!")
# 词分割
> text1<-strsplit(text,"[[:space:]]")
> text1
[[1]]
[1] "The"    "war"    "broke"    "out"    "in"    "1938."
[[2]]
[1] "Many"    "people"    "were"    "killed."
[[3]]
[1] "It"    "was"    "a"    "disaster!!!"
# 改列表为向量
> text2<-unlist(text1)
```

```
> text2
[1] "The"    "war"     "broke"   "out"    "in"    "1938."  "Many"   "people"
[9] "were"   "killed." "It"      "was"    "a"     "disaster!!!"
# 去除标点
> text3<-gsub("[[:punct:]]","",text2)
> text3
[1] "The"    "war"    "broke"  "out"    "in"    "1938"   "Many"   "people"
[9] "were"   "killed" "It"     "was"    "a"     "disaster"
# 去除数字
> text4<-gsub("[[:digit:]]","",text3)
> text4
[1] "The"    "war"    "broke"  "out"    "in"    ""       "Many"   "people"
[9] "were"   "killed" "It"     "was"    "a"     "disaster"
# 利用函数 which()去除作为剔除字符痕迹的空格,再利用索引符"[]"得到结果
> text5<-text4[which(text4!="")]
> text5
[1] "The"    "war"    "broke"  "out"    "in"    "Many"   "people" "were"
[9] "killed" "It"     "was"    "a"     "disaster"
# 利用函数 tolower()把所有字母改为小写
> tolower(text5)
[1] "the"    "war"    "broke"  "out"    "in"    "many"   "people" "were"
[9] "killed" "it"     "was"    "a"     "disaster"
```

我们也可以使用下面更简洁的 R 命令开展形符化:

```
> text1<-tolower(unlist(strsplit(text,"\\W+")))
> text2<-gsub("[[:digit:]]","",text1)
> text2[which(text2!="")]
[1] "the"    "war"    "broke"  "out"    "in"    "many"   "people" "were"
[9] "killed" "it"     "was"    "a"     "disaster"
```

2.4 R 基础包中的正则表达式函数

使用正则表达式的目的是查找和替换。前面一节介绍了不同类型的正则表达式,并且利用 R 基础函数 strsplit( )、grep( )和 gsub( )做了举例说明。除此之外,R 基础包(base)还包括其他六种查找和替换函数,即 grepl( )、sub( )、regexpr( )、gregexpr( )、regexec( )和 gregexec( )。

函数 grepl( )比函数 grep( )在名称上多个一个字母 l。l 是 logical(逻辑符)的简称,

grepl()查找字符向量中的每个元素,对每个元素返还 TRUE(表示匹配)或者 FALSE(表示不匹配)。例如,我们有下面的一个短文本(改编自维基百科:https://en.wikipedia.org/wiki/North_America),要求用正则表达式查找文本中的数字:

```
> text<-c("North America covers an area of about 24,709,000 square kilometres.","It has about 16.5% of Earth's land area.","North America is the third-largest continent by area, following Asia and Africa, and the fourth by population after Asia, Africa, and Europe.","In 2013, its population was estimated at nearly 579 million people in 23 independent states.")
```

这个文本有 4 个句子,其中的 3 个句子包括数字,每个数字包括多个字符,因而正则表达式应为"\\d+"(允许有多个数字符)或"\\d"。本例的 R 命令如下:

```
> grepl("\\d+",text)
[1] TRUE  TRUE  FALSE  TRUE
> grepl("\\d",text)
[1] TRUE  TRUE  FALSE  TRUE
> grep("\\d+",text)
[1] 1 2 4
> grep("\\d",text)
[1] 1 2 4
```

以上结果显示,若句子中有数字匹配,则函数 grepl()返还 TRUE;若没有数字匹配,则返还 FALSE。函数 grep()返还匹配的位置,如 1、2 和 4 表示在第 1、第 2 和第 4 个句子中有匹配的数字。如果我们要在文本(特别是大文本)中找出这些数字所在的语境,可以利用索引符"[ ]"。例如:

```
> text[grepl("\\d+",text)]
[1] "North America covers an area of about 24,709,000 square kilometres."
[2] "It has about 16.5% of Earth's land area."
[3] "In 2013, its population was estimated at nearly 579 million people in 23 independent states."
```

函数 sub()与 gsub()都为替换函数,不同之处在于 sub()只替换第一次出现的匹配模式,而 gsub()则替换所有出现的匹配模式。数字匹配的正则表达式是"\\d+"或"[[:digit:]]"。试比较:

```
> sub("\\d+","?",text)
[1] "North America covers an area of about ?,709,000 square kilometres."
[2] "It has about ?.5% of Earth's land area."
[3] "North America is the third-largest continent by area, following Asia and Africa, and the fourth by population after Asia, Africa, and Europe."
```

[4] "In ?, its population was estimated at nearly 579 million people in 23 independent states."
> gsub("\\d+","?",text)
[1] "North America covers an area of about ?,?,? square kilometres."
[2] "It has about ?.?% of Earth's land area."
[3] "North America is the third-largest continent by area, following Asia and Africa, and the fourth by population after Asia, Africa, and Europe."
[4] "In ?, its population was estimated at nearly ? million people in ? independent states."

以上结果显示，sub()在每个句子中找到第一个数字时，如找到 24,709,000 中的第一个数字 24，便将之替换为"?"，而其他位置上的数字则保持不变。相比之下，gsub()把在每个句子中查找到的每个数字都替换为"?"。我们利用下面的向量 vector 看一下使用 gsub() 的复杂情况：

> vector<-c("p = 0.045","m = 85.23","s = 10.68","n = 50")

如果我们想要从 vector 中提取字母，排除数字和等号，该使用怎样的正则表达式呢？我们可以利用元字符"^"、字符类"[:alpha:]"或"a-z"保留 vector 中的字母，非字母部分用符号""替换。[^]表示匹配不在中括号里的字符。注意，符号""内部没有空格，增加符号内的空格（如" "）会使字母与空格并存。例如：

> gsub("[^[:alpha:]]", "", vector)
[1] "p" "m" "s" "n"
> gsub("[^a-z]", "", vector)
[1] "p" "m" "s" "n"
> gsub("[^a-z]", " ", vector)
[1] "p   " "m   " "s   " "n   "

如果只提取向量 vector 中的数字，可以使用以下 R 命令：

> gsub("[^0-9\\.]", "", vector)
[1] "0.045" "85.23" "10.68" "50"
> gsub("^[^.]|=|\\s","", vector)
[1] "0.045" "85.23" "10.68" "50"

以上命令表示，从向量 vector 中排除字母和等号，只保留数字和标点"."。

函数 regexpr() 用于在字符串中查找匹配的位置。输出的第一个数字表示在字符串中匹配的初始位置，值为 -1 表示没有发现匹配。例如：

> regexpr("America",text)
[1] 7 -1 7 -1
attr(,"match.length")

```
[1] 7  -1   7  -1
attr(,"index.type")
[1] "chars"
attr(,"useBytes")
[1] TRUE
```

以上结果显示,"America" 在文本 text 中出现两次,分别在第一句和第三句中,属性输出中显示匹配长度(match.length)为 7 个字符,索引类型(index.type)是字符(chars),"useBytes" 为 TRUE 表示匹配按字节进行。函数 gregexpr() 同 regexpr() 一样也用于在字符串中查找匹配,但是每个元素被分开查找,返回一个列表而不是一个向量。试比较下面的例子:

```
> example<-"How much do you earn? 1,000 dollars or 2,000 dollars? I earn 1,500 dollars."
> gregexpr("\\d+",example)
[[1]]
[1] 23 25 40 42 62 64
attr(,"match.length")
[1] 1 3 1 3 1 3
attr(,"index.type")
[1] "chars"
attr(,"useBytes")
[1] TRUE
> regexpr("\\d+",example)
[1] 23
attr(,"match.length")
[1] 1
attr(,"index.type")
[1] "chars"
attr(,"useBytes")
[1] TRUE
```

以上结果显示,gregexpr() 输出的结果是一个列表,在一个字符串中查到 6 个数值。第一个数字在第 23 个字符的位置,长度为 1(值为"1"),第二个数字(在"1,000"中的","号之后)在第 25 个字符的位置,长度为 3(值为"000"),其他以此类推。相比之下,regexpr() 输出的结果是一个向量,且只有一个值,即只报告字符串 example 中第一次出现的数字("1")。

函数 regexec() 和 gregexec() 相似,但是前者返还在每一个字符串中第一次出现的匹配位置,以列表形式给出,而后者则返还在每个字符串中出现的所有匹配位置,以矩阵形式给出。例如:

```
> regexec("\\d+",example)
[[1]]
[1] 23
attr(,"match.length")
[1] 1
attr(,"index.type")
[1] "chars"
attr(,"useBytes")
[1] TRUE
> gregexec("\\d+",example)
[[1]]
     [,1] [,2] [,3] [,4] [,5] [,6]
[1,]  23   25   40   42   62   64
attr(,"match.length")
     [,1] [,2] [,3] [,4] [,5] [,6]
[1,]   1    3    1    3    1    3
attr(,"useBytes")
[1] TRUE
attr(,"index.type")
[1] "chars"
```

2.5 字符串操作数据包 stringr

除了 R 基础包之外，R 外置数据包 stringr 也提供常用字符串操作方法，该数据包由哈德利·威克姆开发（Wickham，2022）。哈德利·威克姆也是著名数据包 ggplot2 的开发者（详见第九章）。在 R 操作界面输入 install.packages("stringr")执行自动安装，也可以指定安装路径，如 install.packages('stringr',lib="C:/Program Files/R/R-4.2.2/library/")。如果在调用 stringr 包中出现有关 rlang 的错误，可以先利用以下代码安装 R 数据包 pak：install.packages("pak")，然后加载该数据包：require(pak)，最后执行以下命令安装 R 数据包 rlang：pkg_install("r-lib/rlang")。注意，在执行 R 代码 pkg_install("r-lib/rlang")时，会提示三个选项，我们选择第一项：1. Have pak unload them before the installation. (Safest option.)，即在"Your choice [1]":后输入 1，继续执行命令即可。

本节利用以下简短的文本 text 介绍 R 数据包 stringr 包括的几个重要函数，即 str_extract_all()、str_remove_all()、str_trim()、str_replace_all()、str_split()、str_detect()、str_to_upper()和 str_to_lower()。本节举例使用的文本如下：

```
text<-"   Once there was a little boy who was _hungry_ for success in sports_.   To him, winning was everything and success was measured by _results.   He succeeded in 7 out of 10 races.        "
```

要从 text 中提取所有包括"success"的个案,需要调用函数 str_extract_all(string, pattern, simplify = FALSE),其中变元 string 是字符串向量,pattern 是查找或匹配模式,simplify=FALSE 表示返还一个字符向量列表,simplify = TRUE 表示返还一个字符矩阵。例如:

```
> require(stringr)
> str_extract_all(text," success")
[[1]]
[1] " success"  " success"
```

如果要计算文本中出现"success"的次数,先调用 R 基础函数 unlist()把列表转化为向量,再调用 R 基础函数 length()即可:

```
> length(unlist(str_extract_all(text," success",simplify=TRUE)))
[1] 2
```

如果要提取 text 中的所有数字,可以利用正则表达式"[:digit:]+"或"\\d+"。为什么要在表达式中使用加号(+)呢? 这里,"+"表示一次或多次匹配数字。如果不使用"+",文本 text 中的"10"就会被分割成"1"和"0"。试比较下面的例子:

```
> str_extract_all(text, "[:digit:]+")
[[1]]
[1] "7"   "10"
> str_extract_all(text, "\\d+")
[[1]]
[1] "7"   "10"
> str_extract_all(text, "[:digit:]")
[[1]]
[1] "7"   "1"   "0"
> str_extract_all(text, "\\d")
[[1]]
[1] "7"   "1"   "0"
```

在文本 text 中,有四处出现文本处理不需要的着重号"_"。要去除这些着重号,可以调用函数 str_remove_all(string, pattern),其中变元 string 是字符串向量,pattern 是匹配模式。如果匹配模式是正则表达式"[[:punct:]]",那么 text 中的所有标点符号都会被去除。如果只去除着重号"_",那么只需使用表达式"_":

```
> str_remove_all(text," _ ")
[1] "     Once there was a little boy who was hungry for success in sports.    To him, winning was everything and success was measured by results.    He succeeded in 7 out of 10 races.    "
```

细心的读者不难发现，文本 text 中有多余的空格。要删除多余的空格，则需调用函数 str_trim(string, side=c("both", "left", "right"))，其中变元 string 是字符串向量，side 表示删除字符串开头或结尾多余空格的位置，如左边(left)、右边(right)或两边(both)。在这个例子中，R 命令和统计分析结果如下：

```
> str_trim(text, side="both")
[1] "Once there was a little boy who was _hungry_ for success in sports_.    To him, winning was everything and success was measured by _results.    He succeeded in 7 out of 10 races."
```

以上结果显示，调用函数 str_trim() 删除了字符串开头和结尾多余的空格，但是" To him" 和" He succeeded" 前面多余的空格并未删除。

一个解决方法是在调用函数 str_trim() 之后，再调用替换函数 str_replace_all()，并使用正则表达式" \\s+"。这个替换函数的结构是 str_replace_all(string, pattern, replacement)，其中变元 string 是字符串向量，pattern 是匹配模式，replacement 指替换的字符向量。对 text 的操作命令和结果如下：

```
> text1<-str_trim(text, side="both")
> str_replace_all(text1, "\\s+", " ")
[1] " Once there was a little boy who was _hungry_ for success in sports_. To him, winning was everything and success was measured by _results. He succeeded in 7 out of 10 races."
```

对字符串切分或形符化的函数是 str_split(string, pattern, n=Inf, simplify=FALSE)，其中变元 string 是字符串向量，pattern 是匹配模式，n = Inf 指函数默认分割所有的形符，simplify=FALSE 指函数默认返还一个字符向量列表，simplify=TRUE 则返还一个字符矩阵。我们对文本 text 开展形符化时可以使用前面学过的正则表达式" \\W"，以便分割出所有形符或词。如果要把分割结果由列表形式改为向量形式，那么可以调用 R 基础函数 unlist()。R 命令和分割结果如下：

```
> text_toks<-unlist(str_split(text,"\\W"))
> text_toks
 [1] ""          ""          "Once"      "there"     "was"       "a"
 [7] "little"    "boy"       "who"       "was"       "_hungry_"  "for"
[13] "success"   "in"        "sports_"   ""          ""          ""
[19] "To"        "him"       ""          "winning"   "was"       "everything"
[25] "and"       "success"   "was"       "measured"  "by"        "_results"
[31] ""          ""          ""          "He"        "succeeded" "in"
[37] "7"         "out"       "of"        "10"        "races"     ""
[43] ""          ""
```

以上结果显示,文本中所有的词都被正确地析出,在切分后标点符号和多余的空格以空号("")显示。

在开展形符化时,要删除切分后留下的标点符号和多余的空格痕迹,可以调用 stringr 中的函数 str_split(string, boundary("word")),其中变元 string 是字符串向量,boundary("word") 表示以词(word)切分。例如:

```
> unlist(str_split(text, boundary("word")))
 [1] "Once"      "there"     "was"       "a"         "little"    "boy"
 [7] "who"       "was"       "_hungry_"  "for"       "success"   "in"
[13] "sports_"   "To"        "him"       "winning"   "was"       "everything"
[19] "and"       "success"   "was"       "measured"  "by"        "_results"
[25] "He"        "succeeded" "in"        "7"         "out"       "of"
[31] "10"        "races"
```

如果我们要以句子而非词为单位切割字符串,那么同样可以调用函数 str_split(),只是用 boundary("sentence") 替代 boundary("word")。例如:

```
> unlist(str_split(text, boundary("sentence")))
[1] "    Once there was a little boy who was _hungry_ for success in sports_.   "
[2] "To him, winning was everything and success was measured by _results.   "
[3] "He succeeded in 7 out of 10 races.   "
```

如果要查找字符向量中的某个形符,可以调用 R 函数 str_detect(string, pattern, negate=FALSE),其中变元 string 是字符串向量,pattern 是匹配模式,negate=FALSE 表示返还匹配的元素。若 negate=TRUE,则表示返还非匹配的元素。函数 str_detect() 返还逻辑符 TRUE 和 FALSE。若要提取匹配元素,可以使用索引符[ ]。例如,提取向量 text_toks 中的词"success"可以执行以下 R 命令:

```
> text_toks[str_detect(text_toks,"success")]
[1] "success"  "success"
```

如果要把动词"succeeded"也包含在匹配词中,可以使用以下 R 命令:

```
> text_toks[str_detect(text_toks,"succe(ss|.*d$)")]
[1] "success"   "success"   "succeeded"
```

把所有的字符转化为大写字母或小写字母,可以调用函数 str_to_upper(string) 或 str_to_lower(string),其中的 string 为字符串或字符向量。这两个函数在功能上分别等同于 R 基础函数 toupper() 和 tolower()。例如:

```
> str_to_upper(text_toks)
```

[1] " "	" "	"ONCE"	"THERE"	"WAS"	"A"	
[7] "LITTLE"	"BOY"	"WHO"	"WAS"	"_HUNGRY_"	"FOR"	
[13] "SUCCESS"	"IN"	"SPORTS_"	" "	" "	" "	
[19] "TO"	"HIM"	" "	"WINNING"	"WAS"	"EVERYTHING"	
[25] "AND"	"SUCCESS"	"WAS"	"MEASURED"	"BY"	"_RESULTS"	
[31] " "	" "	" "	"HE"	"SUCCEEDED"	"IN"	
[37] "7"	"OUT"	"OF"	"10"	"RACES"	" "	
[43] " "	" "					

2.6 语境中的关键词检索

语境指与关键词在某个长度范围内共同出现的词语。在语料库研究中,关键词检索的长度范围称作窗口(windows)。例如,窗口长度为 5 表示在关键词(即检索词)左边和右边出现的词语数。语境为关键词提供了关键词使用的语言环境。探索语料库或文本中词语使用的特点和模式需要考察词语使用的语境。语境中的关键词检索(Keyword in Context, KWIC)是常用的自然语言处理方法。本节以我们保存在本地 D:\Rpackage 文件夹中的纯文本文件为例介绍语境中的关键词检索方法,保存的文件名为 story.txt。这个文本是 2019 年全国英语专业四级口语测试中的听后复述故事。故事讲述一个在体育比赛中渴望获胜的小男孩参加三次赛跑,第一次比赛赢了另外两个小男孩,第二次比赛赢了两位长者,第三次比赛和两位长者携手穿过终点线。故事原文如下:

Once there was a little boy who was hungry for success in sports. To him, winning was everything and success was measured by results. One day, the boy took part in a race in his village, and his competitors were two other young boys. A large crowd gathered to watch this sporting spectacle, and a wise old man, upon hearing about the boy, also came from afar to watch. Unsurprisingly the boy was the winner. The crowd cheered and waved at him. The little boy felt proud and important. The wise man, however, remained still and calm, showing no excitement. "Another race, another race!" pleaded the little boy, hoping to impress the wise man. The wise old man stepped forward and presented the little boy with two new competitors, a frail granny and a blind man. This time the boy was the only one who finished the race, the other two being left standing at the starting line. The boy raised his arms in delight. The crowd, however, was silent, with no cheering at all. "What happened? Why is no one cheering my success?" he asked the wise old man. "Race again," replied the wise man. "This time, finish together, all three of you. Finish together." The little boy stood between the blind man and the frail granny and took them by the arm. The race began and the little boy walked slowly, so the other two could also move together toward the finishing line. They crossed it at the same time. Now the crowd cheered like a thunderstorm. The wise man smiled, gently nodding his head. The little boy felt puzzled. "I don't understand, grandpa," asked the little boy. "Who is the crowd cheering for? Which one of us three won?" The wise old man looked into the boy's eyes and replied softly, "Remember, little boy: for this race you have won much more than in any other race you ran before, and for this race, the crowd is cheering not for any winner!"

我们想要知道听力文本如何使用关键词"cheer"表现观众的反应,检索的窗口为 5 个词,即把关键词前后出现的 5 个词作为语境。在开展语境中的关键词检索之前,首先要把文本文件调入 R 操作界面。根据第一章所学,我们利用 R 基础函数 scan( ) 把 story. txt 导入 R:

```
> setwd("D:/Rpackage")
> story<-scan("story.txt", what = "character", sep = "\n", quiet = TRUE)
```

我们已经把文档以一个长字符串的形式调入 R,在检索之前需要开展形符化。利用本章前面所学的函数 strsplit( )、unlist( ) 和正则表达式"\\s"(为了保留标点符号)得到字符串分割后的字符向量(包括词语和标点符号),并用函数 head( ) 显示处理结果的开头部分:

```
> story_w<-unlist(strsplit(story, "\\s"))
> head(story_w)
[1] "Once"    "there"    "was"    "a"    "little"    "boy"
```

检索的关键词"cheer"可以是名词单数或复数,也可以是现在分词、过去式或过去分词,还可以是大写形式,如"cheers""cheered""cheering"和"Cheered"。就本例而言,"cheer"的变化形式包括"cheering"和"cheered",只需检索这两种形式即可。但是在实际研究中,我们使用的语料库可能会很大,因而需要全面检索关键词的各种形式。由于"cheer"词形变化规则,我们只需用表达式"cheer",利用 ignore. case = TRUE 忽略大小写问题。R 代码示例如下:

```
> cheer < - c ("cheer", "cheers", "cheering", "cheered", "Cheer", "Cheers", "Cheering", "Cheered")
> grep("cheer", cheer, value = TRUE, ignore.case = TRUE)
[1] "cheer"    "cheers"    "cheering"    "cheered"    "Cheer"    "Cheers"    "Cheering"    "Cheered"
```

在有些情况下,编写正则表达式要复杂一些。例如,我们有以下一个词串,要检索与"血"有关的词语:"blood""bloodless""bleed""bleeding""bled""stumbled"和"wrinkled"。这些词语中只有"stumbled"和"wrinkled"与"血"没有直接的关系。尝试用以下代码得到结果:

```
> example<-c("blood", "bloodless", "bleed", "bleeding", "bled", "stumbled", "wrinkled")
> grep("bl(oo|ee?)d", example, value=TRUE)
[1] "blood"    "bloodless"    "bleed"    "bleeding"    "bled"    "stumbled"
```

在以上代码中,(oo|ee?)表示 bl 后面紧跟着"oo",或者至少跟一个"e"(因为"?"表示第二个"e"可出现,也可不出现),其后再紧跟 d。这个表达式排除了"wrinkled",但是没有排除"stumbled",因为"stumbled"包括"bled",满足表达式的条件。与"stumbled"相比,其他词均以"b"开头,可以利用元符号"^"。R 代码和统计分析结果如下:

```
> grep("^bl(oo|ee?)d",example,value=TRUE)
[1] "blood"    "bloodless"    "bleed"    "bleeding"    "bled"
```

我们回到对关键词"cheer"的语境检索。先利用函数 grep() 确定关键词（又称节点，node）的位置：

```
> node.positions<-grep("cheer",story_w,ignore.case=TRUE)
```

然后，利用 for 循环检索"cheer"所有语境，R 代码如下：

```
for(i in 1:length(node.positions)){
start<-node.positions[i]-5
end<-node.positions[i]+5
output<-story_w[start:end]
node<-story_w[node.positions[i]]
output[which(output==node)]<-paste("|",node,"|",sep="")
cat(output,"\n")
}
```

执行以上 R 代码，会得到以下结果：

```
was the winner. The crowd |cheered| and waved at him. The
however, was silent, with no |cheering| at all. "What happened? Why
happened? Why is no one |cheering| my success?" he asked the
same time. Now the crowd |cheered| like a thunderstorm. The wise
boy. "Who is the crowd |cheering| for? Which one of us
this race, the crowd is |cheering| not for any winner!" NA
```

在以上 R 代码中，for 循环从第 1 次开始执行大括号（{}）中的命令，直至执行的次数等于节点位置的长度（length(node.positions)）。循环的开始值和结束值根据窗口大小设定为节点位置顺序值减 5 个和加 5 个词符，输出结果（output）是节点窗口长度范围包含的语境词（story_w[start:end]）。在代码主体中利用合并函数 cat() 增加命令 cat(output,"\n") 即可得到检索结果。为了凸显节点词（检索关键词），我们编写了其他命令。具体做法是，利用匹配函数 which() 定位各个节点，利用粘贴函数 paste() 在节点前后添加分割线"|"，且字符间不留空格（sep=""）。最后，利用函数 cat() 把结果打印出来，不同检索行分行显示（\n）。在最后一个输出行中有一个 NA（Not Available）。我们看到 cheering 之后只有 4 个词，没有 5 个词，语境窗口长度超出了文本边界。

检索行显示，小男孩第一次获胜后，观众为之喝彩和挥手致意。但是在小男孩第二次获胜后，观众却鸦雀无声。当小男孩与两位老人携手穿过终点线后，观众却掌声如雷。小男孩不明原因，向智慧老人询问。

第三章 文本基础统计

本章介绍语料库(corpus)研究使用的一个重要数据包 koRpus,主要依据这个数据包介绍文本基础统计,包括计算描述性统计量和词汇密度,制作文本词频分布表和绘制词频分布图形。本章调用的数据包括 koRpus、SnowballC、tm、lattice、RColorBrewer 和 wordcloud。

3.1 数据包 koRpus 的安装与调试

R 数据包 koRpus 由梅克·米夏尔克开发,当前版本号为 0.13-8(Michalke,2021)。这个数据包提供一整套文本分析工具,主要特色是词类(word class,即词性)标注(POS tagging)、词汇多样性(lexical diversity)不同测量方法和文本可读性(readability)多种测量指标。

第一章介绍了 R 数据包的安装方法。既可以在 R 工作界面输入 R 命令 install.packages("koRpus")选择默认安装路径进行安装,也可以指定安装路径进行安装,如 install.packages("koRpus",lib="C:/Program Files/R/R-4.2.2/library"),选择就近的站点在线安装。安装完成后,利用 R 命令 require(koRpus)调用 koRpus 数据包。利用 koRpus 包处理文本时,需要先安装和调用语言支持数据包。如果我们选择英语作为工作语言,在 R 界面中输入命令 install.koRpus.lang("en",lib="C:/Program Files/R/R-4.2.2/library")或采用默认的安装路径。安装完成后利用 R 命令 require(koRpus.lang.en)调用英语数据包。

为了实现数据包的形符化和词类标注功能,需要安装第三方软件 TreeTagger。我们使用的操作系统是 Windows 11。在 Windows 系统上安装 TreeTagger 比较麻烦。具体的安装步骤如下:

1. 安装 Perl 解释器。如果你的系统尚未安装,可以进入网站 http://www.perl.org,下载 Strawberry Perl 软件并安装。

2. 进入网站 https://www.cis.uni-muenchen.de/~schmid/tools/TreeTagger/,下载 TreeTagger 软件,解压 zip 文件,把解压后的文件放在本地硬盘 C:\的根目录下,文件夹名为 TreeTagger。

3. 下载 TreeTagger 软件所需语言的参数文件 english.par.gz 和 english-chunker.par.gz。利用解压软件对其解压,把解压后文件放在子目录 TreeTagger\lib 里。

4. 把路径 C:\TreeTagger\bin 增加到路径(PATH)环境变量中。在本地电脑属性中点击"高级系统设置",在"系统属性"面板中点击"环境变量"。在环境变量面板选中 OneDrive 所在行,点击"新建"栏,在"新建用户变量"窗口的"变量名"栏输入:TreeTagger,

在"变量值"栏输入：C:\TreeTagger\bin,然后点击"确定"。

5. 右击桌面开始菜单,选择"运行(R)"。如果使用快捷键,可以同时按 Windows+R 键启动"运行"。在"打开(O)"栏输入：cmd,然后点击"确定"。在命令执行窗口设置：PATH=C:\TreeTagger\bin;%PATH%,按回车键执行。在新行输入：cd c:\TreeTagger,并按回车键,然后关闭窗口,安装结束。

成功安装后,就可以调用函数 treetag( )对文本开展形符化和词类标注了。用 tagged. text 表示标注后存储的文件名,treetag( )的变元设置如下：

```
tagged. text <- treetag(
  "~/···/. txt",
  treetagger="manual",
  lang="en",
  TT. options=list(
    path="~/···/treetagger",
    preset="en"
  )
)
```

在以上设置中,第一个变元显示包括路径的文本文件名,lang="en"表示调用英语支持包,preset="en"表示指定英语语言作为预设语言,treetagger="manual"表示使标注器的安装位置指向 TT. options 中的位置(即 treetagger 的安装路径)。

为了检查数据包和软件安装是否正确,我们把新建文档 tk. txt 放在 D:\Rpackage 文件夹中,文本内容为：This package aims to be a versatile tool for text analysis. R 命令和执行结果如下：

```
> require(koRpus)
> require(koRpus. lang. en)
> tk. text<-treetag("D:/Rpackage/tk. txt",treetagger
="manual",lang="en",TT. options=list(path= "C:/TreeTagger",preset="en"))
> tk. text
```

	doc_id	token	tag	lemma	lttr	wclass	desc	stop	stem	idx	sntc
1	tk. txt	This	DT	this	4	determiner	NA	NA	NA	1	1
2	tk. txt	package	NN	package	7	noun	NA	NA	NA	2	1
3	tk. txt	aims	VVZ	aim	4	verb	NA	NA	NA	3	1
4	tk. txt	to	TO	to	2	to	NA	NA	NA	4	1
5	tk. txt	be	VB	be	2	verb	NA	NA	NA	5	1
6	tk. txt	a	DT	a	1	determiner	NA	NA	NA	6	1
7	tk. txt	versatile	JJ	versatile	9	adjective	NA	NA	NA	7	1
8	tk. txt	tool	NN	tool	4	noun	NA	NA	NA	8	1
9	tk. txt	for	IN	for	3	preposition	NA	NA	NA	9	1
10	tk. txt	text	NN	text	4	noun	NA	NA	NA	10	1
11	tk. txt	analysis	NN	analysis	8	noun	NA	NA	NA	11	1
12	tk. txt	.	SENT	.	1	fullstop	NA	NA	NA	12	1

显示的结果说明 TreeTagger 安装成功。在输出结果中，doc_id 是文档编号，token 是文本分割后的形符，tag 为标注码，wclass 为词性。例如，"package" 为名词（noun），标注码为 NN。词目（lemma）指词典中的条目，即词条，包括词的形态变化，如词目"run"包括的形符为："run""runs""running"和"ran"。本例的一个句子包括的词目如下："this""package""aim""to""be""a""versatile""tool""for""text""analysis""."，从中可以发现只有词形"aims"转化成了词目"aim"，其他词形均未发生变化。字母组合 lttr 指字符数，如"package"包括 7 个字母。

词类描述（desc）、停用词（stop）和词干（stem）三列没有显示结果。如果我们在函数 treetag( ) 中增加变元 add.desc = TRUE，那么执行 R 命令会得到以下词类描述：

```
> tk.text<-treetag("D:/Rpackage/tk.txt",treetagger
  ="manual",lang="en",TT.options=list(path=
  "C:/TreeTagger",preset="en"),add.desc=TRUE)
> tk.text
```

	doc_id	token	tag	lemma	lttr	wclass	desc	stop	stem	idx	sntc
1	tk.txt	This	DT	this	4	determiner	Determiner	NA	NA	1	1
2	tk.txt	package	NN	package	7	noun	Noun, singular or mass	NA	NA	2	1
3	tk.txt	aims	VVZ	aim	4	verb	3rd person singular present (-s form) of lexical verbs	NA	NA	3	1
4	tk.txt	to	TO	to	2	to	to	NA	NA	4	1
5	tk.txt	be	VB	be	2	verb	Verb, base form of "to be"	NA	NA	5	1
6	tk.txt	a	DT	a	1	determiner	Determiner	NA	NA	6	1
7	tk.txt	versatile	JJ	versatile	9	adjective	Adjective	NA	NA	7	1
8	tk.txt	tool	NN	tool	4	noun	Noun, singular or mass	NA	NA	8	1
9	tk.txt	for	IN	for	3	preposition	Preposition or subordinating conjunction	NA	NA	9	1
10	tk.txt	text	NN	text	4	noun	Noun, singular or mass	NA	NA	10	1
11	tk.txt	analysis	NN	analysis	8	noun	Noun, singular or mass	NA	NA	11	1
12	tk.txt	.	SENT	.	1	fullstop	Sentence ending punctuation	NA	NA	12	1

以上结果表明，在 desc 列，有些词类的描述比 wclass 列更详细，如 aims 为实义动词第三人称单数现在式，为末尾加 s 的形式（3rd person singular present (-s form) of lexical verbs）。停用词指统计分析中被排除在分析之外的词，通常包括冠词、代词、介词和连词等。我们在这里没有设置停用词表。词干通常指除去词的结尾部分，如 leafs 的词干是 leaf，leaves 的词干是 leav。在本例中，如果你安装了数据包 SnowballC，那么可以调用这个数据包中的函数 wordStem( )（在 R 函数 treetag( ) 中增加变元 stemmer = SnowballC::wordStem）得到以下结果："Thi""packag""aim""to""be""a""versatil""tool""for""text""analysi"和"."，其中一些词干，如"Thi"和"packag"，不再是英文词。注意，在 SnowballC::wordStem 中的双冒号"::"是除函数 require( ) 和 library( ) 之外的另外一种数据包调用方式。有些自然语言处理研究使用词干化方法，但是在语言学研究中词目化比词干化更常用。最后两列是索引（idx，表示字符在文本中的位置）和句子序号（sntc）。本例只包含一个句子，所以句子序号都是 1。

3.2 描述性统计

本节和后面各节使用的例子是 2019 年全国英语专业四级口语测试中的听后复述故事(详见第二章)。故事文件名为 story.txt,以纯文本格式保存在本地 D:\Rpackage 文件夹中。

koRpus 通过形符化把字符串转化为字符(包括词语和标点符号),因而使用这个数据包计算文本词汇基本统计量很方便。利用数据包 koRpus 中的函数 treetag() 得到标注文本,再利用函数 describe() 计算基本统计量,其中的主要变元是 koRpus 对象。计算 story.txt 文本基本统计量的 R 命令和统计分析结果(保留两位小数)如下:

```
> require(koRpus)
> require(koRpus.lang.en)
> story.tag<-treetag("D:/Rpackage/story.txt",treetagger="manual",lang="en",TT.options=list(path="C:/TreeTagger",preset="en"))
> describe(story.tag)
$ all.chars
[1] 1847
$ lines
[1] 1
$ normalized.space
[1] 1846
$ chars.no.space
[1] 1512
$ punct
[1] 74
$ digits
[1] 0
$ letters
   all   11   12   13   14   15   16   17   18   19  110  111  112  113  114
  1438    8   30  125   60   33   28   26   16    4    3    2    1    0    1
$ letters.only
[1] 1438
$ char.distrib
                    1          2          3          4          5          6          7
num            80.0000    30.000000  125.00000    60.00000    33.00000    28.000000   26.000000
cum.sum        80.0000   110.000000  235.00000   295.00000   328.00000   356.000000  382.000000
cum.inv       329.0000   299.000000  174.00000   114.00000    81.00000    53.000000   27.000000
pct            19.5599     7.334963   30.56235    14.66993     8.06846     6.845966    6.356968
```

```
cum. pct        19.5599    26.894866   57.45721    72.12714    80.19560    87.041565   93.398533
pct. inv        80.4401    73.105134   42.54279    27.82786    19.80440    12.958435    6.601467
                     8           9          10          11          12          13          14
num             16.000000   4.0000000   3.0000000   2.0000000   1.0000000   0.0000000   1.0000000
cum. sum       398.000000 402.0000000 405.0000000 407.0000000 408.0000000 408.0000000 409.0000000
cum. inv        11.000000   7.0000000   4.0000000   2.0000000   1.0000000   1.0000000   0.0000000
pct              3.911980   0.9779951   0.7334963   0.4889976   0.2444988   0.0000000   0.2444988
cum. pct        97.310513  98.2885086  99.0220049  99.5110024  99.7555012  99.7555012 100.0000000
pct. inv         2.689487   1.7114914   0.9779551   0.4889976   0.2444988   0.2444988   0.0000000
$ lttr.distrib
                     1           2           3           4           5           6           7
num              8.000000   30.000000  125.00000    60.00000    33.000000   28.000000   26.000000
cum. sum         8.000000   38.000000  163.00000   223.00000   256.000000  284.000000  310.000000
cum. inv       329.000000  299.000000  174.00000   114.00000    81.000000   53.000000   27.000000
pct              2.373887    8.902077   37.09199    17.80415     9.792285    8.308605    7.715134
cum. pct         2.373887   11.275964   48.36795    66.17211    75.964392   84.272997   91.988131
pct. inv        97.626113   88.724036   51.63205    33.82789    24.035608   15.727003    8.011869
                     8           9          10          11          12          13          14
num             16.000000    4.000000   3.0000000   2.0000000   1.0000000   0.0000000   1.0000000
cum. sum       326.000000  330.000000 333.0000000 335.0000000 336.0000000 336.0000000 337.0000000
cum. inv        11.000000    7.000000   4.0000000   2.0000000   1.0000000   1.0000000   0.0000000
pct              4.747774    1.186944   0.8902077   0.5934718   0.2967359   0.0000000   0.2967359
cum. pct        96.735905   97.922849  98.8130564  99.4065282  99.7032641  99.7032641 100.0000000
pct. inv         3.264095    2.077151   1.1869436   0.5934718   0.2967359   0.2967359   0.0000000
$ words
[1] 337
$ sentences
[1] 31
$ avg.sentc.length
[1] 10.87097
$ avg.word.length
[1] 4.267062
$ doc_id
[1] "story.txt"
```

在以上结果中,$ all.chars 计算所有字符,包括词、标点符号和空格。$ lines 计算行数,本例只有一个字符串。normalized.space 类似于 all.chars,但是空格字符串(含换行)计作一个字符。$ chars.no.space 计算不含空格的字符数。$ punct 计算标点符号数。$ digits 计算数字数。$ letters 计算字母总数和每个字母长度上的词数,如11表示第一列只有一个字母组成的单词数为8个,即"a"出现6次、"A"和"I"各出现1次。$ letters.only 只计算字母总数(不含数字)。$ char.distrib 和 $ lttr.distrib 分别计算字符分布和字母分布,如由两个字母组成的单词数(num)为30,即"in"出现5次,"at"出现4次,"is"出现3次,"no""to"各出现3次,"by""of"各出现2次,"do""he""it""my""so""To""nt"和"us"各出现1次。由一个或两个字母组成的单词数累计(cum.sum)为38,剩余词数累计为(cum.inv)为299,由两个字母组成的单词数占比(pct)约为8.9%,由一个和两个字母组成的单词数累计占比约为11.3%,剩余词数占比(pct.inv)约为88.7%。$ words 计算词数,$ sentences 计算句子数,$ avg.sentc.length 计算平均句长,$ avg.word.length 计算平均词长,$ doc_id 为文档编号。

3.3 词汇密度

词汇密度(Lexical Density,LD)指文本中的实词(lexical words)数占整个文本词数的比率,是词汇丰富性(lexical richness)指标之一。本节以纯文本格式保存在本地 D:\Rpackage 文件夹中的文本文档 story.txt 为例,介绍词汇密度计算方法。

词汇密度的计算关键在于确定实词。但是英语实词量大,且具有开放性,因而很难列举穷尽。切合实际的做法是列举功能词(function words)。功能词是相对封闭的词类,包括冠词、代词、连词和介词等。R 数据包 tm 中的 stopwords("en")包括174个英语常用停用词表中,如:"i""me""my""myself""we""our""ours""ourselves""than""too"和"very"等。在 R 操作界面中输入命令 tm::stopwords("en")可以看到所有的停用词。注意,执行本命令前需要安装数据挖掘(text mining)包 tm。当然,这个词表不算很完整,包括了一些常用实词,如副词 once 和动词 doing,未包括代词 us 和介词 upon 等。我们现在以该词表作为停用词表计算词汇密度,后面再考虑把代词 us 和介词 upon 等词作为停用词。

在编写 R 代码时,首先要利用函数 treetag()提取文本中的各个字符(包括词和标点符号)。为了在提取过程中自动记录停用词,我们需要在 treetag()的括号中增加变元 stopwords=tm::stopwords("en")。通常情况下,自然语言处理不考虑字母大小写问题,可以利用 R 基础函数 tolower()把大写字母统一转化为小写字母。由于得到的字符包括词和标点符号,需要利用函数 gsub()和正则表达式"[[:punct:]]"除去标点符号和替换字符。再利用 R 基础函数 which()得到词向量,并利用 R 基础函数 length()计算文本总词数。要计算停用词的数量,需利用 R 基础函数 sum()计算输出表(story.tag)中 stop 所在列结果显示为 TRUE 的逻辑符数量。有实际意义词(主要为实词)总数等于文本总词数减去停用词数量。最后,将有实际意义的总词数除以文本总词数得到词汇密度值。本例完整 R 代码如下:

```
require(koRpus)
require(koRpus.lang.en)
story.tag<-treetag("D:/Rpackage/story.txt",treetagger="manual",
```

```
lang="en",TT.options=list(path="C:/TreeTagger",preset="en"),
stopwords=tm::stopwords("en"))
tokens<-tolower(story.tag[,"token"])
rm_puncts<-gsub("[[:punct:]]","",tokens)
words<-rm_puncts[which(rm_puncts!="")]
story_len<-length(words)
stop_w<-sum(story.tag[,"stop"])
content<-story_len-stop_w
LD<-content/story_len;LD
```

执行以上 R 代码,得到词汇密度值为 0.563 798 2。为了使计算尽可能准确,我们每执行一条代码都会查看处理结果。具体而言,在执行 R 对象"words"后查看结果发现,原文中的"don't"转化为"do"和"nt"(词类标注为副词),"boy's"转化为"boy"和"s"(词类标注为所有格)。我们可以接受这些处理方式。

把哪些词视作有实际意义的词计算词汇密度具有一定的主观性,依研究目的而定。如果你对数据包 tm 提供的停用词表不满意,可以对之进行调整或另建词表。例如,把"another""toward""upon"和"us"作为增加的停用词重新计算词汇密度,我们只需把停用词变元重新设置为:stopwords=c(tm::stopwords("en"),"another","toward","upon","us")。本例完整 R 命令和计算结果如下:

```
> story.tag<-treetag("D:/Rpackage/story.txt",treetagger
  ="manual",lang="en",TT.options=list(path="C:/TreeTagger",preset="en"),stopwords=c(tm::
  stopwords("en"),"another","toward","upon","us"))
> tokens<-tolower(story.tag[,"token"])
> rm_puncts<-gsub("[[:punct:]]","",tokens)
> words<-rm_puncts[which(rm_puncts!="")]
> story_len<-length(words)
> stop_w<-sum(story.tag[,"stop"])
> content<-story_len-stop_w
> LD<-content/story_len;LD
[1] 0.5489614
```

以上结果显示,由于增加了几个停用词,词汇密度值略有减小。

我们可以利用数据包 koRpus 中的函数 treetag()提供的词类标注计算不同词类上的词汇密度。treetag()标注准确性很高,但是并非百分之百正确,如在本例文本词类标注中 everything 标为 NN(单数或物质名词),实际为代词(pronoun)。如果接受少量词的标注存疑,使用函数 treetag()就很方便。treetag()把名词标注为 NN 和 NNS(复数名词)。

计算文本中名词密度的 R 命令和统计分析结果如下:

```
> word_tag<-story.tag[,"tag"]
> Nouns<-c(word_tag[which(word_tag=="NN")],
  word_tag[which(word_tag=="NNS")])
> N_LD<-length(Nouns)/story_len; N_LD
[1] 0.2195846
```

另外一种计算名词密度的方法是利用数据包 koRpus 中的函数 summary( ),其中的主要变元是 koRpus 对象。例如,执行 R 命令 summary(story.tag)得到以下统计分析结果(只显示词类分布前 10 排数据):

```
> summary(story.tag)
  Sentences: 31
  Words:    337 (10.87 per sentence)
  Letters:  1438 (4.27 per word)
  Word class distribution:
              num      pct
  noun         74   21.9584570
  verb         63   18.6943620
  determiner   57   16.9139466
  adjective    40   11.8694362
  preposition  29    8.6053412
  adverb       24    7.1216617
  pronoun      20    5.9347181
  conjunction  13    3.8575668
  number       10    2.9673591
  to            4    1.1869436
```

以上结果显示,story.txt 文本长度为 337 个词,包括 31 个句子,平均句长为 10.87,字母总数为 1 438 个,平均词长约为 4 个字母。在词类分布(word class distribution)中,名词数为 74,占比约为 21.96%,与上面的计算结果相同。

3.4 词频表

由于 koRpus 包能够通过形符化把字符串转化为字符,因而使用这个数据包得到文本词频表很方便。本节介绍三种词频表。第一种词频表利用词符。第二种词频表利用停用词表。第三种词频表利用停用词表和词目(lemma,即词条)。

3.4.1 利用词符的词频表

如果我们只是计算文本词符表,那么可以使用 koRpus 数据包提供的函数 tokenize

("~/···/.txt",lang="en"),其中第一个变元显示包括路径的文本文件名,第二个变元 lang="en"指调用英语支持包。我们先利用函数 tokenize()把 story.txt 转化为字符,然后利用函数 tolower()将所有字母转化为小写字母,最后利用函数 gsub()和 which()得到所有的词语。初始 R 代码如下:

```
require(koRpus)
require(koRpus.lang.en)
token.tag<-tokenize("D:/Rpackage/story.txt",lang="en")
tokens<-tolower(token.tag[,"token"])
rm_puncts<-gsub("[[:punct:]]","",tokens)
words<-rm_puncts[which(rm_puncts!="")]
```

执行以上 R 代码,查看结果发现,原文中的"don't"转化为"don"和"t"。一种调整方式是把"I don't"转化为"do"和"n't"。进行以上调整的 R 代码如下:

```
words[274]<-"do"
words[275]<-"n't"
```

执行以上 R 代码,在得到各个词符(word tokens)后,利用 R 基础函数 table()计算每个词符出现的频次,并利用 R 基础函数 data.frame()把表格形式转化为数据框,最后利用 R 基础函数 order()将各个词符按照频次降序排列(按频次排序前词符按字母排序)。本例的 R 操作命令和统计分析部分结果如下:

```
> word.table<-table(words)
> word.freq<-data.frame(word.table)
> word.freq<-word.freq[order(-word.freq$Freq),]
> word.freq[1:20,]
         words    Freq
119        the     37
19         boy     15
7          and     13
71         man     10
69      little     9
94        race     9
148       wise     8
1            a     7
138        was     7
29       crowd     6
40         for     5
61          in     5
123       this     5
13          at     4
25    cheering     4
54         his     4
84         old     4
86         one     4
88       other     4
127         to     4
```

由于词频表较长,这里利用命令 word.freq[1:20,]显示前20排的结果。结果表明,"the" 出现频次最高(37次),其次是"boy"(15次),其他词符以此类推。

下面我们调用数据包 koRpus 中的函数 treetag()计算词频表,操作步骤与上面的方法大致相同。得到词符的 R 操作代码如下:

```
require(koRpus)
require(koRpus.lang.en)
story.tag<-treetag("D:/Rpackage/story.txt",treetagger="manual",
lang="en",TT.options=list(path="C:/TreeTagger",preset="en"))
tokens<-tolower(story.tag[,"token"])
rm_puncts<-gsub("[[:punct:]]","",tokens)
words<-rm_puncts[which(rm_puncts!="")]
```

以上 R 代码合理地从字符串中分割出形符,"don't"被恰当地分割为"do"和"nt",从而避免了调整的麻烦。执行这些命令得到的结果与调用 R 函数 tokenize()开展计算得到的结果相同,在此从略。

3.4.2 利用停用词表的词频表

在上面的词频表中,"the""and""a"等功能词出现频次较高,但是这些功能词只有语法功能,往往没有实际内容方面的意义,因此词频表分析主要集中于有意义的词(通常为实词)。3.3 节利用数据包 koRpus 中的函数 treetag()提供的词类标注和数据包 tm 提供的停用词表计算词汇密度。我们这里也利用这个停用词表(另增 4 个功能词)计算文本词频分布。对 story.txt 文本开展词频分析时,利用 R 基础函数 which()提取所有的非停用词,再利用正则表达式提取所有的词语。本例利用停用词表计算文本词频的 R 代码如下:

```
require(koRpus)
require(koRpus.lang.en)
story.tag<-treetag("D:/Rpackage/story.txt",treetagger="manual",
lang="en",TT.options=list(path="C:/TreeTagger",preset="en"),
stopwords=c(tm::stopwords("en"),"another","toward","upon","us"))
tokens<-tolower(story.tag[,"token"])
stop_w<-story.tag[,"stop"]
tokens<-tokens[which(stop_w==FALSE)]
rm_puncts<-gsub("[[:punct:]]","",tokens)
words<-rm_puncts[which(rm_puncts!="")]
word.table<-table(words)
word.freq<-data.frame(word.table)
word.freq<-word.freq[order(-word.freq$Freq),]
```

执行以上 R 代码会得到一个文本词频分布对象 word.freq。由于词频表较长，这里利用 R 命令 word.freq[1:20,]显示前 20 排的结果：

```
> word.freq[1:20,]
            words  Freq
8            boy    15
46           man    10
44         little    9
61          race    9
97          wise    8
16         crowd    6
13       cheering   4
54           old    4
55           one    4
88           two    4
82       success    3
85          time    3
86       together   3
2           also    2
5          asked    2
7          blind    2
12        cheered   2
14      competitors 2
22          felt    2
23        finish    2
```

以上结果表明，故事主人公"boy"出现频次最高，其次是老人("man")、比赛("race")、人群("crowd")和喝彩("cheering")等词出现的频次也较高，烘托出比赛的热闹场面。

3.4.3 利用词目和停用词表的词频表

前两节介绍的词频表依据词符，即实际出现的词，这些意义相同却屈折变化不同的词符被视作不同的词，可能不利于词表对文本内容和措辞特点的概括。利用词目便于捕捉文本重要信息，探究词汇使用的特点。本节介绍利用词目和 tm 数据包提供的停用词表（另增 4 个功能词）制作 story.txt 文本词频表。与上一节词频表的制作不同的是，利用词目和停用词表的词频表是根据函数 treetag()计算得到的词目。计算文本词目的 R 代码如下：

```
require(koRpus)
require(koRpus.lang.en)
story.tag<-treetag("D:/Rpackage/story.txt",treetagger="manual",
lang="en",TT.options=list(path="C:/TreeTagger",preset="en"),
stopwords=c(tm::stopwords("en"),"another","toward","upon","us"))
```

```
lemmas<-story.tag[,"lemma"]
```

执行以上 R 代码,发现文本处理结果可以接受。接下来,使用索引符[ ]和正则表达式,并且调用函数 which( ),目的是从词目中排除停用词。R 命令如下:

```
stop_w<-story.tag[,"stop"]
lemmas<-lemmas[which(stop_w==FALSE)]
lemmas<-gsub("[[:punct:]]","",lemmas)
lemmas<-lemmas[which(lemmas!="")]
```

最后,利用以下 R 代码得到有意义词的词频分布表:

```
lemma.table<-table(lemmas)
lemma.freq<-data.frame(lemma.table)
lemma.freq<-lemma.freq[order(-lemma.freq$Freq),]
```

执行以上所有 R 代码,得到按词频降序排列的词频表,保存的 R 对象为 lemma.freq。由于词频表较长,这里利用 R 命令 word.freq[1:20,]显示前 20 排的结果:

```
>lemma.freq[1:20,]
         lemmas   Freq
7           boy    16
42          man    10
40        little    9
57         race    9
93         wise    8
9         cheer    6
13        crowd    6
50          old    4
51          one    4
83          two    4
20       finish    3
77      success    3
81         time    3
82     together    3
2          also    2
3           arm    2
4           ask    2
6         blind    2
11    competitor  2
19         feel    2
```

以上结果显示，词目表与词符表有所不同。例如，以词目为单位，"boy"出现16次，而以词符为单位，"boy"出现15次；以词目为单位，"cheer"出现6次，而以词符为单位，"cheering"出现4次，"cheered"出现2次。

3.5 词频分布图

我们在3.4.3节已经计算了文本词频表，利用词频表绘制图形使统计分析结果能够简洁直观地显示出来，词汇使用的特点能够更清晰地被发现。本节依据上一节利用词目和停用词表创建的词频表绘制线图（line plot）、条形图（bat chart）和词云（word cloud）。

3.5.1 词频分布线图

3.4.3节利用词目和停用词表创建的词频表名称为lemma.freq。我们利用线图绘制使用频次最高的10个词。关于图形基本结构，参见《语言学研究统计分析方法》（鲍贵 等，2020b）。在绘图之前，要利用函数par()中的变元设置图形结构。常用变元包括mai和omi。变元mai是一个数值向量，形式为c(bottom, left, top, right)，以英寸为单位确定图形边缘的大小。也可以用变元mar通过边缘线数设定图形边缘的大小，R默认值为c(5，4，4，2)+0.1。变元omi也是一个数值向量，形式为c(bottom, left, top, right)，以英寸为单位设定图形区外部边缘（outer margin）的大小。我们也可以用变元oma通过设定文本线确定外部边缘的大小。由于在图形显示中可以任意改变图形的大小，通过变元设置的数值表示相对大小。除了在参数函数par()中设置mai和omi的值之外，还可以设置其他变元的值。在R操作界面输入?par，按回车键后可以查看其他变元设置。

绘制线图使用的基础函数是plot()。本例图形的横坐标是各个高频词，设置为x=1:10；纵坐标是频次，设置为lemma.freq[1:10,"Freq"]；用空心圆点显示词，用线把点连接起来显示变化模式，设置type="b"。我们不打算设置横坐标名称，因而设置xlab=""（如要设置名称，可在引号内添加）。纵坐标名称为频数，设置为ylab="Frequency"。我们把横坐标值设为1到10，但是实际想要的是各个高频词，因而在plot()变元中设置xaxt="n"，以便抑制数值显示，后面再利用函数axis()设置高频词轴签。本例绘制线图的R命令和结果如下所示：

```
> par(mai=c(0.40,0.8,0,0),omi=c(0,0,0,0.01))
> plot(x=1:10,y=lemma.freq[1:10,"Freq"],type="b",xlab="",
  ylab="Frequency",xaxt="n")
> axis(1,1:10,labels=lemma.freq$lemmas[1:10])
```

图3.1显示，"boy""man""little""race""wise"等词出现的频次较高，说明比赛主要在小男孩和老人之间进行，比赛的教育意义主要通过智慧（"wise"）老人的反应以及其与小男孩的对话来反映。

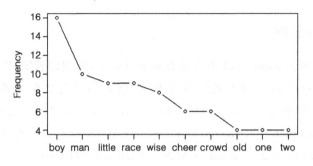

图 3.1 10 个高频词线图

3.5.2 词频分布条形图

我们利用 R 数据包 lattice 中的函数 barchart() 绘制文本中的词频分布条形图。条形图用分离的长方条绘制变量的频数、平均数或其他统计量。函数 barchart() 的主要变元是 y~x, 其中 x 是条件变量, y 是结果变量, y 和 x 位于数据框中。其他变元包括颜色设置 (col)、主标题 (main) 和轴名称。本例将颜色设置为"cadetblue"（军服蓝）, 主标题设为"Top 10 words"（10 个高频词）, 水平轴名称为"Frequency"（频次）。在绘制图形之前, 调用 R 基础函数 sort() 将前面得到的词目表 (lemma.table) 中的词目按照频次降序排列, 然后调用 R 基础函数 data.frame() 把表格转化为数据框。本例 R 命令和生成的条形图 (图 3.2) 如下所示:

```
> lemma.table<-sort(lemma.table,decreasing=TRUE)
> df<-data.frame(lemma.table)
> require(lattice)
> barchart(lemmas~Freq,data=df[1:10,],col="cadetblue",main="Top 10 words",
    xlab="Frequency")
```

图 3.2 显示,"boy""man""little""race""wise"出现的频数较高,特别是"boy"一词。"crowd"和"cheer"均出现 6 次, 而"old""one"和"two"均出现 4 次。

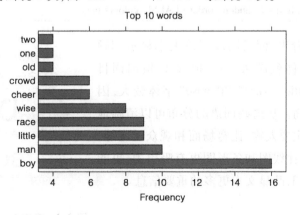

图 3.2 10 个高频词条形图

3.5.3 词频分布词云图

我们调用 R 数据包 wordcloud 中的函数 wordcloud() 绘制词云图,以字体变化显示文本词频分布的特点。这个数据包由费洛斯开发,当前版本号为 2.6(Fellows,2018)。函数 wordcloud() 的基本结构为:wordcloud(words,freq,scale=c(4,0.5),min.freq=3,max.words=Inf,random.order=TRUE,random.color=FALSE,rot.per=0.1,colors="black",ordered.colors=FALSE,use.r.layout=FALSE,fixed.asp=TRUE)。变元 words 指词频数据框中的词语,freq 指词频数据框中词语的频次。scale 是长度为 2 的向量,表示词语显示的大小范围;min.freq=3 是函数默认的最小频数。max.words=Inf 指图形显示的词语最大量无上限。变元 random.order=TRUE 指以随机顺序绘制词语。如果 random.order=FALSE,那么词语按频次从图中心位置向外以降序排列。变元 random.color=TRUE 表示随机选择颜色。如果 random.color=FALSE,那么颜色的选择依据频次。rot.per=0.1 指旋转 90°的词语比率为 10%。变元 colors="black" 指按照词频给词语着色,默认颜色为黑色。变元 ordered.colors=FALSE 指不按顺序给词语着色。变元 use.r.layout=FALSE 表示使用 C++代码进行碰撞检测。变元 fixed.asp=TRUE 表示固定宽高比。本例在变元设置中调整了字体大小的范围,即设置变元 scale=c(5,1),并且设置变元 random.order=FALSE。由于本例显示词语的最小频数为 2,因而设置 min.freq=2。颜色设置调用 R 调色板数据包 RColorBrewer 中的函数 brewer.pal(),其中第一个变元为颜色数,最大值取决于调色板中的颜色数,第二个变元为颜色名称。这个数据包由埃里希·诺维尔什开发,为图形提供色彩方案而设计,当前版本号为 1.1-3(Neuwirth,2022)。在 R 操作界面执行 display.brewer.all() 命令可以查看调色板名称。其他变元接受默认设置。针对本例,选择的调色板是 Dark2,R 命令和词云图如下所示:

```
> require(wordcloud)
> require(RColorBrewer)
> pal<-brewer.pal(8,"Dark2")
> wordcloud(words=lemma.freq $ lemmas,freq=lemma.freq $ Freq,
    scale=c(5,1),min.freq=2,random.order=FALSE,colors=pal)
```

从图 3.3 可以看出,词云图以字体大小显示词频高低,词频越高,字体就越大。在 30 个高频词词目中,"boy""man""little""race"和"wise"字体较大,因而出现的频次也最高。从这些词语的分布可以清晰地了解故事中出现的主要人物、比赛场面和观众反应等。从视觉上看,词云图比线图和条形图更有吸引力,也能够容纳更多的高频词,凸显文本更多的重要信息。

图 3.3 频次至少为 2 的高频词目词云

第四章 文本词汇多样性测量

本章以 koRpus 为数据包,以复述故事文本 story.txt 为例,介绍词汇多样性多种测量方法,包括传统的类符-形符比(TTR)、平均分段类符-形符比(MSTTR)、移动平均类符-形符比(MATTR)、Herdan's C、Guiraud's R、尤伯指数(U)、萨默指数、Yule's K、Maas 指数、HD-D、文本词汇多样性测量(MTLD)和文本词汇多样性移动平均测量(MTLD-MA)。

4.1 传统的类符-形符比

类符-形符比(Type-Token Ratio,TTR)是传统的词汇多样性(lexical diversity; lexical variation)测量方法。这里,type 是词型或类符,指在文本中出现的不相同的词,而 token 是词符或形符,指在文本中实际出现的词,不考虑出现的词是否相同。例如,在"We run."和"They run, too."这两个句子中,"run"作为一个类符出现一次,但是作为一个形符出现两次。这两个句子包括 4 个类符,5 个形符,因而类符-形符比为 0.8。下面以第三章使用的复述故事文本 story.txt 为例,计算文本类符-形符比。文本类符-形符比的计算调用数据包 koRpus 中的函数 TTR(),其中的主要变元是 koRpus 文本,如利用函数 treetag() 得到的标注文本。计算文档 story.txt 类符-形符比的 R 命令和统计分析结果如下:

```
> require(koRpus)
> require(koRpus.lang.en)
> story.tag<-treetag("D:/Rpackage/story.txt",treetagger
    ="manual",lang="en",TT.options=list(path="C:/TreeTagger",preset="en"))
> TTR(story.tag)
Language: "en"

Total number of tokens: 337
Total number of types: 152
Total number of lemmas: 153

Type-Token Ratio
        TTR: 0.45

Note: Analysis was conducted case insensitive.
```

以上结果显示,复述故事 story.txt 类符-形符比为 0.45,使用的形符(词符)总数为 337 个,类符总数为 152 个。注意,这里的 lemmas 不是指我们在第三章介绍的 lemmas(词目)。为了说明这一点,我们利用 story.tag 提供的词目,调用函数 unique() 计算不同词目数,R 命令和统计分析结果如下:

```
> lm<-story.tag[,"lemma"]
> lm<-gsub("[[:punct:]]","",lm)
> lm<-lm[which(lm!="")]
> length(unique(lm))
[1] 145
```

以上结果显示,文档 story.txt 包括的词目数为 145 个,而不是 153 个,这是不同的算法导致的。

我们下面用 R 命令计算文本类符数:

```
> tokens<-story.tag[,"token"]
> tokens<-gsub("[[:punct:]]","",tokens)
> tokens<-tolower(tokens[which(tokens!="")])
> length(unique(tokens))
[1] 152
```

这一结果与函数 TTR() 计算的类符数相同。

对传统的类符-形符比的主要批评是类符-形符比受文本长度的影响,即随着文本长度的增加,重复使用的词就会越多,类符-形符比就会降低。如果我们在研究中用 TTR 比较不同长度文本的形符比,造成结果差异的原因既有可能是不同类符数的增加提高了词汇多样性,又有可能是文本长度对类符-形符比的负面影响。下面我们利用 R 命令绘制文本长度对类符-形符比影响的线图:

```
> story.tag<-treetag("D:/Rpackage/story.txt",treetagger
="manual",lang="en",TT.options=list(path="C:/TreeTagger",preset="en"))
> ttr.res<-TTR(story.tag,char=TRUE,quiet=TRUE)
>plot(ttr.res@TTR.char,type="l",main="TTR degredation over text length")
```

在以上 R 命令中,TTR(story.tag,char=TRUE,quiet=TRUE) 以文本开头的 5 个形符作为文本长度开始计算类符-形符比,每增加 5 个形符作为文本长度依次计算类符-形符比,一共计算出 67 种长度条件下的类符-形符比;ttr.res@TTR.char 是一个数据框,列名称为"token"(形符)和"value"(类符-形符比值)。变元 char=TRUE 表示分析包含标注文本的 koRpus 对象。函数默认 char=FALSE,即默认不分析包含标注文本的 koRpus 对象。变元 quiet=TRUE 表示不显示计算状态提示。图 4.1 以数据框中"token"为横坐标、"value"为纵坐标绘制而成;type="l"指绘制线条。若 type="b",则绘制各个数据点,并把点连成线。变元 main = "TTR degredation over text length"指设定线图的标题"类符-形符比随文本长度的衰减"。线图显示,文本长度与类符-形符比呈现很强的负相关($r = -0.92, p < 0.001$)。这意味着,随着文本长度的增加,类符-形符比呈明显的下降趋势,重复使用的形符量随之增加。

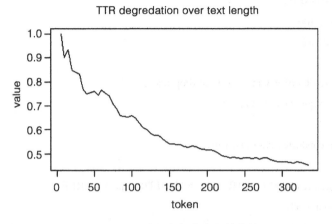

图 4.1　TTR 随文本长度变化的曲线

4.2　平均分段类符-形符比

由于传统的类符-形符比受文本长度的影响,后期研究采用各种测量方法降低或控制文本长度对词汇多样性的影响。本节以 story.txt 文本为例介绍平均分段类符-形符比(Mean Segmental Type-Token Ratio, MSTTR)。后面的几节也以 story.txt 文本为例,依据数据包 koRpus 提供的测量方法介绍其他几种词汇多样性测量方法。

平均分段类符-形符比又称分割类符-形符比(split TTR)。计算时,把形符按照某个长度(如每100个形符作为一个语段)进行分割,文本尾部形符数不足某个长度值的语段舍去,根据每个分割计算类符-形符比,各个类符-形符比的平均值即为平均分段类符-形符比。计算文本平均分段类符-形符比的函数是 MSTTR(txt),其中的 txt 是 koRpus 文本对象,默认的文本分割长度为100个形符。计算平均分段类符-形符比的另外一个通用函数是 lex.div(txt, measure, char, quiet),其中的 txt 是 koRpus 文本对象,measure 设置测量名称,char 设置绘制特征曲线所需数据的测量名称,quiet = TRUE 表示不显示计算状态提示。计算文本 story.txt 的平均分段类符-形符比的两个 R 命令和统计分析结果如下:

> MSTTR(story.tag)
Language:"en"

Total number of tokens:337
Total number of types:152

Total number of lemmas: 153

Mean Segmental Type-Token Ratio
        MSTTR: 0.66
      SD of TTRs: 0.05
    Segment size: 100
  Tokens dropped: 37

Hint: A segment size of 110 would reduce the drop rate to 7.
    Maybe try ? segment.optimizer()

Note: Analysis was conducted case insensitive.

> lex.div(story.tag, measure = "MSTTR", char = "MSTTR", quiet = TRUE)

Total number of tokens: 337

Total number of types: 152

Total number of lemmas: 153

Mean Segmental Type-Token Ratio
        MSTTR 0.66
      SD of TTRs: 0.05
    Segment size: 100
  Tokens dropped: 37

Hint: A segment size of 110 would reduce the drop rate to 7.
    Maybe try ? segment.optimizer()

Note: Analysis was conducted case insensitive.

以上两个结果显示,文本 story.txt 的平均分段类符-形符比为 0.66,函数 MSTTR()和 lex.div()默认的分段长度为 100 个形符。各个分段类符-形符比的标准差(SD of TTRs)为 0.05。由于文本词符数为 337,三个分段总词符数为 300,因而舍弃的词符数(Tokens dropped)为 37,结果提示以 110 个形符作为分割单位会把未包括在分段内的形符数减少到 7 个。分割优化函数是 segment.optimizer(txtlgth, segment = 100, range = 20, favour.min = TRUE),其中 txtlgth 是文本中的形符数,segment = 100 是函数默认的分段初始量(即形符数),range = 20 是函数默认的寻找合适分段量的形符数变化范围,favour.min = TRUE 表示如果结果存疑,那么保留较小的分段量。执行 R 命令 segment.optimizer(txtlgth = 337)得到分段量 110。如果分段量设为 110,那么执行 R 命令 MSTTR(story.tag, segment = 110)得到平均分段类符-形符比为 0.63,比使用默认的分割量(100 个形符)得到的平均分段类符-形符比要小一点。以上输出结果还说明,本次词汇多样性分析对字母大小写不敏感(case insensitive),即把大写字母全部转化为小写字母。

下面我们编写 R 命令计算平均分段类符-形符比,比较结果的一致性。首先,利用函数 treetag( )和 gsub( )等计算 tokens。然后,按照每 100 个词符作为文本长度计算类符-形符比。最后,利用平均数函数 mean( )计算三个类符-形符比的平均值,利用函数 round(x, digits)把结果保留两位小数,其中的第一个变元为数值向量,第二个变元设置小数位数。本例 R 命令和统计分析结果如下:

```
> require(koRpus)
> require(koRpus.lang.en)
> story.tag<-treetag("D:/Rpackage/story.txt",treetagger=
    "manual",lang="en",TT.options=list(path="C:/TreeTagger",preset="en"))
> tokens<-story.tag[,"token"]
> tokens<-gsub("[[:punct:]]","",tokens)
> tokens<-tolower(tokens[which(tokens!="")])
> ttr1<-length(unique(tokens[1:100]))/100
> ttr2<-length(unique(tokens[101:200]))/100
> ttr3<-length(unique(tokens[201:300]))/100
> round(mean(c(ttr1,ttr2,ttr3)),digits=2)
[1] 0.66
```

以上结果与函数 MSTTR( )和 lex.div( )的计算结果相同。

4.3 移动平均类符-形符比

移动平均类符-形符比(Moving-Average Type-Token Ratio,MATTR)按照固定的窗口(即词符数)从文本开头计算类符-形符比,然后在文本中依次移动窗口计算类符-形符比,直至在包括文本最后 1 个词符的窗口中计算类符-形符比,最后计算这些窗口类符-形符比的平均值(Covington et al.,2010)。相对于平均分段类符-形符比,移动平均类符-形符比不会因为文本长度不足舍弃文本末尾的词符。

文本 story.txt 共有 337 个形符,按照每 100 个形符作为一个窗口依次计算类符-形符比,得到 238 个窗口和 238 个类符-形符比。第 1 个窗口的最后一个形符是 excitement,第 2 个窗口的最后一个形符是 another,以此类推,直到第 238 个窗口的最后一个形符是 winner。计算移动平均类符-形符比的函数是 MATTR(txt),其中的 txt 是 koRpus 文本对象,默认的文本分割长度为 100。计算移动平均类符-形符比的另外一个函数是 lex.div(txt,measure,char,quiet),其中的 txt 是 koRpus 文本对象,measure 设置测量名称,char 设置绘制特征曲线所需数据的测量名称,quiet=TRUE 表示不显示计算状态提示。在这个例子中,计算文本 story.txt 词汇多样性的 R 命令和统计分析结果如下:

```
> story.tag<-treetag("D:/Rpackage/story.txt", treetagger = "manual", lang = "en", TT.options =
list(path = "C:/TreeTagger", preset = "en"))
> MATTR(story.tag)
```
Language: "en"

Total number of tokens: 337
Total number of types: 152
Total number of lemmas: 153

Moving-Average Type-Token Ratio
        MATTR: 0.65
     SD of TTRs: 0.04
   Window size: 100

Note: Analysis was conducted case insensitive.

```
> lex.div(story.tag, measure = "MATTR", char = "MATTR", quiet = TRUE)
```
Total number of tokens: 337
Total number of types: 152
Total number of lemmas: 153

Moving-Average Type-Token Ratio
        MATTR: 0.65
     SD of TTRs: 0.04
   Window size: 100

MATTR characteristics:
  Min. 1st Qu. Median Mean 3rd Qu. Max.
0.6127 0.6150 0.6230 0.6268 0.6359 0.6540
  SD
0.0129

Note: Analysis was conducted case insensitive.

以上两种方法计算的移动平均类符-形符比相同(0.65)。第 2 个命令的计算结果除了包括第 1 个命令的结果之外，还包括在不同文本长度条件下移动平均类符-形符比分布的基本统计量，即最小值(Min.)、第 1 个四分位数(1st Qu.)、中位数(Median)、平均数(Mean)、第 3 个四分位数(3rd Qu.)和最大值(Max.)。在上一节中，文本 story.txt 的平均分段类符-形符比为 0.66，本节计算移动平均类符-形符比得到 0.65，这两个结果几乎相同。

4.4 Herdan's C

Herdan's C,又称类符–形符比对数(log TTR),是另外一种词汇多样性测量方法,计算公式为 $C = \dfrac{\log V}{\log N}$,其中 log 是常用对数,$V$ 是类符数,N 是形符数(Herdan,1960;Tweedie et al.,1998)。

计算 Herdan's C 的函数是 C.ld(txt),其中 txt 是 koRpus 文本对象。计算 Herdan's C 的另外一个通用函数是 lex.div(txt,measure,char,quiet),其中的 txt 是 koRpus 文本对象,measure 设置测量名称,char 设置绘制特征曲线所需数据的测量名称,quiet = TRUE 表示不显示计算状态提示。针对本例,计算文本 story.txt 词汇多样性的 R 命令和统计分析结果如下:

```
> C.ld(story.tag)
Language: "en"

Total number of tokens: 337
Total number of types: 152
Total number of lemmas: 153

Herdan's C
              C: 0.86

Note: Analysis was conducted case insensitive.

> lex.div(story.tag, measure = "C", char = "C", quiet = TRUE)
Total number of tokens: 337
Total number of types: 152
Total number of lemmas: 153

Herdan's C
              C: 0.86

C characteristics:
  Min.    1st Qu.   Median    Mean    3rd Qu.    Max.
 0.8641   0.8681    0.8777   0.8921   0.9082    1.0000
  SD
 0.0308

Note: Analysis was conducted case insensitive.
```

以上结果显示,两个命令的计算结果相同,即 Herdan's C 为 0.86。Herdan's C 的计算比较简单。在本例中,形符数为337,类符数为152,利用 R 命令 log 10(152)/log 10(337),即可得到同样的结果。第 2 个命令的计算结果除了包括第 1 个命令的结果之外,还包括在不同文本长度条件下 Herdan's C 分布(参看图 4.2)的基本统计量,即最小值(Min.)、第 1 个四分位数(1st Qu.)、中位数(Median)、平均数(Mean)、第 3 个四分位数(3rd Qu.)和最大值(Max.)。

直觉上使用类符和形符的对数比值测量词汇多样性似乎不太可能消除文本长度对测量结果的影响。我们依据文本 story.txt,以文本长度为 5 个形符数开始计算 Herdan's C,长度依次增加 5,直至文本结尾,由此得到 67 个文本长度和 67 个 Herdan's C。Herdan's C 随文本长度的变化趋势如图 4.2 所示。绘图 R 命令和执行结果如下:

```
> ttr.res<- C.ld(story.tag,char=TRUE,quiet=TRUE)
> plot(ttr.res@C.char,type="l",main="")
```

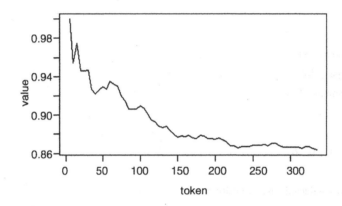

图 4.2　Herdan's C 随文本长度变化的曲线

图 4.2 显示,文本长度与 Herdan's C 呈现很强的负相关($r=-0.89, p<0.001$)。这意味着,随着文本长度的增加,Herdan's C 呈明显的下降趋势。只有在文本长度达到 150 以后,Herdan's C 值才趋于稳定。

4.5　Guiraud's R

Guiraud 平方根类符-形符比(Guiraud's Root TTR,简称 Guiraud's R)用文本长度(N)的平方根代替文本长度计算词汇多样性,以便降低文本长度对类符-形符比的影响,计算公式为 $R = \dfrac{V}{\sqrt{N}}$,其中 V 是类符数,N 是形符数(Tweedie et al.,1998)。

计算 Guiraud's R 的函数是 R.ld(txt),其中 txt 是 koRpus 文本对象。计算 Guiraud's R 的另外一个通用函数是 lex.div(txt, measure, char, quiet),其中的 txt 是 koRpus 文本对象,

measure 设定测量名称,char 设置绘制特征曲线所需数据的测量名称,quiet = TRUE 表示不显示计算状态提示。本例用以上两个函数计算文本 story. txt 词汇多样性的 R 命令和统计分析结果如下:

```
> R. ld( story. tag)
Language："en"


Total number of tokens：337
Total number of types：152
Total number of lemmas：153


Guiraud's R
            R：8.28


Note：Analysis was conducted case insensitive.

> lex. div( story. tag,measure = " R" , char = " R" , quiet = TRUE)
Total number of tokens：337
Total number of types：152
Total number of lemmas：153

Guiraud's R
            R：8.28

R characteristics：
  Min.     1st Qu.    Median    Mean     3rd Qu.    Max.
  2.236    6.217      6.929     6.718    7.678      8.376
  SD
  1.3464

Note：Analysis was conducted case insensitive.
```

以上两个命令得到的结果相同,即 Guiraud's R 为 8.28。Guiraud's R 的计算比较简单。本例中,形符数为 337,类符数为 152,利用 R 命令 152/sqrt(337) 即可得到同样的结果。第 2 个命令的计算结果除了包括第 1 个命令的结果之外,还包括在不同文本长度条件下 Guiraud's R 分布(参看图 4.3)的基本统计量,即最小值(Min.)、第 1 个四分位数(1st Qu.)、中位数(Median)、平均数(Mean)、第 3 个四分位数(3rd Qu.)和最大值(Max.)。

利用文本长度的平方根计算词汇多样性似乎不能很好地控制文本长度对测量结果的影响。采用与上一节同样的方法,我们以文本长度为 5 个形符数开始计算 Guiraud's R,每增加5个形符作为文本长度依次计算Guiraud's R，绘制由此得到的67个文本长度和对应的67个Guiraud's R值的线图，如图4.3所示。

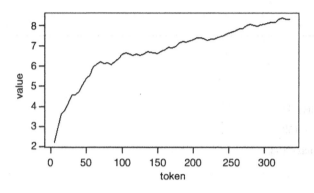

图4.3　Guiraud's R随文本长度变化的曲线

图4.3显示，与Herdan's C测量不同的是，Guiraud's R测量值随着文本长度的增加而增加，正相关关系的强度很大（$r=0.91$，$p<0.001$）。这表明，Guiraud's R不是很好的词汇多样性测量方法。

4.6　尤伯指数（U）

尤伯指数（Uber Index，简称U）通过同时调整文本长度（N）和类符数（V）降低文本长度对词汇多样性测量的影响，计算公式为 $U = \dfrac{(\log N)^2}{\log N - \log V}$，其中log是常用对数，$V$是类符数，$N$是形符数（Tweedie et al.，1998）。

计算尤伯指数的函数是U.ld(txt)，其中txt是koRpus文本对象。另外一个通用函数是lex.div(txt,measure,char,quiet)，其中的txt是koRpus文本对象，measure设定测量名称，char设置绘制特征曲线所需数据的测量名称，quiet=TRUE表示不显示计算状态提示。本例利用以上两个函数计算文本story.txt词汇多样性的R命令和统计分析结果如下：

```
> U.ld(story.tag)
Language："en"

Total number of tokens：337
Total number of types：152
Total number of lemmas：153

Uber Index
          U：18.48
```

```
Note：Analysis was conducted case insensitive.

> lex.div(story.tag,measure="U",char="U", quiet=TRUE)
Total number of tokens：337
Total number of types：152
Total number of lemmas：153

Uber Index
          U：18.48

U characteristics：
  Min.    1st Qu.   Median    Mean    3rd Qu.    Max.    NA's
  17.63   18.34     18.64     20.30   21.03      46.16   1
  SD
  4.139

Note：Analysis was conducted case insensitive.
```

以上两个命令得到的结果相同,即尤伯指数值为 18.48。尤伯指数的计算比较简单,利用 R 命令 log10(337)^2/(log10(337)-log10(152)) 即可得到同样的结果。第2个命令的计算结果除了包括第1个命令的执行结果之外,还包括在不同文本长度条件下尤伯指数分布(参看图 4.4)的基本统计量,即最小值(Min.)、第 1 个四分位数(1st Qu.)、中位数(Median)、平均数(Mean)、第 3 个四分位数(3rd Qu.)、最大值(Max.)和一个 NA(缺失值)。之所以有一个 NA,是因为 lex.div()在以形符数为 5(Once there was a little)作为文本长度计算尤伯指数时类符数和形符数均为 5,所以无法计算尤伯指数值。

采用与上一节同样的方法,我们以文本长度为 10 的形符数开始,每增加 5 个形符作为文本长度依次计算尤伯指数,绘制由此得到的 66 个文本长度和对应的 66 个尤伯指数值的线图,如图 4.4 所示。

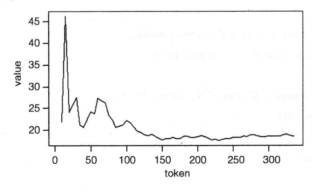

图 4.4　尤伯指数随文本长度变化的曲线

如图 4.4 所示,在文本长度达到 100 之后,尤伯指数值趋于稳定。文本长度达到 150 之后,尤伯指数值波动很小。因此,文本较长时,尤伯指数是较为稳定的词汇多样性测量方法。

4.7 萨默指数(S)

萨默指数(Summer's Index,简称 S)利用常用对数同时调整文本长度(N)和类符数(V),以降低文本长度对词汇多样性测量的影响,计算公式为 $S = \dfrac{\log(\log V)}{\log(\log N)}$,其中 log 是常用对数。

计算萨默指数的函数是 S.ld(txt),其中 txt 是 koRpus 文本对象。另外一个通用函数是 lex.div(txt, measure, char, quiet),其中的 txt 是 koRpus 文本对象,measure 设定测量名称,char 设置绘制特征曲线所需数据的测量名称,quiet = TRUE 表示不显示计算状态提示。本例利用以上两个函数计算文本 story.txt 词汇多样性的 R 命令和统计分析结果如下:

```
> S.ld(story.tag)
Language: "en"

Total number of tokens: 337
Total number of types: 152
Total number of lemmas: 153

Summer's S
            S: 0.84

Note: Analysis was conducted case insensitive.
Warning message:
The implementations of these formulas are still subject to validation:
  S
  Use the results with caution, even if they seem plausible!
  See lex.div(measure = "validation") for more details.

> lex.div(story.tag, measure = "S", char = "S", quiet = TRUE)
Total number of tokens: 337
Total number of types: 152
Total number of lemmas: 153

Summer's S
```

```
           S: 0.84

S characteristics:
   Min.     1st Qu.   Median    Mean     3rd Qu.   Max.      NA's
   0.7880   0.8381    0.8414    0.8462   0.8468    1.0000    1
   SD
   0.0237

Note: Analysis was conducted case insensitive.
```

以上两个命令得到相同的结果,即萨默指数值为0.84。第1个命令的计算结果显示一条警告,提示萨默指数测量需要验证。通过执行以下R命令发现以上结果正确:log 10(log 10(152))/log 10(log 10(337))。第2个命令的计算结果除了包括第1个命令的结果之外,还包括在不同文本长度条件下萨默指数分布(参看图4.5)的基本统计量,即最小值(Min.)、第1个四分位数(1st Qu.)、中位数(Median)、平均数(Mean)、第3个四分位数(3rd Qu.)、最大值(Max.)和一个NA(缺失值)。之所以有一个NA,是因为函数lex.div()在以形符数为10("Once there was a little…")作为文本长度计算萨默指数时,萨默指数公式的分母为0(即log 10(log 10(10))=0),所以无法计算萨默指数值。

采用与上一节同样的方法,我们以文本长度为5个形符数开始,每增加5个形符作为文本长度(排除文本长度为10的情况)依次计算萨默指数,绘制由此得到的66个文本长度和对应的66个萨默指数值的线图,如图4.5所示。

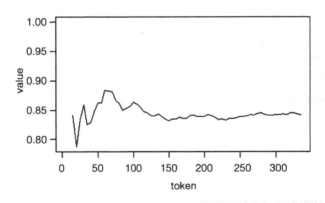

图4.5 萨默指数随文本长度变化的曲线

图4.5显示,在文本长度达到120之后,萨默指数值趋于稳定。因此,文本较长时,萨默指数是较为稳定的词汇多样性测量方法。

4.8 Yule's K

Yule's K(简称K)是较为复杂的词汇多样性测量方法。它利用类符频次控制文本长度对词汇多样性的影响,计算公式为 $K = 10^4 \times \dfrac{(\sum_{X=1}^{X} f_X X^2) - N}{N^2}$,其中 X 是每个类符频次向量,f_X 是每个类符出现的频次,N 是文本长度,即形符数(Tweedie et al. ,1998)。

我们以 story. txt 开头的前两句为例,说明 K 值的计算方法。这两句如下:

> Once there was a little boy who was hungry for success in sports. To him, winning was everything and success was measured by results.

这两个句子的形符总数为 24,其中共有 20 个类符,18 个类符出现 1 次("once""there""a""little""boy""who""hungry""for""in""sports""to""him""winning""everything""and""measured""by""results"),1 个形符出现 2 次("success"),另 1 个类符出现 4 次("was"),则 $\dfrac{(\sum_{X=1}^{X} f_X X^2) - N}{N^2} = \dfrac{(18 \times 1^2 + 1 \times 2^2 + 1 \times 4^2) - 24}{24^2} \approx 0.024\,306$。因此,$K = 10^4 \times 0.024\,305\,56 \approx 243.06$。

计算 Yule's K 的函数是 K. ld(txt),其中 txt 是 koRpus 文本对象。另外一个函数是 lex. div(txt, measure, char, quiet),其中的 txt 是 koRpus 文本对象,measure 设定测量名称,char 设置绘制特征曲线所需数据的测量名称,quiet = TRUE 表示不显示计算状态提示。本例利用以上两个函数计算文本 story. txt 词汇多样性的 R 命令和统计分析结果如下:

```
> K. ld( story. tag)
Language: "en"


Total number of tokens: 337

Total number of types: 152

Total number of lemmas: 153


Yule's K
              K: 206.75


Note: Analysis was conducted case insensitive.
Warning message:
The implementations of these formulas are still subject to validation:
  K
Use the results with caution, even if they seem plausible!
See lex. div( measure = " validation" ) for more details.
```

```
> lex.div(story.tag,measure="K",char="K",quiet=TRUE)

Total number of tokens: 337
Total number of types: 152
Total number of lemmas: 153


Yule's K
         K: 206.75

K characteristics:
  Min.    1st Qu.    Median    Mean    3rd Qu.    Max.
  0.0     174.3      200.0     189.9   220.3      227.7
SD
40.1884

Note: Analysis was conducted case insensitive.
```

以上两个命令得到的结果相同,即 Yule's K 值为 206.75。对其他结果的解释见 4.7 节。

采用与上一节同样的方法,我们以文本长度为 5 个形符数开始,每增加 5 个形符作为文本长度依次计算 Yule's K,绘制由此得到的 67 个文本长度和对应的 67 个 Yule's K 值的线图,如图 4.6 所示。

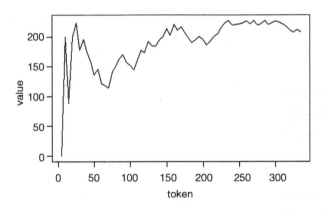

图 4.6 Yule's K 随文本长度变化的曲线

图 4.6 显示,在文本长度达到 150 之后,Yule's K 值趋于稳定。文本较长时,Yule's K 是较为稳定的词汇多样性测量方法。

4.9 Maas 指数

Maas 指数(Maas' Indices)是较为复杂的词汇多样性测量方法,利用文本长度和类符数的对数计算词汇多样性，有三个计算公式。第一个计算公式为 $a^2 = \dfrac{\log N - \log V}{(\log N)^2}$，其中 log 是常用对数，$N$ 是文本长度，V 是类符数。从公式可以看出，a^2 是尤伯指数的倒数。第二个计算公式为 $\log V_0 = \dfrac{\log V}{\sqrt{1-\left(\dfrac{\log V}{\log N}\right)^2}}$。第三个公式本质上等同于第二个公式，只是把常用对数改为自然对数（ln，以 e 为底，数值约为 2.718）。

对于 story.txt 文本，我们已经知道文本长度为 337，类符数为 152，计算词汇多样性 a 和词汇多样性 $\log V_0$ 的 R 命令分别为：

```
sqrt((log10(337)-log10(152))/log10(337)^2);
log10(152)/sqrt(1-(log10(152)/(log10(337)))^2)
```

执行这两个命令分别得到 a 约为 0.23，$\log V_0$ 约为 4.32。若利用自然对数，则执行 R 命令 log(152)/sqrt(1-(log(152)/(log(337)))^2)，得到词汇多样性（函数 maas( ) 的输出结果为 lgeV0）约为 9.95。

利用数据包 koRpus 计算 Maas 指数的简单函数是 maas(txt)，其中 txt 是 koRpus 文本对象。另外一个通用函数是 lex.div(txt, measure, char, quiet)，其中的 txt 是 koRpus 文本对象，measure 设定测量名称，char 设置绘制特征曲线所需数据的测量名称，quiet=TRUE 表示不显示计算状态提示。本例利用以上两个函数计算文本 story.txt 中的 Maas 指数的 R 命令和统计分析结果如下：

```
> maas(story.tag)
Language: "en"

Total number of tokens: 337
Total number of types: 152
Total number of lemmas: 153

Maas' Indices
            a: 0.23
         lgV0: 4.32
        lgeV0: 9.95

Relative vocabulary growth (first half to full text)
            a: 0.76
         lgV0: 5.16
           V': 0.34 (34 new types every 100 tokens)
```

Note: Analysis was conducted case insensitive.

> lex.div(story.tag, measure="Maas", char="Maas", quiet=TRUE)

Total number of tokens: 337
Total number of types: 152
Total number of lemmas: 153

Maas' Indices
        a: 0.23
        lgV0: 4.32
        lgeV0: 9.95

Relative vocabulary growth (first half to full text)
        a: 0.76
        lgV0: 5.16
        V': 0.34 (34 new types every 100 tokens)

Maas Indices characteristics:

Min.	1st Qu.	Median	Mean	3rd Qu.	Max.	
0.0000	0.2175	0.2316	0.2208	0.2335	0.2382	
SD						
0.0317						
Min.	1st Qu.	Median	Mean	3rd Qu.	Max.	NA's
3.191	4.092	4.181	4.200	4.307	5.110	1
SD						
0.2387						
Min.	1st Qu.	Median	Mean	3rd Qu.	Max.	NA's
7.348	9.422	9.627	9.671	9.918	11.766	1
SD						
0.5495						

Note: Analysis was conducted case insensitive.

以上两个 R 命令计算的结果相同。第二个命令比第一个命令多报告了三个 Maas 指数分布(参看图 4.7、图 4.8 和图 4.9)的基本统计量,即最小值(Min.)、第 1 个四分位数(1st Qu.)、中位数(Median)、平均数(Mean)、第 3 个四分位数(3rd Qu.)、最大值(Max.)。这两个命令还计算前半个文本与整个文本比较得到的词汇相对增长指数(relative vocabulary growth)。例如,在 $a = 0.76$ 时,词汇相对增长指数 V' 可以通过执行以下 R 命令计算得到:

$(\log 10(152)/\log 10(0.76*337))\char`\^3*(152/337)$。

采用与上一节同样的方法,我们以文本长度为 5 个形符数开始,每增加 5 个形符作为文本长度依次计算 a、$\lg V_0$ 和 $\lge V_0$,绘制由此得到的 67 个文本长度和对应的 67 个 a 值、$\lg V_0$ 值和 $\lge V_0$ 值的线图,如图 4.7、4.8 和 4.9 所示。

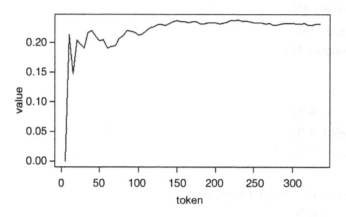

图 4.7 指数 a 随文本长度变化的曲线

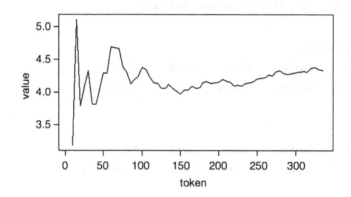

图 4.8 指数 $\lg V_0$ 随文本长度变化的曲线

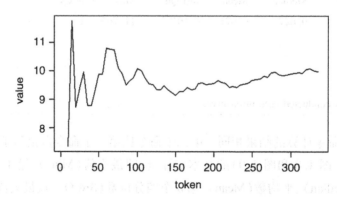

图 4.9 指数 $\lge V_0$ 随文本长度变化的曲线

图 4.7 显示,在文本长度达到 150 之后,a 值趋于稳定。文本较长时,Maas 指数 a 是较为稳定的词汇多样性测量方法。

图 4.8 显示,在文本长度达到 180 之后,$\lg V_0$ 值趋于稳定。文本较长时,Maas 指数 $\lg V_0$ 是较为稳定的词汇多样性测量方法。

图 4.9 显示,在文本长度达到 180 之后,$\lg e V_0$ 值也趋于稳定。文本较长时,Maas 指数 $\lg e V_0$ 是较为稳定的词汇多样性测量方法。

4.10 HD-D

HD-D 是 vocd_D 的理想化形式(McCarthy et al.,2007)。从文档中随机抽取一定量的形符(数据包 koRpus 中的函数 HDD()默认的抽样数量为 42 个形符),利用超几何分布(hypergeometric distribution)计算至少一次得到每个类符的概率,把这些概率相加即为词汇多样性测量 HD-D 值。

利用数据包 koRpus 计算 HD-D 值的简单函数是 HDD(txt),其中 txt 是 koRpus 文本对象。另外一个通用函数是 lex.div(txt,measure,char,quiet),其中的 txt 是 koRpus 文本对象,measure 设定测量名称,char 设置绘制特征曲线所需数据的测量名称,quiet=TRUE 表示不显示计算状态提示。本例利用以上两个函数计算文本 story.txt 中 HD-D 值的 R 命令和统计分析结果如下:

```
> HDD(story.tag)
Language: "en"


Total number of tokens: 337
Total number of types: 152
Total number of lemmas: 153


HD-D
            HD-D: 32.44
            ATTR: 0.77
      Sample size: 42


Note: Analysis was conducted case insensitive.

> lex.div(story.tag,measure="HD-D",char="HD-D",quiet=TRUE)
Total number of tokens: 337
Total number of types: 152
Total number of lemmas: 153
```

```
HD-D
            HD-D: 32.44
            ATTR: 0.77
   Sample size: 42

HD-D characteristics:
   Min.     1st Qu.    Median    Mean    3rd Qu.    Max.
   5.00     31.69      32.08     30.60   32.28      34.29
   SD
   5.369

Note: Analysis was conducted case insensitive.
```

以上两个 R 命令得到相同的 HD-D 值。ATTR 是 HD-D 与样本量的比率（即 32.44/42）。

为了加深对 HD-D 与文本长度之间关系的了解，我们采用与上一节同样的方法，以文本长度为 5 个形符数开始，每增加 5 个形符作为文本长度依次计算 HD-D 值，绘制由此得到的 67 个文本长度和对应的 67 个 HD-D 值的线图，如图 4.10 所示。

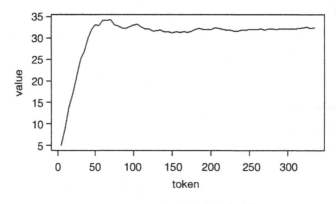

图 4.10　HD-D 随文本长度变化的曲线

图 4.10 显示，在文本长度达到 40 之后，HD-D 值趋于稳定。如果文本长度不是太短，HD-D 是较为稳定的词汇多样性测量方法。

4.11　文本词汇多样性测量（MTLD）

文本词汇多样性测量（Measure of Textual Lexical Diversity，MTLD）是文本词汇多样性的一个指标，其数值等于文本中的连续词串在其类符-形符比小于某个类符-形符比（即因子大小，factor size，默认值为 0.72）时的平均长度（McCarthy et al.，2010）。根据麦卡锡和贾维斯，下面词串中各个词后括号内的值是词串以这个词结束时的类符-形符比：of

(1) the (1.00) people (1.00) by (1.00) the (0.800) people (0.667) for (0.714) the (0.625) people (0.556)(McCarthy et al.,2010)。例如,词串以 for 结尾时,形符总数为 7 个,类符数为 5,因而类符-形符比为 0.714。但是,MTLD 测量设定,在以某个词结束的词串类符-形符比小于 0.72 时,该词串为一个因子。然后,以下一个词开始重新计算因子,如此计算直至文本结束。在上面的这个例子中,当词串达到 by the people 中的 people 时,类符-形符比为 0.667,低于因子阈限值 0.72,因而 of the people by the people 计作一个因子,然后以紧接着的 for 开始重新计算因子,如 for (1.00) the (1.00) people (1.00)。在利用数据包 koRpus 中的函数 MTLD()计算因子数时,部分因子值(即以文本结尾词结束的词串类符-形符比大于因子阈限值)等于 1 与其类符-形符比离差和 1 与因子阈限值离差的比值。MTLD 为形符总数与因子总数的比值。函数 MTLD()采用顺向和逆向方法计算 MTLD 值,最终的 MTLD 值是两次计算 MTLD 值的平均数。

为了帮助了解 MTLD 值的计算方法,我们从文本 story.txt 中选取开头的 4 个句子:

Once there was a little boy who was hungry for success in sports. To him, winning was everything and success was measured by results. One day, the boy took part in a race in his village, and his competitors were two other young boys. A large crowd gathered to watch this sporting spectacle, and a wise old man, upon hearing about the boy, also came from afar to watch.

表 4.1 和表 4.2 列出计算过程。

表 4.1 MTLD 测量顺向计算举例

	once	there	was	a	little	boy	who	was	hungry	for
Type	1	2	3	4	5	6	7	7	8	9
Token	1	2	3	4	5	6	7	8	9	10
类符-形符比	1	1	1	1	1	1	1	0.88	0.89	0.90
	success	in	sports	to	him	winning	was	everything	and	success
Type	10	11	12	13	14	15	15	16	17	17
Token	11	12	13	14	15	16	17	18	19	20
类符-形符比	0.91	0.92	0.92	0.93	0.93	0.94	0.88	0.89	0.89	0.85
	was	measured	by	results	one	day	the	boy	took	part
Type	17	18	19	20	21	22	23	23	24	25
Token	21	22	23	24	25	26	27	28	29	30
类符-形符比	0.81	0.82	0.83	0.83	0.84	0.85	0.85	0.82	0.83	0.83

	in	a	race	in	his	village	and	his	competitors	were
Type	25	25	26	26	27	28	28	28	29	30
Token	31	32	33	34	35	36	37	38	39	40
类符-形符比	0.81	0.78	0.79	0.76	0.77	0.78	0.76	0.74	0.74	0.75
	two	other	young	boys	a	large	crowd	gathered	to	watch
Type	31	32	33	34	34	35	36	37	37	38
Token	41	42	43	44	45	46	47	48	49	50
类符-形符比	0.76	0.76	0.77	0.77	0.76	0.76	0.77	0.77	0.76	0.76
	this	sporting	spectacle	and	a	wise	old	man	upon	hearing
Type	39	40	41	41	41	42	43	44	45	46
Token	51	52	53	54	55	56	57	58	59	60
类符-形符比	0.76	0.77	0.77	0.76	0.75	0.75	0.75	0.76	0.76	0.77
	about	the	boy	also	came	from	afar	to	watch	
Type	47	47	47	48	49	50	51	51	51	
Token	61	62	63	64	65	66	67	68	69	
类符-形符比	0.77	0.76	0.75	0.75	0.75	0.76	0.76	0.75	0.74	

根据表 4.1，顺向计算得到 51 个类符，形符数为 69，但是在词串结束时类符和形符比为 0.74，大于因子阈限值 0.72。部分因子值为 (1−51/69)/(1−0.72) ≈ 0.931 677，即类符和形符比率 0.74 达到从 1 到 0.72 轨迹的 93.167 7%。MTLD 值为形符数与因子值的比率，即 69/0.931 677 ≈ 74.06。

表 4.2 MTLD 测量逆向计算举例

	watch	to	afar	from	came	also	boy	the	about	hearing
Type	1	2	3	4	5	6	7	8	9	10
Token	1	2	3	4	5	6	7	8	9	10
类符-形符比	1	1	1	1	1	1	1	1	1	1

（续表）

	upon	man	old	wise	a	and	spectacle	sporting	this	watch
Type	11	12	13	14	15	16	17	18	19	19
Token	11	12	13	14	15	16	17	18	19	20
类符-形符比	1	1	1	1	1	1	1	1	1	0.95

	to	gathered	crowd	large	a	boys	young	other	two	were
Type	19	20	21	22	22	23	24	25	26	27
Token	21	22	23	24	25	26	27	28	29	30
类符-形符比	0.90	0.91	0.91	0.92	0.88	0.88	0.89	0.89	0.90	0.90

	competitors	his	and	village	his	in	race	a	in	part
Type	28	29	29	30	30	31	32	32	32	33
Token	31	32	33	34	35	36	37	38	39	40
类符-形符比	0.90	0.91	0.88	0.88	0.86	0.86	0.86	0.84	0.82	0.83

	took	boy	the	day	one	results	by	measured	was	success
Type	34	34	34	35	36	37	38	39	40	41
Token	41	42	43	44	45	46	47	48	49	50
类符-形符比	0.83	0.81	0.79	0.80	0.80	0.80	0.81	0.81	0.82	0.82

	and	everything	was	winning	him	to	sports	in	success	for
Type	41	42	42	43	44	44	45	45	45	46
Token	51	52	53	54	55	56	57	58	59	60
类符-形符比	0.80	0.81	0.79	0.80	0.80	0.79	0.79	0.78	0.76	0.77

	hungry	was	who	boy	little	a	was	there	once	
Type	47	47	48	48	49	49	49	50	51	
Token	61	62	63	64	65	66	67	68	69	
类符-形符比	0.77	0.76	0.76	0.75	0.75	0.74	0.73	0.74	0.74	

根据表4.2,逆向计算时,在词串结束时类符和形符比率也为0.74,大于因子阈限值0.72。因此,部分因子值也为0.931 677,MTLD值也为74.06。这两个值的平均数74.06就是最终的MTLD值。

利用数据包 koRpus 计算 MTLD 值的简单函数是 MTLD(txt),其中 txt 是 koRpus 文本对象。另外一个通用函数是 lex.div(txt, measure, char, quiet),其中的 txt 是 koRpus 文本

对象，measure 设定测量名称，char 设置绘制特征曲线所需数据的测量名称，quiet=TRUE 表示不显示计算状态提示。本例利用以上两个函数计算文本 story.txt 中的 MTLD 值的 R 命令和统计分析结果如下：

```
> MTLD(story.tag)
Language："en"

Total number of tokens：337
Total number of types：152
Total number of lemmas：153

Measure of Textual Lexical Diversity
            MTLD：56.55
Number of factors：NA
     Factor size：0.72
  SD tokens/factor：28.63（all factors）
            27.09（complete factors only）

Note：Analysis was conducted case insensitive.

> lex.div(story.tag,measure="MTLD",char="MTLD",quiet=TRUE)
Total number of tokens：337
Total number of types：152
Total number of lemmas：153

Measure of Textual Lexical Diversity
            MTLD：56.55
Number of factors：NA
     Factor size：0.72
  SD tokens/factor：28.63（all factors）
            27.09（complete factors only）

MTLD characteristics：
  Min.    1st Qu.   Median    Mean    3rd Qu.   Max.    NA's
  28.00   44.00     47.57     49.29   51.83     76.22   1
  SD
8.8966

Note：Analysis was conducted case insensitive.
```

以上两个 R 命令得到相同的 MTLD 值(56.55)。输出结果还报告函数默认的因子阈限值(factor size)和每个因子包括形符数的标准差(SD tokens/factor)。第二个函数还包括 MTLD 分布的基本描述性统计量。

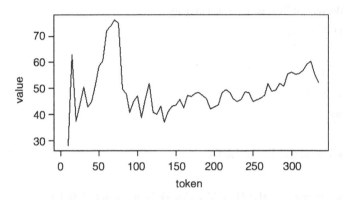

图 4.11　MTLD 随文本长度变化的曲线

我们采用与上一节相似的方法,以文本长度为 10 个形符数开始,每增加 5 个形符作为文本长度依次计算 MTLD,绘制由此得到的 66 个文本长度和对应的 66 个 MTLD 值的线图,如图 4.11 所示。

图 4.11 显示,在文本长度达到 100 之后,MTLD 值随文本长度波动性减小。整体上,MTLD 值与文本长度没有统计显著性相关关系 ($r = 0.06, p = 0.619 > 0.05$)。

4.12　文本词汇多样性移动平均测量(MTLD-MA)

文本词汇多样性移动平均测量(Moving-Average Measure of Textual Lexical Diversity,MTLD-MA)类似于移动平均类符-形符比(MATTR),把因子计算与移动窗口结合在一起。在计算每个因子之后,下一个因子的计算从上一个因子(即词串)起点后移一个形符开始,即把步长(step size)设为 1,如此重复计算直至文本结尾。所有因子长度的平均数即为最终的 MTLD-MA 值。在利用数据包 koRpus 中的函数 MTLD() 计算因子数时,形符长度低于 9 的词串排除在因子计算之外。

利用数据包 koRpus 计算 MTLD-MA 值的简单函数是 MTLD(txt, MA=TRUE),其中 txt 是 koRpus 文本对象,MA=TRUE 表示计算文本词汇多样性移动平均测量值。另外一个通用函数是 lex.div(txt, measure, char, quiet),其中的 txt 是 koRpus 文本对象,measure 设定测量名称,char 设置绘制特征曲线所需数据的测量名称,quiet=TRUE 表示不显示计算状态提示。本例利用以上两个函数计算文本 story.txt 中的 MTLD-MA 值的 R 命令和统计分析结果如下:

```
> MTLD(story.tag, MA=TRUE, quiet=TRUE)
Total number of tokens: 337
```

```
Total number of types: 152
Total number of lemmas: 153

Moving-Average Measure of Textual Lexical Diversity
          MTLD-MA: 57.83
  SD tokens/factor: 21.83
       Step size: 1
      Factor size: 0.72
      Min. tokens: 9

Note: Analysis was conducted case insensitive.

> lex.div(story.tag, measure="MTLD-MA", char=FALSE, quiet=TRUE)
Total number of tokens: 337
Total number of types: 152
Total number of lemmas: 153

Moving-Average Measure of Textual Lexical Diversity
          MTLD-MA: 57.83
  SD tokens/factor: 21.83
       Step size: 1
      Factor size: 0.72
      Min. tokens: 9

Note: Analysis was conducted case insensitive.
```

以上结果显示,文本 story.txt 中的 MTLD-MA 值为 57.83,与 MTLD 值 56.55 稍有差异。

我们以文本长度为 75 个形符数开始,每增加 5 个形符作为文本长度依次计算 MTLD-MA 值,绘制由此得到的 53 个文本长度和对应的 53 个 MTLD-MA 值的线图,如图 4.12 所示。

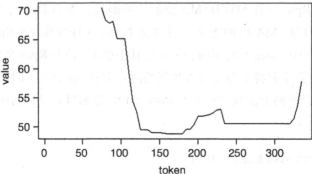

图 4.12　MTLD-MA 随文本长度变化的曲线

在文本长度介于 5~70 之间时,图 4.12 未显示 MTLD-MA 值,说明 MTLD-MA 测量不适用于小样本。在文本长度达到 125 之后,MTLD-MA 值随文本长度有小幅波动,不再随文本长度呈负向变化。

第五章 文本可读性测量

本章利用 R 数据包 koRpus,以复述故事文档 story.txt 为例,介绍文本可读性多种测量方法和一种听力文本难易度的简单测量方法,包括自动化可读性指数(ARI)、科尔曼-廖(Coleman-Liau)指数、戴尔-乔尔(Dale-Chall)可读性新公式、弗莱什(Flesch)阅读难易度、法尔-詹金斯-帕特森(Farr-Jenkins-Paterson)指数、弗莱什-金凯德(Flesch-Kincaid)年级水平、安德森(Anderson)可读性指数、复杂词词频修正指数、复杂词简单测量(SMOG)和听力难易度公式。

5.1 自动化可读性指数(ARI)

自动化可读性指数(Automated Readability Index, ARI)利用文本包含的句子数、词语数和字母数测量文本可读性。ARI 的计算公式为 $ARI = 0.5 \times \frac{W}{St} + 4.71 \times \frac{C}{W} - 21.43$,其中 W 代表文本包含的词数(即文本长度),St 代表文本包含的句子数,C 代表字符数(常指字母数),其他的数字是回归分析得到的常数(Smith et al.,1967)。

下面以第四章使用的文档 story.txt 为例,利用公式编写 R 命令计算 ARI。在编写过程中,我们首先利用数据包 koRpus 中的函数 treetag()对文档开展形符化和词性标注,然后利用函数 describe()计算相关统计量,并利用提取符($)提取公式中包括的变量值,最后利用 ARI 公式计算结果,利用函数 round()使结果保留两位小数。针对本例,计算 ARI 的命令和统计分析结果如下:

```
> require(koRpus)
> require(koRpus.lang.en)
> story.tag<-treetag("D:/Rpackage/story.txt",treetagger=
"manual",lang="en",TT.options=list(path="C:/TreeTagger",preset="en"))
> res<-describe(story.tag)
> W<-res $ words
> St<-res $ sentences
> C<-res $ letters.only
> ARI<-0.5*W/St+4.71*C/W-21.43;round(ARI,digits=2)
[1] 4.1
```

用数据包 koRpus 计算 ARI 值的简单函数是 ARI(txt.file),其中 txt.file 是 koRpus 的文本对象。另外一个计算 ARI 的通用函数是 readability(txt.file,index="ARI"),其中 txt.file 是 koRpus 的文本对象,变元 index 指定使用文本可读性指标的名称。本例利用以上两个函数计算文本 story.txt 中的 ARI 值的 R 命令和统计分析结果如下:

```
> ARI(story.tag)
Automated Readability Index (ARI)
  Parameters: default
      Grade: 4.1


Text language: en

> readability(story.tag,index="ARI")
Automated Readability Index (ARI)
  Parameters: default
      Grade: 4.1


Text language: en
```

以上两个 R 命令计算的结果与自编 R 命令计算的结果相同,即 ARI 值为 4.1。这意味着,美国 4 年级小学生能够读懂文档 story.txt。

5.2 科尔曼-廖指数

科尔曼-廖指数(Coleman-Liau Index)用于近似评估文本的可读性。函数 coleman.liau() 首先估计完形填空测试百分数(cloze percentage),计算公式为 $CL_{ECP} = 141.8401 - 0.214590 \times \frac{100 \times C}{W} + 1.079812 \times \frac{100 \times St}{W}$,然后利用完形填空测试百分数换算对应的年级,计算公式为 $CL_{grade} = -27.4004 \times \frac{CL_{ECP}}{100} + 23.06395$(Coleman et al.,1975)。利用完形填空测试的原理是,随着阅读技能的提升,读者能够更好地填写出缺失的词(DuBay,2004)[27]。根据威廉·杜拜(William DuBay),完形填空测试使用的文本有规律地删除词语,要求读者填写空格,正确填写词的百分比为完形测试得分(DuBay,2004)。CL_{grade} 的简短公式为 $CL_{short} = 5.88 \times \frac{C}{W} - 29.6 \times \frac{St}{W} - 15.8$。在以上公式中,$W$ 代表文本包含的词数,St 代表文本包含的句子数,C 代表字符数(常指字母数)。

下面以 story.txt 为例,利用公式编写 R 命令计算科尔曼-廖指数。上一节已经计算了

W、St 和 C。本例利用公式计算科尔曼-廖指数的 R 命令和统计分析结果如下：

```
> CL_ECP<-141.8401 - 0.214590 * 100 * C/W + 1.079812 * 100 * St/W
> round(CL_ECP,digits=0)
[1] 60
> CL_grade<--27.4004 * CL_ECP / 100 + 23.06395; round(CL_grade,digits=1)
[1] 6.6
> CL_short<-5.88 * C / W - 29.6 * St / W - 15.8; round(CL_short,digits=1)
[1] 6.6
```

利用数据包 koRpus 计算科尔曼-廖指数的函数是 coleman.liau（txt.file），其中 txt.file 是 koRpus 的文本对象。另外一个计算科尔曼-廖指数的通用函数是 readability（txt.file，index="Coleman.Liau"），其中 txt.file 是 koRpus 的文本对象，变元 index 指定使用文本可读性指标的名称。本例利用以上两个函数计算文档 story.txt 中科尔曼-廖指数的 R 命令和统计分析结果如下：

```
> coleman.liau (story.tag)
Coleman-Liau
  Parameters: default
       ECP: 60% (estimted cloze percentage)
       Grade: 6.57
       Grade: 6.57 (short formula)

Text language: en

> readability(story.tag, index = "Coleman.Liau")
Coleman-Liau
  Parameters: default
       ECP: 60% (estimted cloze percentage)
       Grade: 6.57
       Grade: 6.57 (short formula)

Text language: en
```

以上两个 R 命令计算的结果与自编 R 命令计算的结果基本相同，即科尔曼-廖指数值为 6.57。这意味着，美国 7 年级初中学生能够读懂文档 story.txt。由于计算方法的不同，可读性指标值的结果也可能会有差异。例如，story.txt 的 ARI 值为 4.1，而科尔曼-廖指数值为 6.57。

5.3 戴尔-乔尔可读性新公式

戴尔-乔尔可读性新公式（New Dale-Chall Readability Formula）为 $DC_{new} = 64 - 0.95 \times \frac{100 \times W_{-WL}}{W} - 0.69 \times \frac{W}{St}$，其中 W 代表文本包含的词数，St 代表文本包含的句子数，W_{-WL} 代表不在戴尔-乔尔 3 000 常用词词表中的文本词数（Chall et al.，1995）。戴尔-乔尔 3 000 常用词词表下载的网址为：https://countwordsworth.com/download/DaleChallEasyWordList.txt。该词表实际包括 2 950 个词符。

我们利用上面的公式和词表计算戴尔-乔尔可读性新公式。先把词表放在 D:\Rpackage 文件夹中，变量名为 DaleChall。首先，我们利用 R 命令得到文本的形符表，取名称为 tokens，其次利用函数 describe() 和提取符（$）得到文本词数（$W$）和句子数（$St$）。然后，利用以下命令把戴尔-乔尔 3 000 常用词词表加载到 R 操作界面：

```
DaleChall <- scan("D:/Rpackage/DaleChall.txt", what = "char", sep = "", quiet = TRUE)
```

下一步利用 R 逻辑符！和函数 length() 计算不包括在文本中的形符数（W_{WL}）。最后，利用前面的公式计算戴尔-乔尔可读性新公式值。针对文档 story.txt，计算戴尔-乔尔可读性新公式值的 R 命令和统计分析结果如下：

```
> tokens<-story.tag[,"token"]
> tokens<-gsub("[[:punct:]]","",tokens)
> tokens<-tolower(tokens[which(tokens!="")])
> tokens[274]<-"don't"
> tokens[275]<-""
> tokens<-tokens[which(tokens!="")]
> DaleChall<-scan("D:/Rpackage/DaleChall.txt",
what = "char", sep="", quiet = TRUE)
> W<-length(tokens)
> St<-describe(story.tag)$sentences
> W_WL<-length(tokens[!tokens%in%DaleChall])
> DC_new<-64-0.95*100*W_WL/W-0.69*W/St;round(DC_new,digits=2)
[1] 40.41
```

在以上命令中，使用 tokens[274]<-"don't" 和 tokens[275]<-"" 是为了把"do"和"n't"还原为原文中的"don't"。戴尔-乔尔 3 000 常用词词表包括"don't"。执行 R 命令 grep("don't",DaleChall)，得到"don't"在词表中的位置是"752"。

利用数据包 koRpus 计算戴尔-乔尔可读性新公式值的函数是 dale.chall（txt.file,

word.list），其中 txt.file 是 koRpus 的文本对象，word.list 指定要添加的词表。另外一个计算戴尔-乔尔可读性新公式值的通用函数是 readability（txt.file，word.lists＝list（），index），其中 txt.file 是 koRpus 的文本对象，word.lists＝list（）指定要使用的词表，变元 index 指定使用文本可读性指标的名称。本例利用以上两个函数计算文本 story.txt 中的戴尔-乔尔可读性新公式值的 R 命令和统计分析结果如下：

```
> dale.chall（story.tag，word.list＝DaleChall）
Dale-Chall Readability Formula
  Parameters: custom
  Not on list: 17%
    Raw value: 40.43
       Grade: 5-6
         Age: 10-12

Text language: en

> readability（story.tag，word.lists = list（Dale.Chall=DaleChall），index = "Dale.Chall"）
Dale-Chall Readability Formula
  Parameters: New Dale-Chall（1995）
Not on list: 17%
    Raw value: 40.43
       Grade: 5-6
         Age: 10-12

Text language: en
```

以上两个 R 命令计算的结果与自编 R 命令计算得到的可读性指标原始分基本相同，即戴尔-乔尔可读性新公式值为 40.43。这两个函数统计分析的结果还显示，文档 story.txt 适合于美国 5~6 年级的学生阅读，适合的学生年龄为 10~12 岁。

5.4 弗莱什阅读难易度

弗莱什阅读难易度（Flesch Reading Ease）考虑文本长度、句子数和音节数，计算公式为 $F_{EN} = 206.835 - 1.015 \times \dfrac{W}{St} - 84.6 \times \dfrac{Sy}{W}$，其中 W 代表文本包含的词数，St 代表文本包含的句子数，Sy 代表文本中各个词的音节数（Flesch，1948）。

鲁道夫·弗莱什（Rudolph Flesch）提供了文本阅读难易度值与年级的对应关系，如表 5.1 所示。表中显示，弗莱什阅读难易度值越大，文本越容易（Flesch，1949）。

表 5.1 弗莱什阅读难易度等级和年级划分

	弗莱什阅读难易度分值	年级估计
很容易	90~100	5 年级
容易	80~89	6 年级
较容易	70~79	7 年级
正常	60~69	8、9 年级
较难	50~59	10~12 年级
难	30~49	13~16 年级(大学)
很难	0~29	大学研究生

我们下面利用前面的公式计算弗莱什阅读难易度。首先,利用函数 treetag() 对文本开展形符化。然后,利用函数 hyphen() 对词符划分音节,并利用 R 累计函数 sum() 计算总音节数(Sy)。函数 hyphen() 的主要结构是 hyphen(words, hyph.pattern, min.length=4, rm.hyph=TRUE, quiet=FALSE),其中 words 是词向量,hyph.pattern 是命名所用模式语言的有效字符串,hyph.pattern="en" 设置英语连字法。变元 min.length=4 指函数默认使用连字符时单词必须至少包括 4 个字母。变元 rm.hyph=TRUE 表示在模式匹配之前删除单词中出现的连字符。变元 quiet=FALSE 表示不显示状态信息。文本长度(即总词符数)和句子数的计算同上一节一样利用函数 describe()。最后,利用弗莱什阅读难易度公式计算文本难易度。针对文档 story.txt,计算弗莱什阅读难易度的 R 命令和统计分析结果如下:

```
> require(koRpus)
> require(koRpus.lang.en)
> story.tag<-treetag("D:/Rpackage/story.txt",treetagger
  ="manual",lang="en",TT.options=list(path="C:/TreeTagger",preset="en"))
> tokens<-story.tag[,"token"]
> tokens<-gsub("[[:punct:]]","",tokens)
> tokens<-tolower(tokens[which(tokens!="")])
> tokens<-tokens[which(tokens!="")]
> Sy<-sum(hyphen(tokens,hyph.pattern="en",quiet=TRUE)[,1])
> res<-describe(story.tag)
> W<-length(tokens)
> St<-res $ sentences
> F_EN=206.835-1.015*W/St-84.6*Sy/W; F_EN
[1] 85.34399
```

以上结果显示,文档 story.txt 的弗莱什阅读难易度约为 85.34。

利用数据包 koRpus 计算弗莱什阅读难易度的函数是 flesch(txt.file),其中 txt.file 是 koRpus 的文本对象。另外一个计算弗莱什阅读难易度的通用函数是 readability(txt.file, index），其中 txt.file 是 koRpus 的文本对象，变元 index 指定使用文本可读性指标的名称。本例利用以上两个函数计算文档 story.txt 中的弗莱什阅读难易度的 R 命令和统计分析结果如下：

```
> flesch(story.tag,quiet=TRUE)
Flesch Reading Ease
   Parameters: en (Flesch)
          RE: 85.34
       Grade: 6

Text language: en

> readability(story.tag,index="Flesch",quiet=TRUE)
Flesch Reading Ease
   Parameters: en (Flesch)
          RE: 85.34
       Grade: 6

Text language: en
```

以上两个 R 命令计算的结果相同，即弗莱什阅读难易度值为 85.34，与自编 R 命令计算得到的结果一致。根据表 5.1，文档 story.txt 难易度等级为"容易"，适合 6 年级的学生阅读。以上两个函数计算的结果还显示，文档 story.txt 适合于美国 6 年级的学生阅读，与戴尔-乔尔可读性新公式推算的年级大致相同。

5.5 法尔-詹金斯-帕特森指数

法尔、詹金斯和帕特森（Farr, Jenkins & Paterson）对弗莱什提出的阅读难易度公式进行了简化（Flesch, 1948），简化公式为 $FJP = -31.517 - 1.015 \times \frac{W}{St} + 1.599 \times \frac{W^{1Sy}}{W} \times 100$，其中 W 代表文本包含的词数，St 代表文本包含的句子数，$\frac{W}{St}$ 代表平均句长，W^{1Sy} 代表只有一个音节的词数，$\frac{W^{1Sy}}{W} \times 100$ 代表每 100 个词中只包含一个音节的词数（Farr et al., 1951）。

如果要编写 R 命令计算文档 story.txt 中的法尔-詹金斯-帕特森（Farr-Jenkins-Paterson，FJP）指数，那么需要利用函数 hyphen() 对词符划分音节，并利用逻辑符（==）和累计函数 sum() 计算文本中只包括一个音节的单词总数。利用上一节得到的 koRpus 对象 story.tag，计算法尔-詹金斯-帕特森指数的 R 命令和统计分析结果如下：

```
> require(koRpus)
> require(koRpus.lang.en)
> story.tag<-treetag("D:/Rpackage/story.txt",treetagger
    ="manual",lang="en",TT.options=list(path="C:/TreeTagger",preset="en"))
> tokens<-story.tag[,"token"]
> tokens<-gsub("[[:punct:]]","",tokens)
> tokens<-tolower(tokens[which(tokens!="")])
> tokens[274]<-"don't"
> tokens[275]<-""
> tokens[301]<-"boy's"
> tokens[302]<-""
> tokens<-tokens[which(tokens!="")]
> W<-describe(story.tag)$words
> St<-describe(story.tag)$sentences
> Syl<-sum(hyphen(story.tag,hyph.pattern="en",quiet=TRUE)[,1]==1)
> FJP <- -31.517 - 1.015 * W/St + 1.599 * Syl * 100/W; round(FJP,digits=2)
[1] 78.44
```

利用数据包 koRpus 计算法尔-詹金斯-帕特森指数的函数是 farr.jenkins.paterson(txt.file)，其中 txt.file 是 koRpus 的文本对象。另外一个计算该指数的通用函数是 readability(txt.file, index)，其中 txt.file 是 koRpus 的文本对象，变元 index 指定使用文本可读性指数的名称。本例利用以上两个函数计算文档 story.txt 中的法尔-詹金斯-帕特森指数的 R 命令和统计分析结果如下：

```
> farr.jenkins.paterson(story.tag,quiet=TRUE)
Farr-Jenkins-Paterson
  Parameters: default
        RE: 78.44
        Grade: 7

Text language: en

> readability(story.tag,index="Farr.Jenkins.Paterson",quiet=TRUE)
Farr-Jenkins-Paterson
  Parameters: default
        RE: 78.44
        Grade: 7

Text language: en
```

以上两个 R 命令计算的结果与自编 R 命令计算得到的法尔-詹金斯-帕特森指数值相同，即法尔-詹金斯-帕特森阅读难易度值为 78.44。这两个命令还显示，文档 story.txt 适合于美国 7 年级的学生阅读，与根据弗莱什阅读难易度推算的年级差一个级别。

5.6 弗莱什-金凯德年级水平

弗莱什-金凯德年级水平（Flesch-Kincaid Grade Level）的计算公式为 FK = $0.39 \times \frac{W}{St}$ + $11.8 \times \frac{Sy}{W}$ − 15.59，其中 W 代表文本包含的词数，St 代表文本包含的句子数，$\frac{Sy}{W}$ 代表平均每个词包含的音节数（Kincaid et al., 1975）。

我们利用上面的公式计算弗莱什-金凯德年级水平。5.4 节计算弗莱什阅读难易度时，利用函数 treetag() 得到的 koRpus 文本对象 story.tag 和函数 hyphen() 计算出了所有词包含的音节数。执行以下 R 命令即可计算出弗莱什-金凯德年级水平：

```
> require(koRpus)
> require(koRpus.lang.en)
> story.tag<-treetag("D:/Rpackage/story.txt",treetagger
    ="manual",lang="en",TT.options=list(path="C:/TreeTagger",preset="en"))
> tokens<-story.tag[,"token"]
> tokens<-gsub("[[:punct:]]","",tokens)
> tokens<-tolower(tokens[which(tokens!="")])
> tokens[274]<-"don't"
> tokens[275]<-""
> tokens[301]<-"boy's"
> tokens[302]<-""
> tokens<-tokens[which(tokens!="")]
> W<-length(tokens)
> St<-describe(story.tag)$sentences
> Sy<-sum(hyphen(tokens,hyph.pattern="en",quiet=TRUE)[,1])
> FK<-0.39*W/St+11.8*Sy/W-15.59;round(FK,digits=2)
[1] 4.05
```

利用数据包 koRpus 计算弗莱什-金凯德年级水平的函数是 flesch.kincaid(txt.file)，其中 txt.file 是 koRpus 的文本对象。另外一个计算弗莱什-金凯德年级水平的通用函数是 readability(txt.file, index)，其中 txt.file 是 koRpus 的文本对象，变元 index 指定使用文本可读性指标的名称。本例利用以上两个函数计算文档 story.txt 中的弗莱什-金凯德年级水平的 R 命令和统计分析结果如下：

```
> flesch.kincaid(story.tag)
Hyphenation (language: en)

Flesch-Kincaid Grade Level
    Parameters: default
        Grade: 4.06
            Age: 9.06

Text language: en

> readability(story.tag, index = "Flesch.Kincaid")
Hyphenation (language: en)

Flesch-Kincaid Grade Level
    Parameters: default
        Grade: 4.06
            Age: 9.06

Text language: en
```

以上两个 R 命令计算的结果与自编 R 命令计算得到的弗莱什-金凯德年级水平基本相同,即文档 story.txt 的弗莱什-金凯德年级水平为 4。这意味着,该文档适合于美国 4 年级的学生阅读,与根据弗莱什阅读难易度推算的年级差两个级别,与法尔-詹金斯-帕特森指数推算的级别差三个等级。

5.7 安德森可读性指数

安德森可读性指数(Anderson's Readability Index, RIX)的计算公式为 $RIX = \dfrac{W_{7C}}{St}$,其中 W_{7C} 代表文本中至少包括 7 个字母的单词数,St 代表文本包含的句子数(Anderson,1983)。

对于文本 story.txt,要计算安德森可读性指数,需要利用形符化的 koRpus 文本 story.tag 计算文本中至少包括 7 个字母的单词数,再利用前面提到的函数 describe() 计算句子数。针对文本 story.txt,R 命令和统计分析结果如下:

```
> W_7C<-sum(story.tag[,"lttr"]>=7)
> St<-describe(story.tag) $ sentences
> RIX<-W_7C/St; round(RIX, digits = 2)
[1] 1.71
```

根据数据包 koRpus 的介绍,安德森可读性指数值小于 1.8 的文本被认为很容易阅读,安德森可读性指数值在 3.7 左右的文本被认为正常阅读,安德森可读性指数值大于 7.2 的文本被认为很难阅读。本例中安德森可读性指数值小于 1.8,因而文档 story.txt 很容易阅读。

利用数据包 koRpus 计算安德森可读性指数的函数是 RIX(txt.file),其中 txt.file 是 koRpus 的文本对象。另外一个计算安德森可读性指数的通用函数是 readability(txt.file, index),其中 txt.file 是 koRpus 的文本对象,变元 index 指定使用文本可读性指标的名称。本例利用以上两个函数计算文档 story.txt 中安德森可读性指数的 R 命令和统计分析结果如下:

```
> RIX(story.tag)
Readability Index (RIX)
  Parameters: default
        Index: 1.71
        Grade: 5


Text language: en

> readability(story.tag, index = "RIX")
Readability Index (RIX)
  Parameters: default
        Index: 1.71
        Grade: 5


Text language: en
```

以上两个 R 命令计算的结果与自编 R 命令计算得到的结果相同。另外,这两个命令得到的结果中还报告文档 story.txt 适合于美国 5 年级的学生阅读。由此可见,这个文档很容易读懂。

5.8 复杂词词频修正指数

理查德·德·鲍尔斯等人对罗伯特·冈宁提出的文本可读性公式 FOG(Frequency of Gobbledygook,复杂词频率)指数进行了参数调整(Gunning,1952),计算公式为 $FOG_{PSK} = 3.0680 + \left(0.0877 \times \dfrac{W}{St}\right) + \left(0.0984 \times \dfrac{100 \times W_{3Sy}}{W}\right)$,其中 W_{3Sy} 代表文本中至少包括 3 个音节的单词数,W 指文本包含的词数,St 代表文本包含的句子数(Powers et al.,1958)。

要计算文档 story.txt 的复杂词词频修正指数,需要根据函数 treetag() 得到的结果,利用函数 hyphen() 计算每个词包含的音节数,再利用逻辑符>=和累计函数 sum() 计算复杂词词频修正指数值。文本词数和句子数的计算分别调用函数 length() 和 describe()。本例 R 命令和统计分析结果如下:

```
> require(koRpus)
> require(koRpus.lang.en)
> story.tag<-treetag("D:/Rpackage/story.txt",treetagger
  ="manual",lang="en",TT.options=list(path="C:/TreeTagger",preset="en"))
> tokens<-story.tag[,"token"]
> tokens<-gsub("[[:punct:]]","",tokens)
> tokens<-tolower(tokens[which(tokens!="")])
> tokens[274]<-"don't"
> tokens[275]<-""
> tokens[301]<-"boy's"
> tokens[302]<-""
> tokens<-tokens[which(tokens!="")]
> W<-length(tokens)
> St<-describe(story.tag)$sentences
> W_3Sy<-sum(hyphen(tokens,hyph.pattern="en",
  quiet=TRUE)[,1]>=3)
> FOG_PSK<-3.0680+(0.0877*W/St)+(0.0984*100*W_3Sy/W)
> round(FOG_PSK,digits=2)
[1] 4.57
```

利用数据包 koRpus 计算复杂词词频修正指数的函数是 FOG(txt.file),其中 txt.file 是 koRpus 的文本对象。另外一个计算复杂词词频修正指数的通用函数是 readability(txt.file, index="FOG.PSK"),其中 txt.file 是 koRpus 的文本对象,变元 index="FOG.PSK" 指定使用文本可读性指标的名称。本例利用以上两个函数计算文档 story.txt 中的复杂词词频修正指数的 R 命令和统计分析结果如下:

```
> FOG(story.tag,parameters="PSK",quiet=TRUE)

Gunning Frequency of Gobbledygook (FOG)
  Parameters: Powers-Sumner-Kearl
      Grade: 4.55

Text language: en
```

```
> readability(story.tag,index="FOG.PSK",quiet=TRUE)
Gunning Frequency of Gobbledygook (FOG)
  Parameters: Powers-Sumner-Kearl
       Grade: 4.55

Text language: en
```

以上两个 R 命令计算的结果与自编 R 命令计算得到的结果基本一致。结果表明,文档 story.txt 适合于美国 4~5 年级的学生阅读。

5.9 复杂词简单测量(SMOG)

盖·哈利·麦克劳林(G. Harry McLaughlin)认为词长与词义精确性有关系,读者通常需要付出更多的努力确定长词的充分意义,因而提出复杂词简单测量(Simple Measure of Gobbledygook,SMOG),把多音节词数作为文本可读性的一个重要变量。复杂词简单测量利用句子数和多音节词数简化文本可读性测量,计算公式为 $SMOG = 1.043 \times \sqrt{W_{3Sy} \times \dfrac{30}{St}} + 3.1291$,其中 W_{3Sy} 代表文本中至少包括 3 个音节的单词数,St 代表文本包含的句子数(McLaughlin,1969)。

要计算文档 story.txt 的复杂词简单测量值,需要根据函数 treetag() 得到的结果,利用函数 hyphen() 计算每个词音节数,再利用逻辑符 >= 和累计函数 sum() 计算复杂词简单测量值。同上一节一样,文本句子数的计算利用函数 describe()。本例完整的 R 命令和统计分析结果如下:

```
> require(koRpus)
> require(koRpus.lang.en)
> story.tag<-treetag("D:/Rpackage/story.txt",treetagger
  ="manual",lang="en",TT.options=list(path="C:/TreeTagger",preset="en"))
> W_3Sy<-sum(hyphen(story.tag[,"token"],hyph.pattern="en",
  quiet=TRUE)[,1]>=3)
> res<-describe(story.tag)
> St<-res $ sentences
> round(SMOG<-1.043*sqrt(W_3Sy*30/St)+3.1291,digits=2)
[1] 7.6
```

以上结果显示,文档 story.txt 的复杂词简单测量值为 7.6。利用数据包 koRpus 计算复杂词简单测量值的函数是 SMOG(txt.file),其中 txt.file 是 koRpus 的文本对象。另外一个计算复杂词简单测量值的通用函数是 readability(txt.file,index),其中 txt.file 是 koRpus 的文本对象，变元 index 指定使用文本可读性指标的名称。本例利用以上两个函数计算文档 story.txt 中的复杂词简单测量值的 R 命令和统计分析结果如下：

```
> SMOG(story.tag, quiet=TRUE)
Simple Measure of Gobbledygook (SMOG)
   Parameters: default
       Grade: 7.6
           Age: 12.6

Text language: en

> readability(story.tag, index="SMOG")
Hyphenation (language: en)

Simple Measure of Gobbledygook (SMOG)
   Parameters: default
       Grade: 7.6
           Age: 12.6

Text language: en
```

以上两个 R 命令计算的结果相同，即复杂词简单测量值为 7.6（年级水平为 8）。另外，这两个函数得到的结果中还报告文档 story.txt 适合于美国 13 岁左右的学生阅读。

5.10 听力难易度公式

埃尔文·易·方（Irving E. Fang）根据听力文本中词语的多音节数提出听力理解难易度公式（Easy Listening Formula，ELF），即 ELF = $\frac{W_{2Sy}}{St}$，其中 W_{2Sy} 代表文本中至少包括 2 个音节的单词数，St 代表文本包含的句子数（Fang,1966）。这个公式比较简单，只考虑包括 2 个或 2 个以上音节的单词数和句子数。在文档 story.txt 中，计算包括 2 个或 2 个以上音节的单词数可以在利用函数 treetag() 的基础上再利用音节划分函数 hyphen() 和累计函数 sum()。本例 R 命令和统计分析结果如下：

```
> require(koRpus)
> require(koRpus.lang.en)
> story.tag<-treetag("D:/Rpackage/story.txt",treetagger
   ="manual",lang="en",TT.options=list(path="C:/TreeTagger",preset="en"))
> W_2Sy<-sum(hyphen(story.tag[,"token"], hyph.pattern = "en",quiet = TRUE)[,1]>=2)
```

```
> St<-describe(story.tag) $ sentences
> ELF<-W_2Sy/St; round(ELF,digits=2)
[1] 2.65
```

利用数据包 koRpus 计算听力难易度值的函数是 readability(txt.file, index="ELF"),其中 txt.file 是 koRpus 的文本对象,变元 index="ELF" 指定使用文本测量指标的名称为"ELF"。本例利用以上函数计算文档 story.txt 中听力难易度值的 R 命令和统计分析结果如下:

```
> readability(story.tag,index="ELF",quiet=TRUE)
Easy Listening Formula
   Parameters: default
       Exsyls: 82
        Score: 2.65

Text language: en
```

以上结果显示,听力难易度值为 2.65。听力难易度值越大,听力难度就越大。由于听力难易度只考虑多音节词语数和句子数,因而它只是简单化的听力难易度测量方法。

第六章　N元组和关键词

我们在前面几章介绍文本统计分析方法时以单个的形符（tokens）为单位。单个的形符可以纳入N元组（N-grams）这个大类中。N元组是文本中连续出现的长度为n的形符串，其中n是不小于1的整数。长度n为1的元组称作单元组（unigrams），长度n为2的元组称作二元组（bigrams），长度n为3的元组称作三元组（trigrams），以此类推。本章以二元组和三元组为例介绍n元组在文本挖掘（text mining，即文本分析，是自然语言处理的一个重要领域）中的应用。另外，本章还介绍文本比较关键词提取方法。统计分析使用的R数据包为readtext、koRpus、koRpus.lang.en、nsyllable、quanteda、quanteda.textstats、quanteda.textplots、RColorBrewer和DescTools。

6.1　文本数据计量分析数据包安装与初试

R数据包quanteda由肯尼斯·贝诺特等人开发（Benoit et al., 2018）。数据包quanteda提供文本数据计量分析（quantitative analysis of textual data）工具箱。在最初的设计中，这个数据包包括四个主要模块。第一个模块是常见的自然语言处理方法。quanteda允许用户把语料库（corpus，指文本集）快速转化为文档-特征矩阵（Document-Feature Matrix, DFM），并利用得到的矩阵开展常见的自然语言处理，如开展形符化和生成N元组等。第二个模块是文本统计（textual statistics），包括关键词分析、搭配分析和文本相似度分析等。第三个模块是文本可视化（text visualization），包括生成词云图和关键词图。第四个模块对文本数据进行建模和分类。在2021年发布的数据包中，quanteda.textstats、quanteda.textplots以及quanteda.textmodels从原数据包quanteda中分离出来，形成四个彼此联系的数据包，其中quanteda是基础数据包。本章调用前三个数据包。

第一章介绍了R数据包的安装方法。要安装quanteda数据包，若接受默认安装路径，打开R操作界面，输入install.packages("quanteda")，选择就近的站点在线安装。用户也可以自定义安装路径，如：install.packages("quanteda", lib = "C:/Program Files/R/R-4.2.2/library/")。安装完成后，利用R命令require(quanteda)调用quanteda数据包。对于quanteda.textstats和quanteda.textplots包，安装与调用方法相同。

在第一次加载R包quanteda时，你可能会看到以下警告信息：

```
Warning messages:
1: In stringi::stri_info():
```

> Your current locale is not in the list of available locales. Some functions may not work properly. Refer to stri_locale_list( ) for more details on known locale specifiers.
>
> 2: In stringi::stri_info( ) :
>
> Your current locale is not in the list of available locales. Some functions may not work properly. Refer to stri_locale_list( ) for more details on known locale specifiers.

这个警告表明,我们当前使用的语言环境不在函数 stri_locale_list( )提供的列表中。要去除警告,我们可以关闭 R 工作界面并重新打开,输入:stringi::stri_locale_set("zh_Hans_CN"),然后我们再输入 R 命令:require(quanteda),按回车键即可正常运行。

下面利用第三、四章使用的文档 story.txt 初试以上三个数据包的几个函数。我们先利用数据包 quanteda 中的函数 topfeatures( )计算文档中出现频次最高的 10 个词。这个函数的主要变元是 dfm 对象,其他变元为默认方式,如默认变元 $n=10$(即 10 个特征),默认 decreasing=TRUE(即特征频次按照降序排列)。在利用这个函数之前要先调用数据包 readtext 中的函数 readtext( )把 story.txt 读入 R 工作界面,并利用提取符($)提取其中的字符串。然后,调用数据包 quanteda 中的函数 tokens( )对字符串开展形符化。该函数的主要变元是字符串或语料库对象(调用函数 corpus( )),其他变元包括默认的 remove_punct=FALSE 等。我们不打算考虑标点符号,因而设置 remove_punct=TRUE。在文本处理过程中增加了函数 tokens_replace( ),目的是把词符"boy's"改为"boy"。接下来利用函数 dfm( )把文本中的形符转化为文档-特征矩阵(dfm)。该函数默认把词符的大写改为小写(tolower=TRUE)。本例 R 命令和统计分析结果如下所示:

```
> require(readtext)
> story<-readtext("D:/Rpackage/story.txt")$text
> require(quanteda)
> story_token<-tokens(story,remove_punct=TRUE)
> story_replace<-tokens_replace(story_token,pattern = "boy's", replacement = "boy")
> story_dfm<-dfm(story_replace)
> topfeatures(story_dfm)
   the    boy    and    man little   race   wise    was      a  crowd
    37     15     13     10      9      9      8      7      7      6
```

以上结果与第三章 3.4.1 节的词频统计结果相同,如 the 出现 37 次,boy 出现 15 次。

同 R 数据包 koRpus 一样,数据包 quanteda.textstats 也能够利用函数 textstat_lexdiv( )计算文本词汇多样性指标,利用函数 textstat_readability( )计算文本可读性指标。关于各个测量指标,详见数据包 quanteda.textstats 使用说明。例如,我们要测量 MATTR(移动平均类符-形符比),我们需要在函数 textstat_lexdiv( )中添加 dfm 对象或 tokens 对象,设置变元 measure="MATTR",其他变元采用默认的方法,如 remove_punct=TRUE(除去标点符号),MATTR_window=100L(窗口大小为 100 个形符)。计算文本 story.txt 中的词汇多样性测量 MATTR 值的 R 命令和统计分析结果如下：

```
> require(quanteda.textstats)
> textstat_lexdiv(story_replace,measure="MATTR")
  document         MATTR
1   text1        0.6452966
```

函数 textstat_readability()默认使用弗莱什阅读难易度测量。计算 story.txt 中的弗莱什阅读难易度值的 R 命令和统计分析结果如下：

```
> require(quanteda.textstats)
> textstat_readability(story)
  document    Flesch
1   text1   82.86919
```

以上结果与第五章的弗莱什阅读难易度值（85.34）存在少许差异。究其原因，主要是文本中包括的总词数（即文本长度）、句子数和词音节数的计算方法不同。函数 quanteda.textstats 对文本音节数的计算自动调用数据包 nsyllable 中的函数 nsyllable()，文本句子数的计算调用函数 nsentence()。计算文档 story.txt 中包括的音节数、词数和句子数以及弗莱什阅读难易度值的 R 命令和统计分析结果如下：

```
> require(quanteda)
> story_token<-tokens(story,remove_punct=TRUE)
> require(nsyllable)
> Sy<-sum(nsyllable(story_token[[1]]))
> W<-length(story_token[[1]])
> St<-nsentence(story)
> Sy;W;St[[1]]
[1] 446
[1] 335
[1] 30
> 206.835-1.015*W/St-84.6*Sy/W
   text1
82.86919
```

以上结果显示，调用 quanteda 包相关函数得到文本长度为 335 个词，句子数为 30 个，音节数为 446 个。调用数据包 koRpus、koRpus.lang.en 和函数 treetag()等计算弗莱什阅读难易度值时却发现文本中句子数为 31 个，音节数为 440 个，文本长度为 337 个词。第五章的 R 命令和计算结果重复如下：

```
> require(koRpus)
> require(koRpus.lang.en)
> story.tag<-treetag("D:/Rpackage/story.txt",treetagger
  ="manual",lang="en",TT.options=list(path="C:/TreeTagger",preset="en"))
> tokens<-story.tag[,"token"]
> tokens<-gsub("[[:punct:]]",""  ,tokens)
> tokens<-tolower(tokens[which(tokens!="")])
> tokens<-tokens[which(tokens!="")]
> Sy<-sum(hyphen(tokens,hyph.pattern="en",quiet=TRUE)[,1])
> res<-describe(story.tag)
> W<-length(tokens)
> St<-res $ sentences
> Sy;W;St
[1] 440
[1] 337
[1] 31
> F_EN=206.835-1.015 * W/St-84.6 * Sy/W; F_EN
[1] 85.34399
```

最后，我们利用数据包 quanteda.textplots 中的函数 textplot_wordcloud() 绘制词云图。该函数的主要设置方式为：textplot_wordcloud(x, min_size=0.5, max_size=4, min_count=3, max_words=500, color="darkblue", adjust=0, rotation=0.1, random_order=FALSE, random_color=FALSE, ordered_color=FALSE)，其中 x 是 dfm 对象，min_size 是默认的最小词尺寸，max_size=4 是默认的最大词尺寸，min_count=3 指图中显示词的最低频次默认为 3，max_words=500 指图中显示的词总数最大值默认为 500，color="darkblue" 指图形默认颜色为深蓝色。变元 adjust=0 表示默认的词间距为 0，rotation=0.1 表示旋转 90°的词比率默认为 0.1，random_order=FALSE 表示默认由里向外按词频降序排列，ordered_color=TRUE 表示颜色按顺序分配到词，否则随机着色。在这个例子中，我们使用 tokens_select 排除数据包 quanteda 自带的停用词表 stopwords("en") 中的词。本例词云图的绘制选择 min_count=2, min_size=1, max_size=5, adjust=-0.05，颜色调配利用数据包 RColorBrewer 中的函数 brewer.pal()，R 命令和制图结果如下：

```
> require(readtext)
> story<-readtext("D:/Rpackage/story.txt")
> story<-story $ text
> require(quanteda)
> story_token<-tokens(story,remove_punct=TRUE)
```

```
> story_replace<-tokens_replace(story_token,
   pattern="boy's",replacement="boy")
> story_nostop<-tokens_select(story_replace,pattern=
   stopwords("en"),selection="remove")
> story_dfm<-dfm(story_nostop)
> require(quanteda.textplots)
> require(RColorBrewer)
> pal<-brewer.pal(8,"Dark2")
> textplot_wordcloud(story_dfm,min_count=2,min_size=1,
   max_size=5,color=pal,adjust=-0.05)
```

图 6.1　频数至少为 2 的高频词词云

第三章的图 3.3 显示文本 story.txt 中的词目分布,而图 6.1 显示词符分布。从图 6.1 可以看出,词云图以字体大小显示词频高低,词频越高,词体就越大。在 31 个高频词中,"boy""man""little""race""wise"字体较大,出现的频数也最高。

6.2　两个文本的二元组比较

当我们在前面章节对字符串开展形符化时,我们得到的只是单个的词。虽然单个词在文本挖掘中时常也发挥着重要作用,但是其缺点是忽视语境的作用。N 元组,如二元组或三元组,有助于在某种程度上克服这一局限。如果我们想要知道学习者的听后复述在多大程度上准确重复了听力原文的语言表达和措辞,N 元组切分是恰当的方法。N 元组切分指按照文本词出现的顺序以固定的长度依次分割词符串。例如,对词符串"Once there was a little boy"的二元组切分的结果是:"Once there""there was""was a""a little""little boy"。本节通过比较上一节使用的复述故事原文 story.txt 和学生口语复述文本(编号为 st01)介绍二元组在文本挖掘方面所发挥的重要作用。我们的研究问题是,这名学生的复述在多大

程度上准确再现了听力文本中的二元组信息。学生文本st01.txt(存储在D:\Rpackage文件夹中)全文内容如下:

> Once there was a little boy was hungry for success in the sports. One day in a village, he has a sport game with two other young boy. And he is the winner and a wise man came to watch the sports. The man remain calm and still, showing no excitement. So the little boy ask the wise man why and the wise man stop for the two new competitions. This time, the wise man said," This time only one time to finish the other way. " The boy is the winner, but he asked why there is no cheering happened. The old wise man said, " You can have it again, but this time you should finish it together, all of you finish it together. " This time this little young boy, in order to win, so he walk slowly and took them by arm. At the end, the little young boy asked the wise man," Who is cheer for winner?" and the wise man smiled aloud and go ahead. The little young boy asked his grandpa, " I don't understand who is the cheering for winner?" and the grandpa smiled and go ahead. He said," For this race, one of the three man are winner. You know this race. You can know much more than any other match. Cheering is not for any winner. "

计算学生文本st01.txt中重复出现原文本story.txt的二元组数量包括以下六个步骤。

第一步,调用数据包readtext把存放在D:\Rpackage文件夹中的两个文档以字符串形式调入R工作界面,利用提取符($)提取文本字符串。R命令如下:

```
> require(readtext)
> story<-readtext("D:/Rpackage/story.txt")
> story<-story $ text
> st01<-readtext("D:/Rpackage/st01.txt")
> st01<-st01 $ text
```

第二步,调用数据包quanteda,使用函数corpus()把两个文本转化为语料库,并利用函数变元docnames设置文件名,然后利用函数tokens()把语料库中的字符串形符化,设置变元remove_punct=TRUE去除标点符号。R命令如下:

```
> require(quanteda)
> corp<-corpus(c(story,st01),docnames=c("story","st01"))
> texts_tk<-tokens(corp,remove_punct=TRUE)
```

第三步,利用数据包quanteda中的函数tokens_ngrams(x, n=2L, concatenator="_")把得到的形符(即词符)进行二元组切分。在该函数中,x是形符对象,n=2L指函数默认以2个连续形符作为切分单位,concatenator="_"指函数默认使用下划线(_)合并形符。R命令如下:

```
> texts_bi<-tokens_ngrams(texts_tk, n=2L)
```

第四步,利用提取符([[]])提取两个文本的二元组序列,并利用函数length()和查找逻辑符%in%得到 st01.txt 重复 story.txt 中的二元组数。R 命令如下：

```
> results<-texts_bi[[2]][texts_bi[[2]]%in%texts_bi[[1]]]
```

需要指出的是，这种计算方法会导致过多提取重复出现的二元组数。即是说，如果一个二元组在学生文本中重复 5 次，而在对照原文本中出现 1 次，以上命令会得出 5 次重复二元组的结果，而我们则希望得到重复 1 次的结果，多余的重复不应考虑。第五步是解决方案。

第五步，利用查找逻辑符%in%得到 story.txt 重复 st01.txt 中的二元组，再利用函数 table() 和 data.frame() 得到两个重复词词频表，最后利用简单的 for 循环和函数 min() 计算词表中各个成对词频中的低频词（变量名为 count），利用函数 cbind() 把新生成的词频向量添加到数据框中。R 命令和部分统计分析结果如下：

```
> results_2<-texts_bi[[1]][texts_bi[[1]]%in%texts_bi[[2]]]
> freq<-data.frame(table(results),table(results_2))
> count<-NULL
> for(i in 1:length(freq[,"Freq"])){
+ count[i]<-min(freq[,"Freq"][i],freq[,"Freq.1"][i])
+ }
> freq.df<-cbind(freq,count)
> head(freq.df)
    results   Freq   results_2   Freq.1   count
1   a_little   1     a_little    1        1
2   a_wise     1     a_wise      1        1
3   and_a      1     and_a       2        1
4   and_the    3     and_the     2        2
5   and_took   1     and_took    1        1
6   any_other  1     any_other   1        1
```

以上结果显示，二元组"and_a"在 st01.txt 中出现 1 次，在 story.txt 中出现 2 次，只能按 1 次计数；二元组"and_the"在 st01.txt 中出现 3 次，在 story.txt 中出现 2 次，只能按 2 次计数。

如果我们在 for 循环中不使用函数 min()，而使用 if...else 形式，那么执行以下 R 命令会得到同样的结果：

```
> count<-NULL
> for(i in 1:length(freq[,"Freq"])){
if(freq[,"Freq"][i]<freq[,"Freq.1"][i])
count[i]<- freq[,"Freq"][i]
else count[i]<-freq[,"Freq.1"][i]
}
```

第六步，要计算这名学生的复述在多大程度上准确再现了听力文本中的二元组信息，利用函数 sum( ) 计算学生文本再现听力文本中的二元组数，再利用函数 length( ) 计算听力文本二元组数，再现率即为两者的比率。R 命令和统计分析结果如下：

```
> (coverage<-sum(freq.df $ count)/length(texts_bi[[1]]))
[1] 0.1946108
```

这名学生在全国专业四级笔试中的成绩偏低（70 分），说明其英语水平一般，较少地重复原文二元组数（重复率约为 19%）应该与其英语水平低有关。

6.3 两个文本的三元组比较

两个文本三元组的比较与二元组的比较执行的 R 命令相同，只是在三元组比较中把函数 tokens_ngrams(x, n=2L, concatenator="_") 中的变元设置 n=2L 改为 n=3L。接着上一节的第三个步骤，省去上一节命令 head(freq.df)，本例的 R 命令和统计分析结果如下：

```
> texts_tri<-tokens_ngrams(texts_tk, n=3L)
> results<-texts_tri[[2]][texts_tri[[2]]% in% texts_tri[[1]]]
> results_2<-texts_tri[[1]][texts_tri[[1]]% in% texts_tri[[2]]]
> freq<-data.frame(table(results),table(results_2))
> count<-NULL
> for(i in 1:length(freq[,"Freq"])){
+ if(freq[,"Freq"][i]<freq[,"Freq.1"][i])
+ count[i]<- freq[,"Freq"][i]
+ else count[i]<-freq[,"Freq.1"][i]
+ }
> freq.df<-cbind(freq,count)
> (coverage<-sum(freq.df $ count)/length(texts_tri[[1]]))
[1] 0.06606607
```

听力原文使用的三元组数为 333，学生文本中重复的三元组数只有 22，重复率仅约为 7%。二元组和三元组重复率显示，随着 N 元组长度的增加，重复量减少。

6.4 文本比较关键词提取

6.4.1 文本词语关键性检验方法

词语在文本中的重要性是相对而言的。一个文本中的某个词相对于另一个文本凸显重要性，但是相对于其他文本可能不那么重要。在语料库语言学研究中，文本中发挥重要作用的词称作关键词（keywords）。关键词指某些词在一个语料库中出现的频次明显高于在另一个语料库中出现的频次,能够体现文本的主题。关键词的提取依据是词在一个语料库中的关键性(keyness)。词的关键性通常利用卡方检验(χ^2 test)、费希尔精确检验(Fisher's exact test)或似然比检验(likelihood ratio test)。本节简要介绍这三种方法。

令 a 为某个词在目标语料库 A 中出现的频次,b 为该词在参照语料库 B 中出现的频次,c 为语料库 A 中其他词出现的频次,d 为语料库 B 中其他词出现的频次,则两个样本词符总数为 $N=a+b+c+d$。卡方检验统计量的计算公式为 $\chi^2 = \dfrac{N(|ad-bc|-N/2)^2}{(a+b)(c+d)(a+c)(b+d)}$,其中 $|\cdot|$ 为绝对值符号,$N/2$ 是耶茨校正(Yates' correction)(Oakes,1998)。根据 R 数据包 quanteda.textstats,各个单元格值(a、b、c 和 d)满足以下条件,采用耶茨校正卡方值,否则不校正卡方值:abs(a*d-b*c)>=N/2 && ((a+b)*(a+c)/N<5)||(a+b)*(b+d)/N<5||(a+c)*(c+d)/N<5|(c+d)*(b+d)/N<5,其中 abs() 为绝对值函数,&&(和)和 ||(或者)为逻辑符,* 和 / 分别表示乘号和除号。

这里举两个卡方检验的例子。某个词在两个语料库 A 和 B 中出现的频次分别为 $a=18$,$b=14$,其他词在两个语料库中出现的频次依次为 $c=697$,$d=2\,195$。根据以上耶茨校正条件,本例中卡方值无须校正。R 命令和统计分析结果如下:

```
> a=18;b=14;c=697;d=2195;N=a+b+c+d
> chisq<-N*(abs(a*d-b*c))^2/((a+b)*(c+d)*(a+c)*(b+d))
> p<-1-pchisq(chisq,1)
> cat("chisq:",chisq,"df:",1,"p:",p,"\n")
chisq: 17.7076 df: 1 p: 2.575955e-05
```

以上结果表明,某个词在语料库 A 中出现的频次显著高于在语料库 B 中出现的频次($p<0.001$),因而对语料库 A 来说是重要词。在目标语料库中的关键词称作正向关键词(positive keywords),而在参照语料库中出现的关键词称作负向关键词(negative keywords)。

在第二个例子中,某个词在两个语料库 A 和 B 中出现的频次分别为 $a=11$,$b=8$,其他词在两个语料库中出现的频次依次为 $c=704$,$d=2\,201$。根据以上耶茨校正条件,本例中卡方值需要校正。R 命令和统计分析结果如下:

```
> a=11;b=8;c=704;d=2201;N=a+b+c+d
> chisq<-N*(abs(a*d-b*c)-N/2)^2/((a+b)*(c+d)*(a+c)*(b+d))
> p<-1-pchisq(chisq,1)
> cat("chisq:",chisq,"df:",1,"p:",p,"\n")
chisq: 9.827231 df: 1 p: 0.00171947
```

以上结果表明,某个词在语料库 A 中出现的频次显著高于在语料库 B 中出现的频次($p<0.01$),因而对语料库 A 来说是重要词。

卡方检验 R 函数的基本形式是 chisq.test(x, correct=TRUE),其中 x 是矩阵,correct=

TRUE 表示函数默认采用耶茨校正。上面两个例子的 R 命令和统计分析结果如下：

```
> data1<-matrix(c(18,697,14,2195),ncol=2)
> chisq.test(data1,correct=FALSE)

        Pearson's Chi-squared test

data: data1
X-squared = 17.708, df = 1, p-value = 2.576e-05

> data2<-matrix(c(11,704,8,2201),ncol=2)
> chisq.test(data2)

        Pearson's Chi-squared test with Yates' continuity correction

data: data2
X-squared = 9.8272, df = 1, p-value = 0.001719
```

费希尔精确检验用于计算 2×2 列联表中卡方统计量的精确概率。该检验是计算密集型检验，其详细介绍可以参阅《二语习得研究中的常用统计方法（第二版）》一书（鲍贵 等,2020a）。

费希尔精确检验 R 函数的基本形式为 fisher.test(x)，其中 x 是矩阵。对上面两个例子开展弗希尔精确检验的 R 命令和统计分析结果如下：

```
> data1<-matrix(c(18,697,14,2195),ncol=2)
> fisher.test(data1)

        Fisher's Exact Test for Count Data

data: data1
p-value = 0.0001077
alternative hypothesis: true odds ratio is not equal to 1
95 percent confidence interval:
 1.890469 8.840725
sample estimates:
odds ratio
  4.046303


> data2<-matrix(c(11,704,8,2201),ncol=2)
> fisher.test(data2)
```

```
        Fisher's Exact Test for Count Data

data：data2
p-value = 0.001824
alternative hypothesis：true odds ratio is not equal to 1
95 percent confidence interval：
 1.566901 12.355242
sample estimates：
odds ratio
 4.295782
```

以上结果显示,第一个例子的比数比(odds ratio)为 4.046 303,$p<0.001$,某个词在语料库 A 中出现的频次显著高于在语料库 B 中出现的频次。第二个例子的比数比为 4.295 782,$p<0.01$,也说明某个词在语料库 A 中出现的频次显著高于在语料库 B 中出现的频次。顺便提一下,这里的比数比近似等同于样本值的比数比。例如,第一个例子的样本比数比为 4.048 985$\left(\text{即}\dfrac{18/697}{14/2\ 195}\right)$。函数 fisher.test() 计算得到的比数比利用有条件的最大似然估计(maximum likelihood estimate),而不是利用无条件的最大似然估计(即样本比数比)(Desagulier,2017)。

第三个检验是似然比检验。似然比检验,又称独立性 G 检验,与卡方独立性检验同类,但是在样本量小时(如一个单元格计数小于 5)比卡方独立性检验更好。G 检验统计量 G(又称 G^2)服从卡方分布,因而该统计量又称 χ^2。G 检验统计量的计算公式为 $G = 2 \times \sum_{i=1}^{r}\sum_{j=1}^{c} O_{i,j} \times \ln\left(\dfrac{O_{i,j}}{E_{i,j}}\right)$,其中 O 是观测值,E 是期望值,r 是排数,c 是列数,$\ln$ 是自然对数。在上面第一个例子中,边缘汇总和期望值如表 6.1 所示。

表 6.1　2×2 列联表

	目标词频	非目标词频	边缘汇总
观测值	$O_{11} = 18$	$O_{12} = 14$	$O_{11} + O_{12} = 32$
期望值	$E_{11} = \dfrac{(O_{11}+O_{12}) \times (O_{11}+O_{21})}{N}$ $\approx 7.824\ 897$	$E_{12} = \dfrac{(O_{11}+O_{12}) \times (O_{12}+O_{22})}{N}$ $\approx 24.175\ 1$	
观测值	$O_{21} = 697$	$O_{22} = 2\ 195$	$O_{21} + O_{22} = 2\ 892$
期望值	$E_{21} = \dfrac{(O_{21}+O_{22}) \times (O_{11}+O_{21})}{N}$ $\approx 707.175\ 103$	$E_{22} = \dfrac{(O_{21}+O_{22}) \times (O_{12}+O_{22})}{N}$ $\approx 2\ 184.824\ 9$	
边缘汇总	$O_{11} + O_{21} = 715$	$O_{12} + O_{22} = 2\ 209$	$O_{11} + O_{21} + O_{12} + O_{22}$ $= 2\ 924$

根据表 6.1 的结果,利用以下 R 命令得到 G 检验结果:

```
> G<-2 * sum(18 * log(18/7.824897)+14 * log(14/24.1751)
  +697 * log(697/707.175103)+2195 * log(2195/2184.8249))
> p<-1-pchisq(G,1)
> cat("G:",G,"df:",1,"p:",p,"\n")
G: 14.88918 df: 1 p: 0.0001140152
```

以上结果显示,某个词在语料库 A 中出现的频次显著高于在语料库 B 中出现的频次 ($p < 0.001$)。Williams 校正 G 值公式为 $q = 1.0 + (N/(a+b) + N/(c+d) - 1.0) \times (N/(a+c) + N/(b+d) - 1.0)/(6.0 \times N)$,其中 N 是词符总数。如果要采用 Williams 校正,R 命令和统计检验结果如下:

```
> N=2924
> a=18;b=14;c=697;d=2195
> q<-1.0 + (N/(a+b) + N/(c+d) - 1.0) * (N/(a+c) + N/(b+d) - 1.0)/(6.0 * N)
> G<-G/q
> p<-1-pchisq(G,1)
> cat("G:",G,"df:",1,"p:",p,"\n")
G: 14.55459 df: 1 p: 0.0001361562
```

以上结果显示,Williams 校正前后的 G 值稍有变化,但是统计分析结论相同。

开展 G 检验需要调用 R 数据包 DescTools 中的函数 GTest(x),其中 x 是矩阵(Signorell et al.,2022)。如果要对 G 值进行 Williams 校正,那么在函数中增加变元 correct = "williams"。针对上面的例子,R 命令和统计检验结果如下:

```
> require(DescTools)
> data1<-matrix(c(18,697,14,2195),ncol=2)
> GTest(data1)

        Log likelihood ratio (G-test) test of independence without correction

data:  data1
G = 14.889, X-squared df = 1, p-value = 0.000114
```

6.4.2 比较故事复述原文和学生复述文本中的关键词

本节利用关键词技术比较故事复述原文 story.txt 和学生复述文本 st01.txt 中使用的 20 个高频词。这项任务包括以下六个步骤:

第一步,利用与 6.2 节同样的方法,调用数据包 readtext 把存放在 D:\Rpackage 文件夹

中的两个文档以字符串形式调入 R 工作界面,利用提取符($)提取文本字符串。R 命令如下:

```
> require(readtext)
> story<-readtext("D:/Rpackage/story.txt")
> story<-story $ text
> st01<-readtext("D:/Rpackage/st01.txt")
> st01<-st01 $ text
```

第二步,调用本章使用的三个数据包 quanteda、quanteda.textstats 和 quanteda.textplots。使用数据包 quanteda 中的函数 corpus()把两个文本转化为语料库,并利用函数变元 docnames 设置文件名,然后利用函数 tokens()把语料库中的字符串形符化,设置变元 remove_punct=TRUE 去除标点符号。R 命令如下:

```
> require(quanteda)
> require(quanteda.textstats)
> require(quanteda.textplots)
> corp<-corpus(c(story,st01),docnames=c("story","st01"))
> texts_tk<-tokens(corp,remove_punct=TRUE)
```

第三步,调用数据包 quanteda 中的函数 tokens_select()剔除文本中的停用词。停用词表为数据包自带的英语常用停用词表。这里补充解释一下函数 tokens_select()。该函数的主要结构是 tokens_select(x,pattern,selection=c("keep","remove"),valuetype=c("glob","regex","fixed"),case_insensitive=TRUE,padding=FALSE,window=0),其中变元 x 是字符向量、字符向量列表、词典或搭配对象,模式 pattern 要求指定保留("keep")还是排除("remove")与模式匹配的形符。变元 valuetype 有三个选项("glob""regex"和"fixed"),其中"glob"指"glob"式通配符,"regex"指正则表达式,"fixed"指精确匹配。试比较下面的例子:

```
> text<-c("we think this is an idea of importance.","I don't believe it is impressive.","The important thing is to understand it.")
> tks<-tokens(text,remove_punct=TRUE)
> tokens_select(tks,pattern = "imp*", selection = "remove", value = "glob")
Tokens consisting of 3 documents.
text1:
[1] "we"   "think"   "this"   "is"   "an"   "idea"   "of"
text2:
[1] "I"   "don't"   "believe"   "it"   "is"
text3:
[1] "The"   "thing"   "is"   "to"   "understand"   "it"
> tokens_select(tks,pattern = "imp.", selection = "remove", value = "glob")
```

```
Tokens consisting of 3 documents.
text1：
[1] "we"  "think"  "this"  "is"  "an"  "idea"  "of"  "importance"
text2：
[1] "I"  "don't"  "believe"  "it"  "is"  "impressive"
text3：
[1] "The"  "important"  "thing"  "is"  "to"  "understand"  "it"
> tokens_select(tks, pattern = "imp*", selection = "remove", value = "regex")
Tokens consisting of 3 documents.
text1：
[1] "we"  "think"  "this"  "is"  "an"  "idea"  "of"
text2：
[1] "I"  "don't"  "believe"  "it"  "is"
text3：
[1] "The"  "thing"  "is"  "to"  "understand"  "it"
> tokens_select(tks, pattern = "imp.", selection = "remove", value = "regex")
Tokens consisting of 3 documents.
text1：
[1] "we"  "think"  "this"  "is"  "an"  "idea"  "of"
text2：
[1] "I"  "don't"  "believe"  "it"  "is"
text3：
[1] "The"  "thing"  "is"  "to"  "understand"  "it"
```

函数 tokens_select( ) 中的变元 case_insensitive = TRUE 表示函数默认模式匹配忽略大小写。变元 padding（占位）= FALSE 表示在去除匹配形符的地方不留下空白符。变元 window（窗口大小）长度为 1 或 2。window = 0 指函数默认只选择或排除与模式匹配的形符；window = 1 表示保留或排除匹配模式左边和右边的一个形符；window = c(1,2) 表示保留或排除匹配模式左边的一个形符和右边的两个形符，以此类推。试比较下面的例子：

```
> toks<-as.tokens(list(letters[1:10]))
> tokens_select(toks, c("c", "g"), selection = "keep", window = 1)
Tokens consisting of 1 document.
text1：
[1] "b"  "c"  "d"  "f"  "g"  "h"
> tokens_select(toks, c("c", "g"), selection = "keep", window = 2)
Tokens consisting of 1 document.
text1：
```

```
[1] "a"   "b"   "c"   "d"   "e"   "f"   "g"   "h"   "i"
> tokens_select(toks, c("c","g"), selection="remove", window=c(1,2))
Tokens consisting of 1 document.
text1 :
[1] "a"   "j"
```

回到比较两个文本的例子,R 命令如下:

```
> texts_sw<-tokens_select(texts_tk, pattern=stopwords("en"),
selection="remove")
```

第四步,利用数据包 quanteda 中的函数 dfm()把各个去除标点符号和停用词之后的形符转化为文档-词项矩阵。R 命令如下:

```
> dfm<-dfm(texts_sw)
```

第五步,我们利用数据包 quanteda.textstats 中的函数 textstat_keyness()计算文档-词项矩阵中各个词的关键性统计量。函数 textstat_keyness()包括的主要变元有 x、target 和 measure,其中 x 为文档-词项矩阵,target 用于设置目标文件名,文档-词项矩阵中的其他文档默认为参照文档,measure 用于设置关键性检验的名称,函数默认使用卡方检验,也可以通过此变元设置费希尔精确检验(measure="exact")或似然比检验(measure="lr")。

```
> results<-textstat_keyness(dfm, target="st01")
```

第六步,我们利用数据包 quanteda.textplots 中的函数 textplot_keyness()绘制文本比较关键词对照图。该函数的主要变元包括 x、n=20L、min_count=2L、margin=0.05、color=c("darkblue","gray")、labelcolor="gray30" 和 labelsize=4,其中 n=20L 表示图中显示 20 个特征(即词),min_count=2L 表示图中显示的最少特征计数为 2(即目标文档和参照文档中的累计词符频数不少于 2)。我们在这个例子中设置 min_count=3。变元 margin=0.05 指默认的图中特征边缘大小为 0.05。我们在本例中设置 margin=0.03。变元 color=c("darkblue","gray")是函数默认的目标词和参照词卡方值显示条颜色。我们在本例中设置 color=c("darkred","gray40")。变元 labelcolor="gray30" 是特征名的默认颜色,变元 labelsize=4 是函数默认的特征条和特征名的大小。

```
> textplot_keyness(results, n=10L, min_count=3, margin=0.03,
color=c("darkred","gray40"), labelsize=4)
```

为了便于比较,我们利用第五步得到的结果制作如表 6.2 所示的 20 个词符及其统计量表。

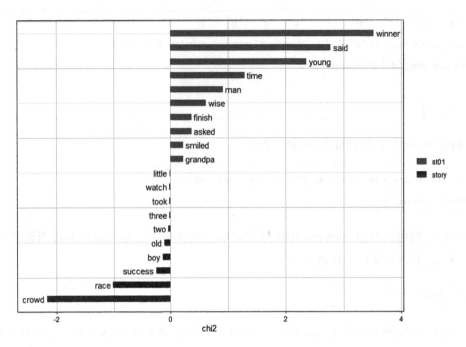

图 6.2 两个文本 20 个关键词比较

表 6.2 20 个关键词词频统计量

特征(词)	χ^2	p	目标特征频次	参照特征频次
winner	3.52	0.061	6	2
said	2.77	0.096	3	0
young	2.35	0.125	4	1
time	1.29	0.256	5	3
man	0.91	0.340	9	10
wise	0.62	0.432	7	8
finish	0.36	0.548	3	2
asked	0.36	0.548	3	2
smiled	0.22	0.643	2	1
grandpa	0.22	0.643	2	1
little	−0.01	0.911	5	9
watch	−0.02	0.891	1	2
took	−0.02	0.891	1	2
three	−0.02	0.891	1	2
two	−0.04	0.846	2	4
old	−0.11	0.740	1	4

(续表)

特征(词)	χ^2	p	目标特征频次	参照特征频次
boy	−0.14	0.709	7	14
success	−0.26	0.613	1	3
race	−1.01	0.314	2	9
crowd	−2.17	0.140	0	6

表6.2显示,由于本例样本较小,两个文本中词的关键性均没有达到统计显著性。例如,said在学生文本st01中出现3次,在文本story.txt中未出现,但是卡方值(χ^2)为2.77, $p=0.096>0.05$。race在文本story.txt中出现9次,在学生文本st01中出现2次,但是$\chi^2=1.01$(表中的卡方值前加个负号仅表示目标文本中的观测值低于期望值),$p=0.314>0.05$。当文本较短时,卡方检验的结果仅供参考。表6.2和图6.2显示出两个文本高频词频数的差异。学生文本中用词偏多的三个词是winner,said和young。例如,winner在学生文本中用了6次,而在复述故事原文只用了2次。我们利用数据包quanteda中的函数kwic()检索关键词winner。这个函数的主要结构是kwic(x, pattern, window = 5, valuetype, separator=" ",case_insensitive=TRUE),其中x是字符、形符或语料库对象,pattern用于设置检索的字符或搭配,window=5是函数默认的语境长度,即检索词前后的文本长度均为5个词。变元valuetype有三个选项("glob" "regex"和"fixed"),其中"glob"指"glob"式通配符,"regex"指正则表达式,"fixed"指精确匹配。变元separator=" "指函数默认用空格分开词符,case_insensitive=TRUE指函数默认模式匹配忽略字母大小写。

对winner检索的R命令和检索结果如下:

```
> kwic(texts_tk,pattern="winner")
Keyword-in-context with 8 matches.
  [story,75]   Unsurprisingly the boy was the | winner | The crowd cheered and waved
  [story,335]              is cheering not for any | winner |
  [st01,33]                    boy And he is the | winner | and a wise man came
  [st01,91]                   way The boy is the | winner | but he asked why there
  [st01,159]                man Who is cheer for | winner | and the wise man smiled
  [st01,184]              who is the cheering for | winner | and the grandpa smiled and
  [st01,203]                of the three man are | winner | You know this race You
  [st01,222]                  Cheering is not for any | winner |
```

以上结果显示,复述原文只在故事开头和结尾使用了"winner",而学生在其他4个地方使用了"winner",且有语法和时态错误,如"The boy is the winner"应为"The boy was the winner"。

再看故事原文使用较多的词"crowd"和"race"。原文使用"crowd"和"race"的频次分别为6次和9次,而学生文本中竟然一次都未出现"crowd","race"也只出现了2次,因而漏掉了较多的原文信息。

鉴于函数 kwic( ) 在文本检索中的有用性,我们依据以上两个文本的比较得到的 R 对象再举三个例子。我们先利用正则表达式"cheer*"在两个文本中检索相关句子。R 命令和检索结果如下:

```
> kwic(texts_tk,pattern="cheer*",valuetype="glob",window=5)
Keyword-in-context with 10 matches.
  [story,78]       was the winner The crowd |  cheered  | and waved at him The
 [story,173]    however was silent with no |  cheering | at all What happened Why
 [story,182]       happened Why is no one  |  cheering | my success he asked the
 [story,256]      same time Now the crowd  |  cheered  | like a thunderstorm The wise
 [story,285]         boy Who is the crowd  |  cheering | for Which one of us
 [story,331]      this race the crowd is   |  cheering | not for any winner
  [st01,99]         asked why there is no  |  cheering | happened The old wise man
 [st01,157]           the wise man Who is  |   cheer   | for winner and the wise
 [st01,182]      don't understand who is the |  cheering | for winner and the grandpa
 [st01,217]      more than any other match |  Cheering | is not for any winner
```

以上结果显示,学生文本(st01)中 3 次使用了"cheering",1 次使用了"cheer"。就语法结构而言,只有 1 个句子"Cheering is not for any winner"正确。故事原文(story)2 次使用了"cheered",4 次使用了"cheering"。从内容上看,学生文本忽略了观众两次反应。即,小男孩第一次比赛获胜时观众欢呼喝彩;小男孩第三次比赛与两位老人携手穿过终点线时观众报以雷鸣般的掌声。当我们要检索"cheer"和"for"的搭配时,需要包括"cheer"的各种变化形式,如"cheer"和"cheering",正则表达式用到元字符"*",并且把"cheer"和"for"两个搭配词放在短语函数 phrase( )中。R 命令和检索结果如下:

```
> kwic(texts_tk,pattern=phrase("cheer* for"),window = 5)
Keyword-in-context with 3 matches.
 [story,285:286]         boy Who is the crowd | cheering for | Which one of us three
  [st01,157:158]           the wise man Who is |   cheer for  | winner and the wise man
  [st01,182:183]   don't understand who is the | cheering for | winner and the grandpa smiled
```

如果要检索由两个以上词语构成的短语或语块(chunks),同样需要调用函数 phrase( )。例如:

```
> kwic(texts_tk,pattern=phrase("the little boy"),window=3)
Keyword-in-context with 8 matches.
  [story,83:85]              waved at him | The little boy | felt proud and
 [story,106:108]        another race pleaded | the little boy | hoping to impress
 [story,123:125]        forward and presented | the little boy | with two new
 [story,207:209]           you Finish together | The little boy | stood between the
 [story,229:231]              race began and | the little boy | walked slowly so
 [story,268:270]              nodding his head | The little boy | felt puzzled I
 [story,278:280]       understand grandpa asked | the little boy | Who is the
   [st01,53:55]              no excitement So | the little boy | ask the wise
```

以上结果显示,学生文本(st01)中只使用了"the little boy"1次,而故事原文(story)则使用了该短语7次。这一对比说明这名学习者在口头复述时不太注重修饰语或语义表达的准确性,也有可能说明学习者短语加工能力不足。

6.4.3 比较总统就职演说文本中的关键词

本节比较唐纳德·特朗普(Donald Trump)2017年的总统就职演说与约瑟夫·拜登(Joseph Biden)2021年的总统就职演说在关键词使用方面的差异。这两个演说的文字来自数据包 quanteda 中的语料库 data_corpus_inaugural。这个语料库包括自乔治·华盛顿(George Washington)于1789年担任第一届美国总统以来的各届总统就职演说,现有59个文档。文档变量名(docvars)包括 Text(文本类别)、Types(类符数)、Tokens(形符数)、Sentences(句子数)、Year(年份)、President(总统姓氏)、First Name(总统名字)和 Party(党派),其中 Text 名称由年份和总统姓组成,如 2017-Trump 和 2021-Biden。Types 指文档的类符数,如特朗普2017年的总统就职演说文本包括582个类符,拜登2021年的总统就职演说文本包括811个类符。Tokens 指文档的形符数,如特朗普2017年的总统就职演说文本包括1 660个形符,拜登2021年的总统就职演说文本包括2 766个形符。Sentences 指文档包含的句子数,如特朗普2017年的总统就职演说文本包括88个句子,拜登2021年的总统就职演说文本包括216个句子。Year 指总统就职演说的年份,如特朗普于2017年就任美国总统,拜登于2021年就任美国总统。President 指总统的姓氏。First Name 指总统的名字。Party 指就职总统隶属的党派,如特朗普属于共和党(Republican),拜登属于民主党(Democratic)。

要计算特朗普2017年与拜登2021年的总统就职演说中的关键词,需要利用 quanteda 等三个数据包开展关键词分析,包括以下六个步骤。

第一步,利用数据包 quanteda 中的函数 corpus_subset(x, subset, drop_docid=TRUE)从语料库 data_corpus_inaugural 中选择相关的文档。在这个函数中,x 指语料库对象,subset 指保留的文档,drop_docid=TRUE 指在执行 subset 之后函数默认排除的文档。本例的 R 命令如下:

```
> require(quanteda)
> corp<-corpus_subset(data_corpus_inaugural,
    President%in%c("Trump","Biden"))
```

第二步,利用函数 tokens()把语料库中的字符串形符化,设置变元 remove_punct=TRUE 去除标点符号。本例的 R 命令如下:

```
> corp<-tokens(corp, remove_punct=TRUE)
```

第三步,利用 quanteda 中的函数 tokens_select()剔除文本中的停用词。停用词表为数据包自带的英语常用停用词表。本例的 R 命令如下:

```
> corp_sw<-tokens_select(corp,pattern=stopwords("en"),
selection="remove")
```

第四步,利用 quanteda 中的函数 dfm()把各个去除标点符号和停用词之后的形符转化为文档-词项矩阵。本例的 R 命令和部分处理结果如下:

```
> dfm<-dfm(corp_sw)
> head(dfm)
Document-feature matrix of: 2 documents, 931 features (41.30% sparse) and 4 docvars.
            features
docs         chief justice roberts president carter clinton bush obama fellow americans
  2017-Trump     1       1       1         5       1       1    1     3      1         4
  2021-Biden     1       5       1         7       1       0    0     0      5         9
[ reached max_nfeat ... 921 more features ]
```

第五步,利用 quanteda.textstats 中的函数 textstat_keyness()计算文档-词项矩阵中各个词的关键性统计量。在本例中,我们把变元 target 设置为"2021-Biden",默认采用关键性卡方检验。本例的 R 命令如下:

```
> require(quanteda.textstats)
> results<-textstat_keyness(dfm,target="2021-Biden")
```

第六步,利用 quanteda.textplots 中的函数 textplot_keyness()绘制文本比较关键词对照图。本例采用函数默认设置,R 命令如下:

```
> require(quanteda.textplots)
> textplot_keyness(results)
```

图 6.3 显示特朗普 2017 年与拜登 2021 年的总统就职演说中的关键词对照。

考虑到本例中两个文档的词符数不是很大,我们选择卡方检验结果中 p 值小于或等于 0.1 的词符,R 命令为 results[which(results $ p<=0.1),]。表 6.3 为 16 个相关关键词及其统计量。

表 6.3 16 个关键词词频统计量

特征(词)	χ^2	p	目标特征频次	参照特征频次
us	12.56	0.0004	27	2
can	7.79	0.005	16	1
democracy	4.81	0.028	10	0
story	4.19	0.041	9	0
know	3.56	0.059	8	0

(续表)

特征(词)	χ^2	p	目标特征频次	参照特征频次
history	2.95	0.086	7	0
war	2.95	0.086	7	0
across	−3.35	0.067	1	5
bring	−4.03	0.045	0	4
everyone	−4.03	0.045	0	4
wealth	−4.03	0.045	0	4
every	−4.29	0.038	2	7
country	−5.15	0.023	4	9
dreams	−5.56	0.018	0	5
protected	−5.56	0.018	0	5
back	−7.12	0.008	0	6

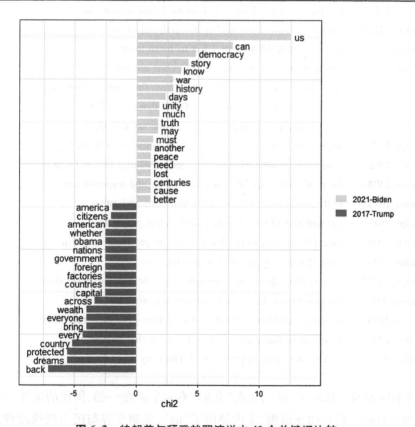

图 6.3　特朗普与拜登就职演说中 40 个关键词比较

图 6.3 和表 6.3 显示,拜登的演讲中使用关键性 p 值小于或等于 0.1 的关键词数为 7 个,最突出使用的 3 个词是"us""can"和"democracy",而特朗普的演讲中使用关键性 p 值小于或等于 0.1 的关键词数为 9 个,最突出使用的 3 个词是"back""protected"和"dreams"。使用不同的关键词体现两位总统执政理念的差异。限于篇幅,我们下面利用数据包

quanteda 自带函数 kwic() 简要分析演说中使用差异显著的两个词"us"和"back"。

执行 R 命令 kwic(corp,"us") 得到以下检索结果：

```
                        Keyword-in-context with 29 matches.
[2017-Trump,1036]    for prejudice The Bible tells | us | How good and pleasant it
[2017-Trump,1399]    and love will forever guide | us | along the way Together we
[2021-Biden,230]     but who cannot be with | us | today but whom we salute
[2021-Biden,270]     not on any one of | us | not on some of us
[2021-Biden,275]     us not on some of | us | but on all of us
[2021-Biden,280]     us but on all of | us | On We the People who
[2021-Biden,429]     years in the making moves | us | The dream of justice for
[2021-Biden,701]     know the forces that divide | us | are deep and they are
[2021-Biden,747]     and demonization have long torn | us | apart The battle is perennial
[2021-Biden,786]     of these moments enough of | us | came together to carry all
[2021-Biden,793]     together to carry all of | us | forward And we can do
[2021-Biden,928]     and in this place let | us | start afresh All of us
[2021-Biden,933]     us start afresh All of | us | Let us listen to one
[2021-Biden,935]     afresh All of us Let | us | listen to one another Hear
[2021-Biden,1171]    our democracy and to drive | us | from this sacred ground That
[2021-Biden,1207]    faith you have placed in | us | To all those who did
[2021-Biden,1215]    those who did not support | us | let me say this Hear
[2021-Biden,1339]    objects we love that define | us | as Americans I think I
[2021-Biden,1362]    weeks and months have taught | us | a painful lesson There is
[2021-Biden,1383]    for profit And each of | us | has a duty and responsibility
[2021-Biden,1614]    in the work ahead of | us | we will need each other
[2021-Biden,1845]    can and should be Let | us | say a silent prayer for
[2021-Biden,1909]    would be enough to challenge | us | in profound ways But the
[2021-Biden,1938]    must step up All of | us | It is a time for
[2021-Biden,2063]    prayers of centuries have brought | us | to this day What shall
[2021-Biden,2097]    my best to you Let | us | add our own work and
[2021-Biden,2130]    children's children will say of | us | they gave their best They
[2021-Biden,2256]    be the story that guides | us | The story that inspires us
[2021-Biden,2261]    us The story that inspires | us | The story that tells ages
```

特朗普的演说两次使用了"us"，表明"我们"有幸生活在一起，国民的勇气、善良和"爱"使"我们"勇往直前。拜登的演讲则 27 次使用了"us"，强调平等与民主的执政理念，美国的辉煌不是依靠某个人、某些人，而是依靠全体的"我们"，只有铭记历史赋予"我们"的启示，只有"我们"坚忍不拔和团结一致，"我们"才能直面当前的困境和挑战，"我们"才能战胜困难，实现"我们"的理想。

我们再来看一下"back"的使用。拜登的演讲中没有出现"back"一词，而特朗普的演讲中则出现了 6 次。利用 R 命令 kwic(corp,"back") 得到以下检索结果：

```
> kwic(corp,"back")
Keyword-in-context with 6 matches.

[2017-Trump,159]    Washington DC and giving it | back | to you the people For
[2017-Trump,828]    never before We will bring | back | our jobs We will bring
[2017-Trump,834]    our jobs We will bring | back | our borders We will bring
[2017-Trump,840]    our borders We will bring | back | our wealth And we will
[2017-Trump,847]    wealth And we will bring | back | our dreams We will build
[2017-Trump,879]    people off of welfare and | back | to work rebuilding our country
```

  特朗普的演说使用"giving...back"1次,"bring...back"4次,"get...back"1次,强调权力移交到"我们"手上,"我们"要拿回属于"我们"的"工作",拿回属于"我们"的"财富",重拾"我们"的梦想,收回"我们"的"边界"。"back"的反复使用体现出特朗普执政的务实性。

第七章 搭配和搭配构式

第六章介绍了 N 元组。N 元组按照设定的词符长度 n 提取文本中的词语组合,包括我们通常所说的搭配(collocations)。搭配构式(collostruction)是对搭配的拓展。本章继续使用 R 数据包 quanteda 和 quanteda.textstats 开展搭配分析,并利用斯蒂芬·托马斯·格赖斯(Stefan Thomas Gries)开发的 R 数据包 Coll. analysis 开展搭配构式分析(Gries,2007)。本章调用的数据包有 quanteda、quanteda.textstats、Coll. analysis 和 DescTools。

7.1 搭配和搭配构式分析方法

英国语言学家约翰·鲁珀特·弗斯(John Rupert Firth)说:"You shall know a word by the company it keeps."(观相伴,知词意。)这句话道出了词语搭配在语言产出中的重要性(Firth,1957)。搭配体现文本语境中词语之间的共现关系,是母语者的习惯性表达。在自然语言处理中,搭配和搭配构式分析利用基于语料库的关联度测量(association measures)技术,有多种测量方法。根据格赖斯和埃利斯,搭配和搭配构式分析有以下五种常用测量方法:点互信息(Pointwise Mutual Information,PMI)、z 值、t 值、G^2 和 $-\log 10\ p_{费希尔-耶茨精确检验}$(Gries et al.,2015)[237]。数据包 quanteda.textstats 计算搭配值 λ 和对应的 z 值。这些方法不仅可以用于搭配分析,而且还可以用于搭配构式分析(collostructional analysis)(Stefanowitsch et al.,2003;Gries et al.,2004a,2004b;Gries et al.,2015)。下面用一个例子介绍格赖斯和埃利斯提到的搭配和搭配构式分析方法,用另外一个例子介绍 λ 值的计算(Gries et al.,2015)。

7.1.1 点互信息

点互信息测量一个词的出现在多大程度上影响另一个词(或构式)的出现,计算公式为 $PMI = \log_2 \dfrac{a}{a_{\exp}}$,其中 a 是共现频次,$a_{\exp}$ 是期望频次。期望频次利用 2×2 列联表计算得到。下面利用"offer"+双及物(ditransitive)构式介绍点互信息等测量,数据来自 Gries 网站(https://stgries.info/teaching/groningen/index.html)的 Excel 文档 1.csv。这个构式文档从包括 138 664 个词的语料库中得到(N = 138 664)。在这个语料库中,"offer"出现 198 次,双及物构式出现的总频次是 1 035,其中"offer"+双及物构式出现的频次为 43。根据以上数据,我们得到如表 7.1 所示的列联表。

表 7.1 "offer"+双及物构式列联表

	+offer	−offer
+双及物	$a=43$	$b=1\,035-43=992$
−双及物	$c=198-43=155$	$d=138\,664-43-155-992=137\,474$

$a_{\exp}$ 是单元格 a 所在排总数和所在列总数的乘积与语料库容量(N)的商。在本例中，$a_{\exp}=\dfrac{(43+992)\times(43+155)}{138\,664}\approx 1.477\,889$，因而 $PMI=\log_2\dfrac{43}{1.477\,889}\approx 4.862\,727$。点互信息值大于 3 可以认为搭配有统计显著性关联(Hunston,2002)[71]。本例中点互信息值约为 5,说明"offer"与双及物构式之间有显著吸引力。点互信息测量的缺点是点互信息值会随着低期望频次(即词配对的稀有性)的出现而增大(Gries et al.,2015)[237]。即是说，一个词在语料库中低频出现，但是多与某个邻近词共现，则点互信息值会很大。

表 7.1 显示,"offer"在语料库中出现 198 次,双及物构式出现 1 035 次。这意味着,双及物构式除了与"offer"共现之外,还与其他词共现。换言之,"offer"对双及物构式的预测力可能比双及物构式对"offer"的预测力更强。如果一个词与另一个词(或构式)彼此有相同的预测力,它们被称作对称搭配(symmetric collocation),否则称作不对称搭配(asymmetric collocatio)。根据格赖斯,在已知 W_1 的情况下 W_1 预测 W_2 的条件概率测量值为 $\Delta P_{(W_2\mid W_1)}=\dfrac{a}{a+b}-\dfrac{c}{c+d}$，其中 W 指词(Gries,2013)。在已知 W_2 的情况下 W_2 预测 W_1 的条件概率测量值为 $\Delta P_{(W_1\mid W_2)}=\dfrac{a}{a+c}-\dfrac{b}{b+d}$。这两个公式对搭配构式的不对称测量同样适用。

在"offer"+双及物构式的例子中,双及物构式预测"offer"的条件概率测量值为 $P_{(\text{offer}\mid\text{ditransitive})}=\dfrac{43}{43+992}-\dfrac{155}{155+137\,474}\approx 0.040\,42$；"offer"预测双及物构式的条件概率测量值 $P_{(\text{ditransitive}\mid\text{offer})}=\dfrac{43}{43+155}-\dfrac{992}{992+137\,474}\approx 0.210\,008$。这两个值存在一定的差异,$P_{(\text{ditransitive}\mid\text{offer})}-P_{(\text{offer}\mid\text{ditransitive})}=0.169\,588>0$,因而"offer"对双及物构式更有预测力,即是说"offer"多使用在双及物构式中。

7.1.2 z 值和 t 值

z 值的计算公式为 $z=\dfrac{a-a_{\exp}}{\sqrt{a_{\exp}}}$，而 t 值的计算公式为 $t=\dfrac{a-a_{\exp}}{\sqrt{a}}$。$t$ 值的计算公式与 z 值的计算公式相似,区别体现在标准化方法上。z 值公式中的分母使用共现频次的期望值,而 t 值公式中的分母使用观测到的共现频次。

我们仍用上一节关于"offer"+双及物构式的例子,根据 z 值公式,得到 $z=\dfrac{43-1.477\,889}{\sqrt{1.477\,889}}\approx 34.155\,33$。根据 t 值公式,得到 $t=\dfrac{43-1.477\,889}{\sqrt{43}}\approx 6.332\,063$。词语共

现观测频次和期望频次差异越大，z 值和 t 值的差异就越大。t 值大于 2 通常认为搭配词具有统计显著性关联（Hunston, 2002）[72]。

7.1.3 G^2

第六章 6.4.1 节在介绍文本词语关键性检验方法时详细介绍了 G 检验。G 检验也适用于搭配构式分析。G 检验方法已在 6.4.1 节做了介绍，不再赘述。我们仍然以 "offer" +双及物构式为例，计算搭配强度。本例 R 命令和统计分析结果如下：

```
> offer<-matrix(c(43,992,155,137474),ncol=2)
> require(DescTools)
> GTest(offer)

        Log likelihood ratio (G-test) test of independence without correction

data: offer
G = 218, X-squared df = 1, p-value<2.2e-16
```

以上结果表明，G 检验得到 G^2 为 218，$p < 0.001$，说明 "offer" 与双及物构式之间有强烈吸引力。G^2 是很有用的强度测量指标。正是因为 $-\log 10\, p_{\text{费希尔-耶茨精确检验}}$ 包括精确检验，能够对每个 x 和 y 配对元素开展数百万次的概率计算，对数似然比统计量时常用作合理的近似值（Gries et al., 2015）[237]。

7.1.4 $-\log 10\, p_{\text{费希尔-耶茨精确检验}}$

费希尔-耶茨精确检验 p 值的计算基于超几何分布（hypergeometric distribution），统计计算利用 R 密度函数 dhyper(x, m, n, k)，其中 x 是分位数向量，表示从一个包括黑球和白球的瓮（urn）中抽取白球的数量，m 指瓮中白球的数量，n 指瓮中黑球的数量，k 指从瓮中抽取的球数。若观测频次 a 大于期望频次 a_{exp}，则利用 R 命令 sum(dhyper(a:(a+c), (a+c), (sum(c(a, b, c, d))-(a+c)), (a+b)))计算费希尔-耶茨精确检验 p 值。若观测频次 a 小于或等于期望频次 a_{exp}，则利用 R 命令 sum(dhyper(0:a, (a+c), (sum(c(a, b, c, d))-(a+c)), (a+b)))计算费希尔-耶茨精确检验 p 值。具体到 "offer" +双及物构式例子，"offer" 与双及物构式共现的观测频次 43 大于期望频次 1.477 889，因而参照表 7.1，x 为 a:(a+c)，m 为 (a+c)，n 为 (sum (c (a, b, c, d))-(a+c))，k 为 (a+b)。计算 $-\log 10\, p_{\text{费希尔-耶茨精确检验}}$ 的 R 命令和统计分析结果如下：

```
> a=43;b=992;c=155;d=137474
> pfye<-sum(dhyper(a:(a+c), (a+c), (sum(c(a, b, c, d))-(a+c)), (a+b)))
> -1*log10(pfye)
[1] 48.48033
```

以上结果显示,$-\log 10\, p_{费希尔\text{-}耶茨精确检验}$值约为 48.48,说明"offer"与双及物构式有较大的搭配强度。格赖斯和埃利斯认为$-\log 10\, p_{费希尔\text{-}耶茨精确检验}$是最有用的测量之一(Gries et al.,2015)[237],因而格赖斯开发的 R 数据包 Coll. analysis 在搭配构式分析中默认使用$-\log 10\, p_{费希尔\text{-}耶茨精确检验}$测量关联强度(Gries,2007)。

7.1.5 λ

R 数据包 quanteda. textstats 利用唐·布拉赫塔和马克·约翰逊(Don Blaheta and Mark Johnson)介绍的技术,计算 λ 值和对应的 z 值(Blaheta et al.,2001)。我们以两个词的搭配为例通过下面一个简单的例子理解搭配值 λ 和 z 的计算:"He is hungry for sports and hungry for success."我们关注的搭配是"hungry for"。这两个词的出现包括四种情形。第一种情形是两个词在各自的位置上均没有出现,计作 c_0,单元格计数为 n_1。本例中,$n_1 = 6$,即这两个词在六个搭配中均未出现在正确的位置上(即"hungry"出现在"for"之前):"He_is""is_hungry""for_sports""sports_and""and_hungry""for_success"。第二种情形是第一个词出现,而第二个词未出现,记作 c_1,单元格计数为 n_2。本例中,$n_2 = 0$,因为本例没有"hungry"出现而"for"没有随后出现的情况。第三种情形是第一个词未出现,而第二个词出现,记作 c_2,单元格计数为 n_3。本例中,$n_3 = 0$,因为本例没有"hungry"未出现而"for"却随后出现的情况。第四种情形是第一个词和第二个词均出现,记作 c_3,单元格计数为 n_4。本例中,$n_4 = 2$,即"hungry for"出现 2 次。我们由此得到如表 7.2 所示的 2×2 列联表,表中的 w 代表词。

表 7.2 搭配"hungry for"列联表

	$w \neq$ for	$w =$ for
$w \neq$ hungry	$n_1 = 6$	$n_3 = 0$
$w =$ hungry	$n_2 = 0$	$n_4 = 2$

在只考虑由两个词构成的搭配时,统计量 λ 值本质上是对数比数比(log odds ratio)。实际的计算在每个观测频次上加上 0.5,作为对小值计数的连续校正,也是为了避免观测值为"0"带来的问题(Blaheta et al.,2001)。这意味着,我们在计算 λ 值时,每个观测值应加上 0.5,如 6 变成 6.5。本例的比数比 OR(odds ratio)为 $\dfrac{6.5 \times 2.5}{0.5 \times 0.5} = 65$,则 $\lambda = \ln(OR) \approx 4.174\,387$,其中 ln 为自然对数。统计量 λ 值对应的 Wald 检验统计量 $z = \dfrac{\lambda}{(\sum_{i=1}^{M} n_i^{-1})^{(1/2)}}$,其中 $M = 2^K$(K 是搭配词数)。如果 z 值小于 2,那么两个词语的搭配未达到统计显著性($p > 0.05$)。本例中,$z = \dfrac{4.174\,387}{\left(\dfrac{1}{6.5} + \dfrac{1}{0.5} + \dfrac{1}{0.5} + \dfrac{1}{2.5}\right)^{(1/2)}} \approx 1.956\,156$。

7.2 文本中的搭配分析案例

本节调用 R 数据包 quanteda.textstats 中的函数 textstat_collocations(x, method = "lambda", size = 2, min_count = 2, smoothing = 0.5, tolower = TRUE) 统计分析数据包 quanteda 自带语料库 data_corpus_inaugural 中自 1945 年第二次世界大战结束之后的美国总统就职演说，计算搭配强度最大的前 20 个二元组。在上面的函数变元中，x 是字符、语料库或形符对象。如果是形符对象，那么它必须包括标点符号。变元 method = "lambda" 是函数默认的关联度测量（见本章 7.1.5 节）。变元 size = 2 是函数默认的搭配长度（即 2 个搭配词）。变元 min_count = 2 是函数默认的最低搭配频次。变元 smoothing = 0.5 是函数默认的加到观测计数的平滑参数（smoothing parameter）。变元 tolower = TRUE 指函数默认按词语小写字母构成搭配。

我们首先利用以下 R 命令调用数据包 quanteda 中的语料库 data_corpus_inaugural，根据要求利用函数 corpus_subset() 筛选所需的文档，并调用数据包 quanteda.textstats 中的函数 textstat_summary() 查看文档基本信息：

```
> require(quanteda)
> corp<-corpus_subset(data_corpus_inaugural,Year>1945)
> require(quanteda.textstats)
> textstat_summary(corp)
```

	document	chars	sents	tokens	types	puncts	numbers	symbols	urls	tags	emojis
1	1949-Truman	13687	116	2504	745	232	1	0	0	0	0
2	1953-Eisenhower	13968	119	2743	865	289	9	0	0	0	0
3	1957-Eisenhower	9213	92	1907	591	248	0	0	0	0	0
4	1961-Kennedy	7642	52	1541	542	175	2	0	0	0	0
5	1965-Johnson	8205	93	1710	535	221	3	0	0	0	0
6	1969-Nixon	11644	103	2416	714	292	0	0	0	0	0
7	1973-Nixon	10007	68	1995	515	193	1	0	0	0	0
8	1977-Carter	6878	52	1369	501	145	3	0	0	0	0
9	1981-Reagan	13743	129	2780	850	348	1	0	0	0	0
10	1985-Reagan	14572	123	2909	876	345	11	0	0	0	0
11	1989-Bush	12529	141	2673	756	356	2	0	0	0	0
12	1993-Clinton	9113	81	1833	605	235	0	0	0	0	0
13	1997-Clinton	12262	111	2436	726	279	0	0	0	0	0
14	2001-Bush	9054	97	1806	592	222	1	0	0	0	0
15	2005-Bush	11923	99	2312	734	241	0	0	0	0	0
16	2009-Obama	13460	110	2689	900	299	0	0	0	0	0
17	2013-Obama	11917	88	2317	786	220	5	0	0	0	0
18	2017-Trump	8433	88	1660	547	215	2	0	0	0	0
19	2021-Biden	13133	216	2766	743	394	6	0	0	0	0

以上结果显示,本例筛选的子语料库包括 19 个文档,即 13 位总统就职演说。这些文档的基本信息是字符数(chars)、句子数(sents)、形符数(tokens)、类符数(types)、标点符号数(puncts)、数字数(numbers)、符号数(symbols)、网址数(urls)、标注数(tags)和表情符数(emojis)。从字符数和形符数来看,1985 年里根(Reagan)总统就职演说最长,包括 14 572 个字符和 2 909 个形符,而 1977 年卡特(Carter)总统就职演说最短,包括 6 878 个字符和 1 369 个形符。从类符数来看,1977 年卡特总统就职演说包括的类符数依然最少,2009 年奥巴马(Obama)总统就职演说包括的类符数最多(900 个)。从句子数来看,2021 年拜登(Biden)总统就职演说用了 216 个句子,而 1961 年肯尼迪(Kennedy)和 1977 年卡特总统就职演说只使用了 52 个句子。

接下来,我们利用函数 textstat_collocations( ) 计算搭配统计量。计算这个子语料库中搭配强度最大的前 20 个二元组的 R 命令和统计分析结果如下:

```
> result<-textstat_collocations(corp)
> result[1:20,]
```

	collocation	count	count_nested	length	lambda	z
1	we will	137	0	2	3.156705	27.49070
2	those who	51	0	2	5.731088	26.43299
3	let us	75	0	2	5.891611	26.42591
4	we have	114	0	2	3.263336	25.51279
5	we are	105	0	2	2.884539	23.17779
6	it is	64	0	2	3.562696	22.64430
7	the world	127	0	2	3.407258	22.25931
8	of our	180	0	2	1.926627	21.58022
9	we must	77	0	2	3.524694	21.48390
10	a new	64	0	2	3.616221	21.22504
11	will be	57	0	2	3.344588	20.98980
12	has been	29	0	2	5.333571	20.65551
13	fellow citizens	24	0	2	6.948951	20.56573
14	my fellow	29	0	2	6.814065	19.94774
15	in the	177	0	2	1.776050	19.78270
16	must be	39	0	2	3.831427	19.63223
17	fellow americans	18	0	2	6.103786	18.40686
18	there is	37	0	2	4.197727	18.39179
19	more than	20	0	2	5.346884	18.25784
20	do not	29	0	2	4.141011	18.13134

以上结果显示具体的搭配(collocation)、搭配计数(count)、嵌套计数(count_nested,即计算出现在高阶搭配中的低阶搭配数)、搭配长度(length)、搭配强度(lambda)和 Wald 检验统计量(z)。结果表明,1945 年之后到 2021 年的总统就职演说较多地使用了以"we"开头的句子,搭配形式为"we will""we have""we are"和"we must",而且还较多地使用了与

"我们"相关的祈使句"let us"。这些演说还频繁使用了"a new""will be""has been""must be"和"do not"。这些表达既体现了演讲的口语特色,又强调了新一届政府的执政理念和面临的主要任务。为了吸引观众和听众的注意力,就职演说频繁使用了带有号召性的词语,如"fellow citizens"和"fellow americans"。以上搭配以功能词之间的搭配(如"we are"和"must be")为主,功能词与实词的搭配(如"the world")以及实词之间的搭配(如"fellow citizens")明显偏少。

如果我们对"we are"这类功能词搭配不感兴趣,只考虑实词搭配的特点,可以先对子语料库开展形符化,调用数据包 quanteda 自带的英语常用停用词表,通过函数 tokens_select()从形符化的子语料库中选择停用词表以外的词,再利用函数 textstat_collocations()计算搭配强度最大的前 20 个二元组。本例 R 命令和统计分析结果如下:

```
> corp1<-tokens(corp)
> corp1<-tokens_select(corp1, pattern = stopwords("en"),
selection = "remove", padding = TRUE)
> result1<-textstat_collocations(corp1)
> result1[1:20,]
```

	collocation	count	count_nested	length	lambda	z
1	let us	75	0	2	5.891611	26.42591
2	fellow citizens	24	0	2	6.948951	20.56573
3	fellow americans	18	0	2	6.103786	18.40686
4	united states	21	0	2	8.576742	17.93428
5	one another	20	0	2	6.416949	17.44485
6	years ago	10	0	2	7.185139	15.09171
7	go forward	9	0	2	6.808241	14.72697
8	god bless	16	0	2	8.969963	14.22490
9	two centuries	8	0	2	8.500532	13.71153
10	united nations	9	0	2	5.158312	13.01172
11	first time	10	0	2	5.034751	13.00550
12	human dignity	7	0	2	5.974751	12.81802
13	every american	11	0	2	4.363102	12.80788
14	four years	9	0	2	7.539344	12.65530
15	created equal	6	0	2	7.384724	12.58439
16	move forward	6	0	2	7.276387	12.26143
17	american people	12	0	2	3.848492	11.91721
18	nuclear weapons	5	0	2	8.339773	11.68565
19	centuries ago	5	0	2	7.586788	11.64757
20	free men	7	0	2	4.905336	11.49351

利用停用词表得到的搭配除了有些与不使用停用词表得到的搭配相同之外,如"let us""fellow citizens"和"fellow americans",以上结果还显示其他一些与演讲主题有关的实词搭

配,如"human dignity"和"nuclear weapons"。从这些搭配可以看出,总统就职演说中搭配的使用体现了就职演说作为一个特殊的语体所表现出来的特点,如演说对象(如"fellow americans"和"every american")、美国历史和总统制(如"centuries ago"和"four years")以及美国人的信仰(如"human dignity"和"created equal")。

7.3 文本中的搭配构式分析案例

格赖斯开发的 R 数据包 Coll. analysis 包括三类分析:搭配词位分析(collexeme analysis)、区别性搭配词位分析(distinctive collexeme analysis)和共变搭配词位分析(covarying collexeme analysis)(Gries,2007)。搭配词位分析测量词位(lexeme)与构式之间关联的强度,如 7.1.1 节关于"offer"与双及物构式的例子所示。区别性搭配词位分析测量交替构式配对与词位之间的关联强度。词位"give"可以与两种构式使用。一种构式是双及物构式,如"Mary gave him a pen."另一种构式是"to+与格"构式,如"Mary gave a pen to him"。格赖斯和斯特凡诺维奇的语料库研究表明,在这两个交替构式中,"give"更倾向于和双及物构式连用(见表 7.3)(Gries et al.,2004a)。共变搭配词位分析是搭配词位分析的拓展,测量同一个构式的两个插槽(slots)中的词位之间的搭配强度。格赖斯和斯特凡诺维奇利用英国《卫报》(*The Guardian*)语料库探讨了英语"into-致使"(into-causative)构式中"into"前后两个词位搭配的特点(Gries et al.,2004b)。该研究发现,具有统计显著性的词位搭配配对包括 torture-confess、force-make 和 mislead-buy 等。根据关联强度排序,在构式中 confess 的搭配词位是:torture、beat、intimidate、trap、coerce 和 lure。这些搭配词都与暴力或欺诈有关,特别是与暴力有关。

三类搭配构式测量的统计分析方法相同。本节利用 R 数据包 Coll. analysis 3.2a 开展区别性搭配词位分析,数据来源于 Gries 搭配构式分析网站(https://zenodo.org/records/10068088)中的区别性搭配词位分析文档 2a. csv。我们把该文件保存在 D:\Rpackage 文件夹中,默认文件名为 2a. csv。该文件提供语料库中与双及物构式或与介词+与格构式共现的 1 772 个词位形符,其中 958 个词位形符与双及物构式共现,814 个词位形符与介词+与格构式共现。我们同时在 D:\Rpackage 文件夹中创建一个文本文档 2a_output. txt,以便保存统计分析结果。

在开展统计分析之前,我们首先要下载和安装 R 软件 Coll. analysis 3.2a。在上述网站下载 Coll. analysis 3.2a,另存在 D:\Rpackage 文件夹中,默认软件名为 coll. analysis. r。然后,打开 R 工作界面,输入以下两行代码调用搭配构式分析软件:

```
> setwd("D:/Rpackage")
> source("coll. analysis. r")
```

该软件以问答式开展搭配构式分析。R 工作界面提示按回车键继续(Press 〈Enter〉 to continue ...)。接下来,R 界面出现以下三个执行选项:

> 1: collocational/collostructional strength, i. e. collexeme analysis (cf. 〈1 *. txt〉 for an example)
> 2: (multiple)distinctive collocates or distinctive collexeme analysis (cf. 〈2 *. txt〉 for an example)
> 3: co-varying collexeme analysis (cf. 〈3 *. txt〉 for an example)

我们在界面提示"Selection:"后输入"2",并按回车键执行区别性搭配词位分析的后续操作。

接下来的 R 界面提示选择区别性类别数,可以是两个类别(2 alternatives)或两个以上的类别(3+ alternatives)。我们在界面提示"Selection:"后输入"1",并按回车键执行两个构式比较的后续分析。

接下来,R 界面提示分析结果要保留的小数位数。输入"4",并按回车键执行下一个步骤。

R 界面提示两种计算搭配强度的方法。一是区别性搭配构式分析利用单尾费希尔-耶茨精确检验计算 p 值,再把 p 值进行对数转化得到搭配强度值(见本章7.1.4节)。二是使用对数似然比(log-likelihood ratio)作为搭配强度值。我们这里选择软件默认的 $-\log 10\, p_{费希尔-耶茨精确检验}$($-\log 10$(Fisher-Yates exact, one-tailed))。在界面提示"Selection:"后输入"1",并按回车键执行下一个步骤。

接下来的 R 界面询问以何种方式对搭配构式分析结果排序,提供四种排序方式:按字母排序(alphabetically)、按第一个构式中的第一个搭配词词频排序(frequency with W1/in C1)、按第二个构式中的第二个搭配词词频排序(frequency with W2/in C2)以及根据搭配构式强度排序(collostruction strength)。我们在界面提示"Selection:"后输入"4",以便按照搭配构式强度对结果排序。

接下来的 R 界面询问输入文档的格式是原始的所有形符表(raw list of all tokens)还是编辑过的词频表(edited list with frequencies)。我们在界面提示"Selection:"后输入"1"(即输入格式为所有形符表),并按回车键执行下一个步骤。

接下来的 R 界面提示按回车键继续(Press 〈Enter〉 to continue),以便加载文档2a. csv。

接下来的 R 界面要求输入双及物构式在语料库中的频次。我们输入"958",再按回车键继续。

接下来的 R 界面要求输入介词+与格构式在语料库中的频次。我们输入"814",再按回车键继续。

接下来的 R 界面询问统计分析结果保存在文本文档(text file)还是保存在终端(terminal)。我们输入"1",并按回车键,把结果保存在事先创建的 2a_output. txt 文档中。按回车键,区别性搭配词位分析至此结束。

表7.3 列出 2a_output. txt 文档中给出的前10个搭配强度分析表。

表 7.3 双及物构式和介词+与格构式与 10 个词的搭配强度

words	obs. freq. 1	obs. freq. 2	exp. freq. 1	exp. freq. 2	pref. occur	delta. p. constr. to. word	delta. p. word. to. constr	coll. strength
give	461	146	328.163 7	278.836 3	DITRANSITIVE	0.301 8	0.332 9	41.429 3
tell	128	2	70.282 2	59.717 8	DITRANSITIVE	0.131 2	0.479 1	32.687 4
cost	20	1	11.353 3	9.646 7	DITRANSITIVE	0.019 6	0.416 7	4.369 4
show	49	15	34.600 5	29.399 5	DITRANSITIVE	0.032 7	0.233 4	3.878 9
teach	15	1	8.650 1	7.349 9	DITRANSITIVE	0.014 4	0.400 5	3.128 2
offer	43	15	31.356 7	26.643 3	DITRANSITIVE	0.026 5	0.207 5	2.940 2
wish	9	1	5.406 3	4.593 7	DITRANSITIVE	0.008 2	0.361 4	1.699 2
promise	7	1	4.325 1	3.674 9	DITRANSITIVE	0.006 1	0.335 9	1.247 9
ask	12	4	8.650 1	7.349 9	DITRANSITIVE	0.007 6	0.211 3	1.134 9
deny	8	3	5.947	5.053	DITRANSITIVE	0.004 7	0.187 8	0.760 3

我们下面以"give"为例简要解释输出结果。在这个例子中,"give"在双及物构式中出现 461 次(obs. freq. 1),在介词+与格构式中出现 146 次(obs. freq. 2)。已知 $N = 1 772$,$N_1 = 958$,$N_2 = 814$。因此,我们构建两个 2×2 列联表,如表 7.4 和表 7.5 所示。

表 7.4 "give"与双及物构式列联表

	+give	-give
+双及物	$a=461$	$b=958-461=497$
-双及物	$c=146$	$d=1 772-461-497-146=668$

表 7.5 "give"与介词+与格构式列联表

	+give	-give
+介词+与格构式	$a=146$	$b=814-146=668$
-介词+与格构式	$c=461$	$d=1 772-146-668-461=497$

由表 7.4 得到期望频次 exp. freq. 1 = $\frac{(461+497) \times (461+146)}{1\,772}$ ≈ 328.163 7。由表 7.5 得到期望频次 exp. freq. 2 = $\frac{(146+668) \times (146+461)}{1\,772}$ ≈ 278.836 3。鉴于"give"在双及物构式中出现的观测频次(461)大于期望频次(328.163 7),因而相对于介词+与格构式,"give"更被双及物构式所吸引(pref. occu,偏向性)。

表 7.3 中,delta. p. constr. to. word 指构式预测词的条件概率测量值。根据本章 7.1.1 节的公式,$\Delta P_{(give|ditransitive)} = \frac{461}{461+497} - \frac{146}{146+668}$ ≈ 0.301 8。表中的 delta. p. word. to. constr 指词预测构式的条件概率测量值。根据本章 7.1.1 节的公式,$\Delta P_{(ditransitive|give)} = \frac{461}{461+146} - \frac{497}{497+668}$ ≈ 0.332 9。这两个测量值表明,"give"对双及物构式的预测力比双及物构式对"give"的预测力要稍强一点。

表 7.3 的最后一列是搭配构式强度值(coll. strength)。利用以下 R 命令得到与表 7.3 显示的同样结果:

```
> a=461; b=497; c=146; d=668
> pfye<-sum(dhyper(a:(a+c),(a+c),(sum(c(a, b, c, d))-(a+c)),
   (a+b)))
> round(-1 * log 10(pfye), digits=4)
[1] 41.4293
```

以上结果表明,"give"与双及物构式有很强的关联性。

第八章 文本相似度和距离测量

文档或文本相似度（document/text similarity）和距离（distance）测量属于聚类（clustering）算术的范畴，是文本挖掘和信息检索的主要手段。文本相似度测量包括文本词汇相似度（text lexical similarity）测量和文本语义相似度（text semantic similarity）测量。传统的文本相似度测量以文档中词频向量为基础。本章先介绍几种常用的文本词汇相似度和距离测量，统计分析辅助数据包是 quanteda 和 quanteda.textstats，后面介绍语义相似度测量，统计分析辅助数据包是 LSAfun。

8.1 词汇相似度测量

8.1.1 皮尔逊相关系数

皮尔逊相关分析（Pearson correlation analysis）是线性相关分析，目的是检验两个变量或向量之间线性关系的强度和方向。皮尔逊相关系数（r）是测量文本相似度较为直接的方法。

皮尔逊相关系数的基本计算公式为 $r = \dfrac{\sum (Z_X Z_Y)}{n-1}$，其中 Z_X 是某个观测值在变量 X 中的 Z 分数（标准分），Z_Y 是某个观测值在变量 Y 中的 Z 分数，n 是样本量。由于 Z 分数是由观测值与平均数的离差除以标准差计算而来，r 的计算公式中又用到两个变量 Z 分数的乘积，因而皮尔逊相关系数又称皮尔逊积差相关系数（Pearson product-moment correlation coefficient）。从公式可以看出，r 的计算利用两个变量的标准化值，因而 r 不会随变量测量尺度的改变而改变。皮尔逊相关系数基本计算公式的变化形式是 $r = \dfrac{\text{cov}(X,Y)}{S_X S_Y} = \dfrac{\sum (X_i - \bar{X})(Y_i - \bar{Y})}{\sqrt{\sum (X_i - \bar{X})^2 \sum (Y_i - \bar{Y})^2}}$，其中，$\text{cov}(X,Y)$ 是变量 X 和 Y 的协方差，S_X 和 S_Y 分别是变量 X 和 Y 的标准差，$\bar{X}$ 是各个 X_i 值的平均数，$\bar{Y}$ 是各个 Y_i 值的平均数。

根据关系的方向，变量或向量之间的线性关系分为正相关（positive correlation）、零相关（zero correlation）和负相关（negative correlation）三种。在正相关关系中，系数 r 大于 0，一个变量的值增大，另一个变量的值也随之增大。同样，随着一个变量的值减小，另一个变量的值也减小。在零相关关系中，系数 r 为 0，一个变量值增加或减小与另一个变量没有关系。在负相关关系中，系数 r 小于 0，一个变量的值增大，另一个变量的值却减小。系数 r 值介于-1和$+1$之间。根据科恩论述,不考虑r的符号,$r=0.10$、$r=0.30$和$r=0.50$分别代表小(small)、中(medium)、大(large)效应量(Cohen, 1988)。

本节和后续各节以下面的两个短文本为例,介绍相似度和距离测量的计算方法:

doc1:
Once there was a little boy who was hungry for success in sports. To him, winning was everything and success was measured by results. One day, the boy took part in a race in his village, and his competitors were two other young boys.

doc2:
Once there was a boy. He wanted to achieve success in all sports. One day he has taken part in the race. His competitors are another two young boys.

以上第一个文本是2019年英语专业四级口试听力原文的开头部分,第二个文本是学生复述的开头部分。在计算文本相似度时,我们首先去除标点符号,把大写字母改为小写字母。然后,计算每个文本的词频。如果在一个文本中出现的词在另一个文本中出现n次(n为正整数),那么计频数为n。如果在一个文本中出现的词在另一个文本中未出现,那么计频数为0。表8.1概括doc1和doc2两个文本中出现的词频。

表8.1 两个文本词频分布比较

	once	there	was	a	little	boy	who	hungry	for	success
doc1	1	1	4	2	1	2	1	1	1	2
doc2	1	1	1	1	0	1	0	0	0	1
	in	sports	to	him	winning	everything	and	measured	by	results
doc1	3	1	1	1	1	1	2	1	1	1
doc2	2	1	1	0	0	0	0	0	0	0
	one	day	the	took	part	race	his	village	competitors	were
doc1	1	1	1	1	1	1	2	1	1	1
doc2	1	1	1	0	1	1	0	0	1	0
	two	other	young	boys	he	wanted	achieve	all	has	taken
doc1	1	1	1	1	0	0	0	0	0	0
doc2	1	0	1	1	2	1	1	1	1	1
	are	another								
doc1	0	0								
doc2	1	1								

表8.1显示,这两个文本共有42个词(类符),未出现在文本doc1中的8个词是:"he""wanted""achieve""all""has""taken""are""another",未出现在文本doc2中的15个词是:"little""who""hungry""for""him""winning""everything""and""measured""by""results" "took""village""were""other"。有较多的词在文本 doc2 未出现主要是因为学生复述漏掉了"To him, winning was everything and success was measured by results."这句话以及用"He wanted to achieve success in all sports"改编了原文从句"who was hungry for success in sports."。

最后,我们把表 8.1 得到的两个文本中的词频分别转化为向量 doc1Freq 和 doc2Freq,利用 R 自带函数 cor( )计算两个向量的皮尔逊相关系数。R 命令和统计分析结果如下:

```
> doc1Freq<-c(1, 1, 4, 2, 1, 2, 1, 1, 1, 2, 3, 1, 1, 1, 1, 1, 2, 1, 1, 1, 1, 1, 1, 1, 1, 2, 1, 1,
1, 1, 1, 1, 1, 0, 0, 0, 0, 0, 0, 0, 0)
> doc2Freq<-c(1, 1, 1, 1, 0, 1, 0, 0, 0, 1, 2, 1, 1, 0, 0, 0, 0, 0, 0, 0, 1, 1, 1, 0, 1, 1, 1, 0, 1,
0, 1, 0, 1, 1, 2, 1, 1, 1, 1, 1, 1, 1)
> round(cor(doc1Freq, doc2Freq),2)
[1] 0.03
```

以上结果表明,从皮尔逊相关系数来看,两个文本的相似度很低。

对于表 8.1 词频的计算和开展皮尔逊相关分析,我们可以调用 quanteda 包中的函数 corpus( )、tokens( )、tokens_tolower( )和 dfm( )。R 命令和操作结果如下:

```
> doc1<-" Once there was a little boy who was hungry for success in sports. To him, winning was everything and success was measured by results. One day, the boy took part in a race in his village, and his competitors were two other young boys."
> doc2<-" Once there was a boy. He wanted to achieve success in all sports. One day he has taken part in the race. His competitors are another two young boys."
> require(quanteda)
> corp<-corpus(c(doc1,doc2),docnames=c("doc1","doc2"))
> token<-tokens(corp, remove_punct = TRUE)
> token<-tokens_tolower(token)
> dfm<-dfm(token)
> doc1.v<-as.vector(dfm[1,])
> doc2.v<- as.vector(dfm[2,])
> round(cor(doc1.v, doc2.v),2)
[1] 0.03
```

8.1.2 余弦相似度

余弦相似度(cosine similarity)测量两个向量之间夹角(θ)的余弦值($\cos \theta$),计算公式为 $\cos \theta = \dfrac{\vec{a} \cdot \vec{b}}{\|\vec{a}\| \, \|\vec{b}\|} = \dfrac{\sum_{i=1}^{n} a_i b_i}{\sqrt{\sum_{i=1}^{n} a_i^2} \sqrt{\sum_{i=1}^{n} b_i^2}}$,其中 $\vec{a}$ 和 $\vec{b}$ 是向量,$\vec{a} \cdot \vec{b}$ 是两个向量的点乘

(dot product)，$\vec{a} \cdot \vec{b} = \sum_{i=1}^{n} a_i b_i = a_1 b_1 + a_2 b_2 + \cdots + a_n b_n$。$\vec{a} \cdot \vec{b}$ 体现向量的方向。如果 $\vec{a} \cdot \vec{b}$ 的值大于 0，那么 θ 是锐角。如果 $\vec{a} \cdot \vec{b}$ 的值为 0，那么 θ 是直角，两个向量垂直，即正交（orthogonal）。如果 $\vec{a} \cdot \vec{b}$ 的值小于 0，那么 θ 是钝角。$\|\vec{a}\|$ 和 $\|\vec{b}\|$ 是欧式范数（Euclidean norms）。$\sqrt{\sum_{i=1}^{n} a_i^2}$ 和 $\sqrt{\sum_{i=1}^{n} b_i^2}$ 体现向量长度。$\cos\theta$ 的取值范围在 -1 和 +1 之间，值越接近 1，文本相似度就越大。

我们在此利用二维向量空间理解余弦相似度的概念。图 8.1 显示两个向量 $\vec{a}(3,1)$ 和 $\vec{b}(2,5)$ 的二维空间。

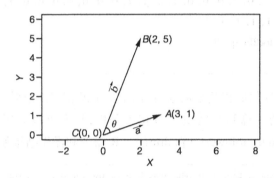

图 8.1　向量二维空间

如图 8.1 所示，原点 C 的坐标为 (0,0)，向量 $\vec{a}$ 和 $\vec{b}$ 的夹角为 θ。点 A 和点 C 的欧式距离为 $d_{AC} = \sqrt{(3-0)^2 + (1-0)^2} = \sqrt{3^2 + 1^2} \approx 3.1623$，点 B 和点 C 的欧式距离为 $d_{BC} = \sqrt{(2-0)^2 + (5-0)^2} = \sqrt{2^2 + 5^2} \approx 5.3852$。$\vec{a} \cdot \vec{b} = 3 \times 2 + 1 \times 5 = 11$。因此，$\cos\theta = \dfrac{11}{3.1623 \times 5.3852} \approx 0.65$。

在文本 doc1 和文本 doc2 的例子中，计算余弦相似度的 R 命令和统计分析结果如下：

```
> costheta<-sum(doc1Freq * doc2Freq)/sqrt(sum(doc1Freq^2) * sum(doc2Freq^2))
> round(costheta,2)
[1] 0.64
```

从余弦相似度值来看，两个文本有某种程度的相似性，似乎比皮尔逊相关系数更可信。当我们调用 R 基础函数 scale() 把两个文本词频标准化（即计算标准分或 z 分数）后计算余弦相似度时，我们得到以下结果：

```
> costheta<-sum(scale(doc1Freq) * scale(doc2Freq))/sqrt(sum(scale(doc1Freq)^2) * sum(scale(doc2Freq)^2))
> round(costheta,2)
[1] 0.03
```

这个值等同于我们前面计算得到的皮尔逊相关系数(r)。因此,对于标准化变量或向量,$r = \cos\theta$。

8.1.3 杰卡德相似度和杰卡德拓展相似度

杰卡德(Jaccard)相似度只考虑两个文本都出现的词,不考虑出现的频次,即某个词在文本中出现,计数为 1,未出现则计数为 0。这是简单的文本相似度测量方法。杰卡德相似度的计算公式为 $J(A,B) = \dfrac{|A \cap B|}{|A \cup B|}$,其中 A 和 B 是两个文本中词的向量,$|A \cap B|$ 表示两个词向量的交集,即两个文本中都出现的词总数;$|A \cup B|$ 表示两个词向量的并集,即两个文本中出现的所有词数量。杰卡德相似度的取值范围在 0 和 1 之间,值越大,文本相似度就越高。

在 doc1 和 doc2 两个文档的例子中,我们首先调用 R 数据包 quanteda 中的重要函数 corpus() 把两个文本转化为语料库对象,其次调用函数 tokens(),同时设置变元 remove_punct=TRUE 把语料库对象形符化,去除标点符号,然后利用函数 tokens_tolower() 把所有的形符转化为小写字母。R 命令操作如下:

```
> doc1<-" Once there was a little boy who was hungry for success in sports. To him, winning was everything and success was measured by results. One day, the boy took part in a race in his village, and his competitors were two other young boys. "
> doc2<-" Once there was a boy. He wanted to achieve success in all sports. One day he has taken part in the race. His competitors are another two young boys. "
> require(quanteda)
> corp<-corpus(c(doc1,doc2),docnames=c("doc1","doc2"))
> token<-tokens(corp, remove_punct = TRUE)
> token<-tokens_tolower(token)
```

接下来,利用索引符[[ ]]提取 R 对象 token 中每个文本的词符。R 命令如下:

```
> doc1.tk<-token[[1]]
> doc2.tk<-token[[2]]
```

最后,利用 R 基础函数 unique() 计算每个文本中出现的不同词,利用 R 基础函数 length() 和 intersect() 计算两个文本中均出现的词总数,利用 union() 计算两个文本中出现的所有词数。R 命令和统计分析结果如下:

```
> doc1.u<-unique(doc1.tk)
> doc2.u<-unique(doc2.tk)
> common<-intersect(doc1.tk,doc2.tk)
> all<-union(doc1.u,doc2.u)
```

```
> round(length(common)/length(all),2)
[1] 0.45
```

以上结果表明,两个文本的杰卡德相似度为 0.45。

与杰卡德相似度不同的是,杰卡德拓展相似度(Extended Jaccard Similarity)考虑词频,计算公式为 $J_e(A,B) = \dfrac{\sum_{i=1}^{n} a_i b_i}{\sum_{i=1}^{n} a_i^2 + \sum_{i=1}^{n} b_i^2 - \sum_{i=1}^{n} a_i b_i}$。针对 doc1 和 doc2 的例子,R 命令计算杰卡德拓展相似度如下:

```
> doc1Freq<-c(1,1,4,2,1,2,1,1,1,2,3,1,1,1,1,1,2,1,1,1,1,1,1,1,1,2,1,1,
1,1,1,1,0,0,0,0,0,0,0,0)
> doc2Freq<-c(1,1,1,1,0,1,0,0,0,1,2,1,0,0,0,0,0,0,1,1,0,1,1,1,0,1,
0,1,0,1,1,2,1,1,1,1,1,1,1)
> ej<-sum(doc1Freq*doc2Freq)/(sum(doc1Freq^2)+sum(doc2Freq^2)-sum(doc1Freq*doc2Freq))
> round(ej,2)
[1] 0.42
```

以上结果表明,两个文档的杰卡德拓展相似度为 0.42,比杰卡德相似度(0.45)稍低。

8.1.4 戴斯相似度和戴斯拓展相似度

戴斯相似度系数(Dice similarity coefficient)同杰卡德相似度一样不考虑文本词出现的频次,计算公式为 $D(A,B) = \dfrac{2a}{2a+b+c}$,其中 a 指两个文本或向量共现的词数,b 指第一个文本独有的词数(即只在第一个文本中出现的词),c 指第二个文本独有的词数(即只在第二个文本中出现的词)(Choi et al.,2010)。根据以上定义,杰卡德相似度的计算公式可以记作 $J(A,B) = \dfrac{a}{a+b+c}$(Choi et al.,2010)。

在 doc1 和 doc2 两个文档比较的例子中,利用交集函数 intersect( )和长度函数 length( )计算两个文本共现词数的 R 命令操作如下:

```
> require(quanteda)
> corp<-corpus(c(doc1,doc2),docnames=c("doc1","doc2"))
> token<-tokens(corp, remove_punct = TRUE)
> token<-tokens_tolower(token)
> doc1.tk<-token[[1]]
> doc2.tk<-token[[2]]
```

```
> doc1.u<-unique(doc1.tk)
> doc2.u<-unique(doc2.tk)
> (a<-length(intersect(doc1.u, doc2.u)))
[1] 19
```

以上结果表明,两个文本中同时出现的词(忽略频次)总数为 19。

利用查找符%in%、逻辑符!和长度函数 length( )计算 doc1 中独有词总数和 doc2 中独有词总数,R 命令操作如下:

```
> (b<-length(doc1.u[!doc1.u%in%doc2.u]))
[1] 15
> (c<-length(doc2.u[!doc2.u%in%doc1.u]))
[1] 8
```

以上结果表明,两个文本中不同时出现的词总数分别为 15 和 8。

根据公式计算戴斯相似度系数的 R 命令和统计分析结果如下:

```
> d<-2*a/(2*a+b+c)
> round(d,2)
[1] 0.62
```

以上结果表明,两个文本的戴斯相似度系数为 0.62,比杰卡德相似度(0.45)大。除非两个文本词完全相同,戴斯相似度系数大于杰卡德相似度。

与戴斯相似度系数不同,戴斯拓展相似度考虑词频,与杰卡德拓展相似度紧密联系。在杰卡德拓展相似度公式的分子和分母上同时加上 $\sum_{i=1}^{n}a_ib_i$ 便可得到戴斯拓展相似度,即戴斯拓展相似度的计算公式为 $D_e(A,B) = \dfrac{2\sum_{i=1}^{n}a_ib_i}{\sum_{i=1}^{n}a_i^2 + \sum_{i=1}^{n}b_i^2}$。

在 doc1 和 doc2 的例子中,计算戴斯拓展相似度的 R 命令和统计分析结果如下:

```
> ed<-2*sum(doc1Freq*doc2Freq)/(sum(doc1Freq^2) + sum(doc2Freq^2))
> round(ed,2)
[1] 0.59
```

8.1.5 简单匹配相似度

简单匹配相似度(simple matching similarity)与杰卡德相似度一样采用二分法,不考虑词频(即词出现记作 1,词未出现记作 0,不重复计数),测量上等同于两个文本共同出现的词数(交集)与所有出现词总数的比率,计算公式为 $SM = \dfrac{a}{a+b+c}$,其中 a 指两个文本或向量共现的词数,b 指只在第一个文本中出现的词,c 指只在第二个文本中出现的词(Batyrshin,2019)。简单匹配相似度的取值范围在 0 和+1 之间。简单匹配相似度与杰卡德相似度计算方法相同。重复 8.1.3 节的 R 命令计算 doc1 和 doc2 的简单匹配相似度:

```
> require(quanteda)
> corp<-corpus(c(doc1,doc2),docnames=c("doc1","doc2"))
> token<-tokens(corp, remove_punct = TRUE)
> token<-tokens_tolower(token)
> doc1.tk<-token[[1]]
> doc2.tk<-token[[2]]
> doc1.u<-unique(doc1.tk)
> doc2.u<-unique(doc2.tk)
> a<-length(intersect(doc1.u, doc2.u))
> b<-length(doc1.u[!doc1.u%in%doc2.u])
> c<-length(doc2.u[!doc2.u%in%doc1.u])
> sm<-a/(a+b+c)
> round(sm,2)
[1] 0.45
```

以上结果表明,两个文本的简单匹配相似度为 0.45。

8.1.6 欧式距离和闵可夫斯基距离

欧式距离(Euclidean Distance, ED)是最被广泛使用的文本不相似测量方法,是连续性变量的测量。假如有两个向量 A 和 B,每个向量各包括 n 个元素,向量 A 的元素为 $a_1, a_2, \cdots, a_n$,向量 B 的元素为 $b_1, b_2, \cdots, b_n$,则欧式距离的计算公式为 $ED = \sqrt{\sum_{i=1}^{n}(a_i - b_i)^2}$。

8.1.1 节计算了两个文档 doc1 和 doc2 的词频向量 doc1Freq 和 doc2Freq。利用欧式距离的计算公式编写 R 命令得到以下结果:

```
> doc1Freq<-c(1,1,4,2,1,2,1,1,1,2,3,1,1,1,1,1,2,1,1,1,1,1,1,1,1,2,1,1,
1,1,1,1,0,0,0,0,0,0,0,0)
> doc2Freq<-c(1,1,1,1,0,1,0,0,0,1,2,1,1,0,0,0,0,0,0,0,1,1,1,0,1,1,1,0,1,
0,1,0,1,1,2,1,1,1,1,1,1,1)
> eu<-sqrt(sum((doc1Freq-doc2Freq)^2)); round(eu,2)
[1] 6.56
```

以上结果显示,文本 doc1 和 doc2 的欧式距离为 6.56。

欧式距离是更为一般性的参数化距离家族——闵可夫斯基距离(Minkowski Distance,

MD) 的特例 (Cichosz, 2015)[315]。欧式距离适用于二维向量 ($p = 2$) 空间距离的测量, 而闵可夫斯基距离则推广到多维向量空间距离的测量 ($p \geq 2$)。针对两个向量之间距离的测量, 闵可夫斯基距离的计算公式为 $MD = \left(\sum_{i=1}^{n} |a_i - b_i|^p \right)^{1/p}$。我们可以利用 R 基础函数 dist(x, method = "minkowski", p) 计算闵可夫斯基距离, 其中 x 是矩阵或数据框, 参数 p 是距离计算的幂, p = 2 等同于计算欧式距离。执行以下 R 命令可以得到同样的结果: dist(rbind(doc1Freq, doc2Freq), method = "minkowski", 2)。

8.1.7 曼哈顿距离和堪培拉距离

在闵可夫斯基距离公式中参数 $p = 1$ 时, 闵可夫斯基距离等同于曼哈顿距离 (Manhattan Distance, MD), 即 $MD = \sum_{i=1}^{n} |a_i - b_i|$。堪培拉距离 (Canberra Distance, CD) 是对曼哈顿距离的改进, 把属性 (这里指词频) 差异与属性本身相比较, 计算公式为 $CD = \sum_{i=1}^{n} \frac{|a_i - b_i|}{|a_i| + |b_i|}$ (Cichosz, 2015)[316]。在这个公式中, 分子表示 A 和 B 两个向量每对对应元素差异的绝对值, 分母表示两个向量每对对应元素绝对值之和, 所有元素分子和分母的商之和即为堪培拉距离。

在文本 doc1 和 doc2 的例子中, 计算曼哈顿距离和堪培拉距离的 R 命令和统计分析结果如下:

```
> manhattan <-sum(abs(doc1Freq-doc2Freq))
> manhattan
[1] 33
> canberra <-sum(abs(doc1Freq - doc2Freq)/(abs(doc1Freq) + abs(doc2Freq)))
> round(canberra, 2)
[1] 25.13
```

在以上命令中, abs() 表示取绝对值, sum() 表示求和, round() 表示四舍五入, 保留小数位数。结果显示, 文本 doc1 和 doc2 的曼哈顿距离为 33, 堪培拉距离为 25.13。

8.2 语义相似度测量

文档或文本语义相似度的一种测量方法是利用潜在语义分析 (Latent Semantic Analysis, LSA)。潜在语义分析把向量空间模型和奇异值分解 (Singular Value Decomposition, SVD) 结合在一起, 将文本词汇表征映射到体现语义结构的向量空间。在语义空间中, 两个词的词向量嵌入 (即表征词义的序列数值) 越近似, 词义就越接近。例如, 我们利用下面介绍的 TASA 潜在语义空间计算冠词 "the" 和 "a" 的语义相似度, 发现它们的余弦相似度值高达 0.74。这是因为它们都属于冠词类别,一个表示特指,一个表示泛指,在句子中的位置相同。语义相似度计算的关键是创建潜在语义空间。研究者可以利用语料库自建潜在语义空间,也可以利用其他研究者已经创建的潜在语义空间。

弗里茨·冈瑟等人创建了 LSAfun 包和潜在语义分析空间(https://sites.google.com/site/fritzgntr/software-resources/semantic_spaces)(Günther et al.,2015)。LSAfun 包用于开展语义空间分析,当前版本号为 0.6.3。该网址提供了多个可下载的潜在语义分析空间。例如,TASA 潜在语义空间利用 TASA(Touchstone Applied Science Associates, Inc.,试金石应用科学同仁公司)语料库创建。这个语料库由不同类型的 37 651 个文档构成,涵盖文学、艺术、科学、经济学和社会学等(Günther et al.,2015)。TASA 语义空间包括 92 393 个词的 300 个维度。EN_100k_lsa 利用包含 20 亿个词的语料库构建。语料库构成包括英国国家语料库(British National Corpus, BNC)、英国网络归档联盟语料库(the UKWAC corpus)和 2009 年维基百科转储(2009 Wikipedia dump),涵盖 5 386 653 个文档。EN_100k_lsa 潜在语义空间包括 100 000 个词的 300 个维度。下面的例子使用 TASA 潜在语义空间计算文本语义相似度。

我们这里援引格罗斯曼和弗里德在介绍潜在语义索引(latent semantic indexing)一节使用的三个简短文本说明如何利用潜在语义空间计算文本相似度(Grossman et al.,2004)。三个文本内容如下:

D1:Shipment of gold damaged in a fire.
D2:Delivery of silver arrived in a silver truck.
D3:Shipment of gold arrived in a truck.

直觉上文本 D1 和 D3 最形似。由这三个文本得到词项-文档频数矩阵(term-document matrix),如表 8.2 所示。

表 8.2 词项-文档矩阵

词项	D1	D2	D3
a	1	1	1
damaged	1	0	0
fire	1	0	0
gold	1	0	1
in	1	1	1
of	1	1	1
shipment	1	0	1
arrived	0	1	1
delivery	0	1	0

(续表)

词项	D1	D2	D3
silver	0	2	0
truck	0	1	1

表8.2显示,"a"在三个文档中均出现一次,"damaged"在第一个文档中出现一次,在其他两个文档中均未出现,其他词的词频分布以此类推。

表8.2中的词频是文本词的原始频次。某个词对于一个文档的重要性不仅与它在该文档中出现的频率有关,而且与它在其他文本或语料库中出现的频率有关。如果一个词在多个文档中高频出现,那么这个词对某个文本语义的重要性就会削弱。词项频率-逆文档频率(Term Frequency-Inverse Document Frequency,TF-IDF)是常规的加权方法,计算上等于原始词项频率与逆文档频率的乘积。逆文档频率的计算公式为 $idf_i = \log_2\left(\frac{n}{df(i)}\right) + 1$,其中 n 表示文档数,$df(i)$ 表示词项 i 在语料库中至少出现一次的文档数(Gefen et al.,2017)。在这个例子中,各个词项的逆文档频率如表8.3所示。

表8.3 逆文档频率

a	damaged	fire	gold	in	of	shipment	arrived	delivery	silver	truck
1	2.584 96	2.584 96	1.584 96	1	1	1.584 96	1.584 96	2.584 96	2.584 96	1.584 96

在这个例子中,"a"在三个文本中均出现一次,$idf_a = \log_2 \frac{3}{3} + 1 = 1$,"silver"在第二个文档中出现两次,在其他两个文档中均未出现,则 $df(i) = 1$,$idf_{silver} = \log_2 \frac{3}{1} + 1 \approx 2.584\,96$,其他词的计算以此类推。

把表8.2的原始词项-文档矩阵与表8.3的逆文档频率向量相乘得到如表8.4所示的词项频率-逆文档频率。

表8.4 词项频率-逆文档频率

词项	D1	D2	D3
a	1	1	1
damaged	2.584 96	0	0
fire	2.584 96	0	0
gold	1.584 96	0	1.584 96
in	1	1	1
of	1	1	1
shipment	1.584 96	0	1.584 96

（续表）

词项	D1	D2	D3
arrived	0	1.584 96	1.584 96
delivery	0	2.584 96	0
silver	0	5.169 93	0
truck	0	1.584 96	1.584 96

观察表 8.4 所示的加权值发现，"a"的权重为 1，"damaged"只在第一个文档中出现一次，因而权重值为 2.584 96，"silver"只在第二个文档中出现两次，权重最高，达到 5.169 93。

对上面得到的词项频率-逆文档频率矩阵（$m \times n$ 矩阵）开展奇异值分解。对矩阵进行奇异值分解限定了潜在语义分析，而且也使这种分析不仅仅是词语共现分析（Gefen et al.，2017）[453]。奇异值分解的计算公式为 $A = U\Sigma V^T$，其中 A 是词项频率-逆文档频率矩阵；U 是 $m \times r$ 正交矩阵，包括作为排向量的词向量；V 是 $n \times r$ 正交矩阵，包括作为排向量的文档向量，V^T 是 V 的转置矩阵；Σ 是 $r \times r$ 对角矩阵，包括作为对角元素的奇异值。应用这种奇异值分解能够使潜在语义分析捕捉更深层次和更基本的语义维度，凸显词语间接共现的重要性（Günther et al.，2015）[931]。本例对词项频率-逆文档频率矩阵开展奇异值分解，得到以下三个矩阵：

$$U = \begin{bmatrix} -0.215\,973 & 0.189\,188 & 0.091\,196 \\ -0.085\,346 & 0.454\,253 & -0.441\,619 \\ -0.085\,346 & 0.454\,253 & -0.441\,619 \\ -0.121\,621 & 0.405\,890 & 0.241\,031 \\ -0.215\,973 & 0.189\,188 & 0.091\,196 \\ -0.215\,973 & 0.189\,188 & 0.091\,196 \\ -0.121\,621 & 0.405\,890 & 0.241\,031 \\ -0.289\,980 & 0.021\,332 & 0.415\,320 \\ -0.359\,928 & -0.172\,933 & -0.157\,367 \\ -0.719\,855 & -0.345\,866 & -0.314\,735 \\ -0.289\,980 & 0.021\,332 & 0.415\,320 \end{bmatrix},$$

$$\Sigma = \begin{bmatrix} 6.683\,2 & 0.000\,0 & 0.000\,0 \\ 0.000\,0 & 4.890\,4 & 0.000\,0 \\ 0.000\,0 & 0.000\,0 & 2.700\,1 \end{bmatrix},$$

$$V = \begin{bmatrix} -0.220\,66 & 0.859\,38 & -0.461\,28 \\ -0.930\,56 & -0.327\,16 & -0.164\,37 \\ -0.292\,17 & 0.392\,98 & 0.871\,89 \end{bmatrix}。$$

上面的奇异值矩阵 Σ 表明，对角线值由左上向右下方向依次减小。潜在语义分析的目的是降维，即维度由 r 降至 k（$k \leqslant r$）。在语料库研究中，k 通常维系在大约 300 个维度。本例

语料库容量很小,只是出于解释潜在语义空间概念的目的。我们选择两个维度得到潜在语义空间,即 $k=2$。 因此,我们有:

$$U \approx U_k = \begin{bmatrix} -0.215\,973 & 0.189\,188 \\ -0.085\,346 & 0.454\,253 \\ -0.085\,346 & 0.454\,253 \\ -0.121\,621 & 0.405\,890 \\ -0.215\,973 & 0.189\,188 \\ -0.215\,973 & 0.189\,188 \\ -0.121\,621 & 0.405\,890 \\ -0.289\,980 & 0.021\,332 \\ -0.359\,928 & -0.172\,933 \\ -0.719\,855 & -0.345\,866 \\ -0.289\,980 & 0.021\,332 \end{bmatrix},$$

$$\Sigma \approx \Sigma_k = \begin{bmatrix} 6.683\,2 & 0.000\,0 \\ 0.000\,0 & 4.890\,4 \end{bmatrix},$$

$$V \approx V_k = \begin{bmatrix} -0.220\,66 & 0.859\,38 \\ -0.930\,56 & -0.327\,16 \\ -0.292\,17 & 0.392\,98 \end{bmatrix}。$$

根据二维空间,我们确定文档向量新坐标。这些新坐标值等同于 V_k 矩阵特征向量值,即 D1 = (-0.220 66,0.859 38),D2 = (-0.930 56,-0.327 16),D3 = (-0.292 17,0.392 98)。最后,我们根据 8.1.2 节的余弦相似度的计算公式,得到:

D1 和 D2 余弦相似度值

$$\cos\theta = \frac{(-0.220\,66) \times (-0.930\,56) + 0.859\,38 \times (-0.327\,16)}{\sqrt{(-0.220\,66)^2 + 0.859\,38^2}\sqrt{(-0.930\,56)^2 + (-0.327\,16)^2}} \approx -0.087,$$

D1 和 D3 余弦相似度值

$$\cos\theta = \frac{(-0.220\,66) \times (-0.292\,17) + 0.859\,38 \times 0.392\,98}{\sqrt{(-0.220\,66)^2 + 0.859\,38^2}\sqrt{(-0.292\,17)^2 + 0.392\,98^2}} \approx 0.926,$$

D2 和 D3 余弦相似度值

$$\cos\theta = \frac{(-0.292\,17) \times (-0.930\,56) + 0.392\,98 \times (-0.327\,16)}{\sqrt{(-0.292\,17)^2 + 0.392\,98^2}\sqrt{(-0.930\,56)^2 + (-0.327\,16)^2}} \approx 0.297。$$

因此,文档 D1 和 D3 最相似,其次是 D2 和 D3,D1 和 D2 之间的相似度最低。

8.3 文本相似度测量举例

本节利用第六章介绍的两个文档 story.txt 和 st01.txt,调用数据包 quanteda、quanteda.textstats 和 LSAfun 计算文本相似度。

8.3.1 基于词频的余弦相似度测量

计算文本词汇相似度测量值的函数是来自数据包 quanteda.textstats 中的 textstat_simil(x,margin=c("documents","features"),method="cosine"),其中 x 是 dfm(文档-特征矩阵)对象,margin=c("documents","features")指定文档-特征矩阵边缘项计算相似度。我们这里选择 margin="documents"。变元 method="cosine"指定相似度测量方法为余弦相似度测量,也可以选择 8.1 节介绍的方法,如"correlation""jaccard""ejaccard""dice""edice"或"simple matching"。

针对文档 story.txt 和 st01.txt 的例子,我们首先调用数据包 readtext,把存储在 D:\Rpackage 文件夹中的两个文档调入 R 控制台。然后,利用数据包 quanteda 中的函数 corpus()、tokens()和 dfm()得到文档-特征矩阵(corp.dfm)。最后,调用数据包 quanteda.textstats 中的函数 textstat_simil()计算文本相似度。本例 R 命令和统计分析结果如下:

```
> require(readtext)
> story<-readtext("D:/Rpackage/story.txt")$text
> st01<- readtext("D:/Rpackage/st01.txt")$text
> require(quanteda)
> corp<-corpus(c(story,st01),docnames=c("doc1","doc2"))
> corp.tk<-tokens(corp,remove_punct=TRUE)
> corp.dfm<-dfm(corp.tk)
> require(quanteda.textstats)
> textstat_simil(corp.dfm,method="cosine",margin="documents")
textstat_simil object; method = "cosine"
      doc1  doc2
doc1  1.00  0.87
doc2  0.87  1.00
```

以上结果表明,两个文本词汇相似度余弦值为 0.87。

如果在计算余弦相似度时排除对意义贡献不大的停用词,那么可以利用函数 tokens_select(),排除停用词。我们这里利用 quanteda 包自带的停用词表 stopwords("en")。本例完整的 R 命令和统计分析结果如下:

```
> require(readtext)
> story<-readtext("D:/Rpackage/story.txt")$text
> st01<- readtext("D:/Rpackage/st01.txt")$text
> require(quanteda)
> corp<-corpus(c(story,st01),docnames=c("doc1","doc2"))
> corp.tk<-tokens(corp,remove_punct=TRUE)
```

```
> corp.tk<-tokens_select(corp.tk,pattern=stopwords("en"),selection="remove")
> corp.dfm<-dfm(corp.tk)
> require(quanteda.textstats)
> textstat_simil(corp.dfm,method="cosine",margin="documents")
textstat_simil object; method = "cosine"
       doc1  doc2
doc1  1.000 0.749
doc2  0.749 1.000
```

以上结果表明,在排除常用停用词后,两个文本的余弦相似度值约为 0.75。

8.3.2 基于语义空间的余弦相似度测量

R 数据包 LSAfun 提供多个有用的函数计算词、词串和文本的语义相似度。计算句子或文档之间余弦值的一个方便函数是 costring(x,y,tvectors = tvectors,split = " ",remove.punctuation=TRUE),其中 x 和 y 是字符向量,tvectors 指定计算使用的语义空间。语义空间是一个数值矩阵,矩阵中的每一排都是一个词向量。变元 split=" "指函数默认使用空格把字符串切分为词;remove.punctuation=TRUE 指函数默认去除向量 x 和 y 中的标点符号。TASA 潜在语义空间存储在 D:\Rpackage 文件夹中。在加载 TASA 潜在语义空间之前,先利用数据包 readtext 中的函数 readtext()调用文档 story.txt 和 st01.txt。然后,加载数据包 LSAfun 和 TASA 潜在语义空间,利用函数 costring()计算文本语义相似度余弦值。本例 R 命令和统计分析结果如下:

```
> require(readtext)
> story<-readtext("D:/Rpackage/story.txt") $ text
> st01<-readtext("D:/Rpackage/st01.txt") $ text
> require(LSAfun)
> setwd("D:/Rpackage")
> load("TASA.rda")
> costring(story,st01,tvectors=TASA)
Note: not all elements in x were found in rownames(tvectors)
Note: not all elements in y were found in rownames(tvectors)
[1] 0.9038474
```

以上结果表明,这两个文本语义相似度余弦值约为 0.9。

如果要使用 quanteda 包自带的停用词表,那么在利用 readtext 包把两个文档读入 R 之后,调用 quanteda 包中的函数 corpus 把文档转化为语料库格式,然后调用函数 tokens()和 tokens_select()得到两个文本在排除停用词之后的词向量。最后,调用 LSAfun 包中的函数 costring()计算文本语义相似度余弦值。本例 R 命令和统计分析结果如下:

```
> require(readtext)
> story<-readtext("D:/Rpackage/story.txt")$text
> st01<-readtext("D:/Rpackage/st01.txt")$text
> require(quanteda)
> corp<-corpus(c(story,st01),docnames=c("doc1","doc2"))
> corp.tk<-tokens(corp,remove_punct=TRUE)
> corp.tk<-tokens_select(corp.tk,pattern=stopwords("en"),
  selection="remove")
> require(LSAfun)
> setwd("D:/Rpackage")
> load("TASA.rda")
> costring(corp.tk[[1]],corp.tk[[2]],tvectors=TASA)
Note: not all elements in x were found in rownames(tvectors)
Note: not all elements in y were found in rownames(tvectors)
[1] 0.8924608
```

以上结果表明,在排除停用词之后,这两个文档语义相似度余弦值约为0.89,与未排除停用词得到的余弦值近似相同。

在结束本节之前,我们就LSAfun包中的几个有用函数举几个例子。计算两个词之间余弦相似度值的函数是Cosine(x,y,tvectors=tvectors),其中x和y是单个词,tvectors指定计算使用的语义空间。例如:

```
> require(LSAfun)
> setwd("D:/Rpackage")
> load("TASA.rda")
> Cosine("husband","wife",tvectors=TASA)
[1] 0.8561325
> Cosine("man","woman",tvectors=TASA)
[1] 0.2055876
```

以上结果显示,"husband"和"wife"之间的余弦值明显高于"man"和"woman"之间的余弦值。函数multicos(x,y=x,tvectors=tvectors)根据词向量计算余弦值矩阵,其中x和y是词向量,函数默认y=x,tvectors指定计算使用的语义空间。

```
> multicos("tiger lion cow","cat bird",tvectors=TASA)
            cat        bird
tiger  0.3696037   0.04184345
lion   0.4732016   0.01308054
cow    0.1460171  -0.07406220
```

以上结果显示，同"tiger"和"cow"相比，"lion"与"cat"之间的语义最接近，"tiger""lion"和"cow"都与"bird"语义相差甚远。

```
> multicos("tiger lion cow",tvectors=TASA)
         tiger         lion          cow
tiger  1.0000000    0.6285499    0.2193815
lion   0.6285499    1.0000000    0.1505321
cow    0.2193815    0.1505321    1.0000000
```

以上结果显示，在三个词中，"tiger"和"lion"之间语义最接近，"lion"和"cow"之间语义最不接近。

函数 neighbors(x,n,tvectors=tvectors)用于查找意义最接近的词，其中 x 是字符向量，n 指需要计算的意义最接近的词数，tvectors 指定计算使用的语义空间。例如：

```
> neighbors("breakfast",n=10,tvectors=TASA)
 breakfast     morning      toast       bacon     pancakes
 1.0000000   0.7516852   0.6714482   0.5341568   0.5280214
 porringer    oatmeal     coffee      poach     tangerine
 0.5181402   0.5062574   0.5044122   0.5025095   0.5025095
```

以上结果显示，在9个词（除了"breakfast"词本身）中，与"breakfast"意义最近的三个词是"morning""toast"和"bacon"，意义最远的两个词是"poach"和"tangerine"。

第九章 基于整洁文本的自然语言处理

哈德利·威克姆(Hadley Wickham)是 R 语言编程方面的杰出"怪才",创建了多个 R 数据包,如 reshape2 和 stringr。本章介绍的整洁工具大部分来自哈德利·威克姆创建的 R 集成包 tidyverse,包括 ggplot2 和 dplyr,当前版本号为 1.3.2。如果要把数据包安装在系统默认的路径,可以在 R 工作界面输入简单的命令:install.packages("tidyverse")。用户也可以自定义,输入带有指定路径的 R 命令,如:install.packages("tidyverse", lib="C:/Program Files/R/R-4.2.2/library"),按回车键对数据包进行自动在线安装(要确保网络通畅,且安装不会被防火墙或代理服务器阻拦)。R 活动期间首次调用 R 数据包 tidyverse 时,界面提示以下信息:

```
> require(tidyverse)
Loading required package: tidyverse
── Attaching packages ─────────────────────── tidyverse 1.3.2 ──
✓ ggplot2 3.4.0     ✓ purrr   1.0.1
✓ tibble  3.1.8     ✓ dplyr   1.0.10
✓ tidyr   1.2.1     ✓ stringr 1.5.0
✓ readr   2.1.3     ✓ forcats 0.5.2
── Conflicts ───────────────────────── tidyverse_conflicts() ──
✗ dplyr::filter()  masks stats::filter()
✗ dplyr::lag()     masks stats::lag()
```

加载的数据包包括 ggplot2、tibble、tidyr、readr 和 dplyr 等。R 界面还提示 tidyverse 包与 R 基础包(即 stats)相冲突的函数。如果在加载 tidyverse 之后仍希望使用基础包中相冲突的函数,如函数 filter(),那么我们可以通过 stats::filter() 的输入方式调用这个函数。本章使用的 tidyverse 包包括 ggplot2、tibble、tidyr 和 dplyr。除此之外,本章还简要介绍数据包 tidytext。这些数据包如需单独安装,安装方式与 tidyverse 的安装相同。数据包的更新可以使用 R 命令 tidyverse_update()。在介绍完这些整洁工具之后,本章重点介绍如何开展情感分析。本章使用的数据包包括:readtext、tidyverse、tibble、dplyr、tidyr、tidytext、stopwords、ggplot2、RColorBrewer、syuzhet、janeaustenr、stringr、gutenbergr、wordcloud2、wordcloud 和 reshape2。

9.1 作为新型数据框的 tibble

2002 年全国英语专业四级测试中的复述故事讲述史密斯先生入住一家酒店的不愉快经历。故事梗概如下:史密斯先生是一家大酒店的老顾客,与经理很熟,每次都入住在同一间景观房。最近一次入住时,经理告诉他可以照例入住此房间,不过酒店在扩建,房间可能有点吵。史密斯先生在入住的第一天,几乎没有听到噪声。第二天下午他在楼上看书时听到有人在砸墙,并且砸墙声越来越大,房间里满是灰尘,史密斯先生随即找经理投诉。他和经理回到房间时却听不到一丝声响,弄得史密斯先生很尴尬。突然,砸墙声又出现了,一块大砖头落到了地板上,在床的正上方屋顶上出现了一个大窟窿。故事的英语原文如下:

Whenever Mr Smith goes to Westgate, he stays at the Grand Hotel. In spite of its name, it is really not very "grand," but it is cheap, clean, and comfortable. Since he knows the manager well, he never has to go to the trouble of reserving a room. He always gets the same room. It is situated at the far end of the building and overlooks a beautiful bay.

On his last visit, Mr Smith was told that he could have his usual room, but the manager added apologetically that it might be a little noisy. So great was the demand for rooms, the manager said, that the hotel had decided to build a new wing. Mr Smith said he did not mind.

During the first day Mr Smith hardly noticed the noise at all. The following afternoon, he borrowed a book from the hotel library and went upstairs to read. No sooner had he sat down than he heard someone hammering loudly at the wall. At first he paid no attention, but after a while he began to feel very uncomfortable. His clothes were slowly being covered with fine white powder. Soon there was so much dust in the room that he began to cough. The hammering was now louder than ever and bits of plaster were coming away from the walls. It looked as though the whole building was going to fall. Mr Smith went immediately to complain to the manager. They both returned to the room, but everything was very quiet. As they stood there looking at each other, Mr Smith felt rather embarrassed for having dragged the manager upstairs for nothing. Suddenly, the hammering began again and a large brick landed on the floor. Looking up, they saw a sharp metal tool had forced its way through the wall, making a very large hole right above the bed!

我们把这则故事保存为一个文本文件,文件名为 hotel0.txt,编码格式为 UTF-8,保存在 D:\Rpackage 文件夹中。利用 R 数据包 readtext 中的函数 readtext( ) 把文件读入 R,用第二章学到的方法把字符串转化为形符,再利用函数 table( ) 和 data.frame( ) 把形符表转化为词频表。本例 R 命令和部分统计分析结果如下:

```
> library(readtext)
> hotel0<-readtext("D:/Rpackage/hotel0.txt")
> hotel0_tok<-unlist(strsplit(hotel0 $ text, "\\W"))
> hotel0_tok<-tolower(hotel0_tok[which(hotel0_tok! ="")])
> hotel0_df<-data.frame(table(hotel0_tok))
> hotel0_freq<-hotel0_df[order(-hotel0_df[,2]),]
> head(hotel0_freq)
```

	hotel0_tok	Freq
146	the	26
74	he	12
151	to	11
1	a	9
12	at	6
99	mr	6

我们下面换一种统计分析方法：利用 R 数据包 tidytext 中的函数 unnest_tokens( )对文档 hotel0.txt 开展形符化，并调用数据包 dplyr 中的函数 count( )开展词频统计。函数 unnest_tokens( )的基本结构是 unnest_tokens(tbl,output,input,token="words")，其中 tbl 是 tibble 缩写，是一种数据框；output 是被创建的输出列，默认输出 word；input 是字符串被切分的输入列，默认为 text；token 指形符化单位，默认单位是词(words)，也可以自定义形符化单位，如 N 元组(ngrams)、句子(sentences)、行(lines)或段落(paragraphs)。我们以词为切分单位，利用以下 R 命令分割字符串中的形符：

```
> library(readtext)
> hotel0<-readtext("D:/Rpackage/hotel0.txt")
> require(tidytext)
> hotel0_toks<-unnest_tokens(hotel0,word,text,token="words")
```

然后，我们调用数据包 dplyr 中的函数 count( )对各个形符(即词)开展词频统计。函数 count( )的主要结构是 count(x,word,sort=FALSE)，其中 x 是数据框，包括 tibble 格式，word 是计数变量，也可以是其他变量，sort=FALSE 表示函数默认不排序。我们在这里设定 sort=TRUE，即表示对各个形符频次进行排序。R 命令和统计分析结果如下：

```
> require(dplyr)
> hotel_df<-count(hotel0_toks,word,sort=TRUE)
> hotel_df
readtext object consisting of 174 documents and 0 docvars.
# Description: df [174 × 3]
  word    n    text
  <chr>  <int> <chr>
1 the    26   "\"\"..."
2 he     12   "\"\"..."
3 to     11   "\"\"..."
4 a       9   "\"\"..."
5 at      6   "\"\"..."
6 mr      6   "\"\"..."
# ... with 168 more rows
```

对比以上两个频次计算结果可以发现，hotel0_freq 是常规数据框。我们输入 R 命令 head(hotel0_freq)使结果只显示前 6 排。如果我们输入 R 命令 hotel0_freq，那么会冗长地显示全部的词符及其频次，甚至在界面看不到变量名。第二个结果 hotel_df 也是数据框，但是提供了更丰富的信息，如 hotel_df 只提供前 6 排信息，提示输出结果是一个 174×3 的数据框，而且还提供变量属性信息，如变量 word 属性是字符(〈chr〉, chr 代表字符[character])，频次(n)属性是整数(〈int〉, int 代表整数[integers])。这意味着，第二个结果虽和第一个结果一样都是数据框，但是它更接近一种新型的数据框格式，即 tibble 格式。

我们利用 R 数据包 tibble 中的函数 tibble()或 as_tibble()把 readtext()读入的数据框格式文档 hotel0 转化为 tibble 格式，再利用函数 count()计算词频。R 命令和统计分析结果如下：

```
> require(tibble)
> hotel0_tbl<-tibble(hotel0)
> hotel0_tokens<-unnest_tokens(hotel0_tbl, word, text, token=" words")
> hotel0_counts<-count(hotel0_tokens, word, sort=TRUE)
> hotel0_counts
# A tibble: 174 × 2
   word     n
   <chr>  <int>
 1 the     26
 2 he      12
 3 to      11
 4 a        9
 5 at       6
 6 mr       6
 7 smith    6
 8 was      6
 9 and      5
10 it       5
# ... with 164 more rows
```

以上结果显示，hotel0_counts 是 tibble 格式，与前面第二个数据框一样显示所有的列变量和列变量属性信息，自动显示数据框的部分信息，但是不像第二个数据框那样有一个多余的列(text)，且显示前 10 排而非前 6 排构成的数据框。

至此，我们对 tibble 格式有了较为深入的了解。tibble 是对传统数据框 data.frame 的技术改进，是整洁的数据框(tidy data frame)，为的是更方便地使用 tidyverse 数据包。在 R 界面输入 vignette("tibble")，再按回车键可以看到有关 tibbles 的网页介绍。根据网页介绍，tibble 和传统的数据框有三方面的差异：打印(printing)、提取子集(subsetting)和循环(recycling)。在打印方面，前面的例子已经说明，在 R 界面打印 tibble 对象时，结果显示前 10 排和所有适合屏显的列变量数据，并且提供列变量属性信息。这对大数据研究很有利。

如果我们要了解数据框全貌，可以利用函数 View( ) 查看，其中的变元是 tibble 对象。在提取子集方面，用索引符"[ ]"提取 tibble 对象总会返还另外一个 tibble 对象，而对传统数据框的提取结果可以是另外一个数据框，也可以是一个向量，如 hotel0_counts[,2] 返还一个 tibble，hotel_df[,2] 则返还一个向量。另外，tibble 不会改变输入类型（如不会把字符串转换为因素），不会改变变量名，也不会创建排名称（row names）（Wickham et al.，2017）[120]。在循环方面，在构建一个 tibble 时，只有长度为 1 的值可以被循环；长度不等于 1 的第一列决定了 tibble 的排数。如果列长度不相容，就会报错。试比较下面的例子：

```
> tibble(a = 1:4, c = 5:8)
# A tibble: 4 × 2
      a     c
  <int> <int>
1     1     5
2     2     6
3     3     7
4     4     8

> tibble(a = 1, c = 4:9)
# A tibble: 6 × 2
      a     c
  <dbl> <int>
1     1     4
2     1     5
3     1     6
4     1     7
5     1     8
6     1     9

> tibble(a = 1:2, c = 4:9)
Error:
! Tibble columns must have compatible sizes.
• Size 2: Existing data.
• Size 6: Column 'c'.
  Only values of size one are recycled.
Run 'rlang::last_error()' to see where the error occurred.

> data.frame(a = 1:2, c = 4:9)
```

```
    a c
1   1 4
2   2 5
3   1 6
4   2 7
5   1 8
6   2 9
```

9.2 整洁数据和数据包 tidyr

tidyverse 中的所有数据包都处理整洁数据。上一节对文档 hotel0.txt 开展词频统计得到的数据框 hotel0_counts 是 tibble 格式,就是整洁数据(tidy data)。当然,不是所有的 tibble 数据都是整洁数据。威克姆和格罗勒芒德提出整洁数据遵循三个原则:(1)每个变量必须有它自己的列;(2)每个观测值必须有它自己的排;(3)每个值必须有它自己的单元格(Wickham et al.,2017)[149]。简单地说,整洁数据的特点是变量按列、观测值按排。不整洁数据可能有两种情形:(1)一个变量可能在多个列上展开;(2)一个观察值可能分散在多个排(Wickham et al.,2017)[152]。R 数据包 tidyr 中的函数 gather( )和 spread( )分别针对这两种情形。

函数 gather( )把数据宽格式调整为长格式,即把多个列合并为两个列——新建 key(键)列(自变量)和新建 value(值)列(因变量),key 列是离散性变量(discrete variable),其水平或值取自原有的列名,value 列包括原有列的匹配值(Mailund,2019)[46-47]。该函数的主要结构是 gather(data, key, value, ...),其中 data 是数据框,key(键)列和 value(值)列是输出变量的新名称,...指需要合并的原有变量名。

假如我们要考察 6 名学生(编号为 ID)在三个背景音乐条件下的言语记忆力,记忆力的测量操作定义为正确回忆出的阅读文本中的意义单位数(idea units)。记忆测量结果如下所示:

```
wide<-data.frame(
  ID=c(1:6),
  music.1=c(45,66,52,75,65,61),
  music.2=c(48,70,53,80,66,68),
  music.3=c(50,69,57,83,70,71)
)
```

把以上结果输入 R 界面,再输入对象名 wide,按回车键得到以下结果:

```
ID music.1 music.2 music.3
1  1  45  48  50
2  2  66  70  69
3  3  52  53  57
4  4  75  80  83
5  5  65  66  70
6  6  61  68  71
```

在以上结果中,ID 在一列,但是记忆测量(recall)却被分成三列,因而数据不整洁。我们想要把 music.1、music.2 和 music.3 三个背景音乐条件变成一列(变量名为 music),把记忆测量结果归为一个新变量(变量名为 recall)。本例 R 命令和合并结果如下:

```
> require(tidyr)
> long<-gather(wide,music,recall,music.1:music.3)
> long
   ID  music   recall
1  1   music.1  45
2  2   music.1  66
3  3   music.1  52
4  4   music.1  75
5  5   music.1  65
6  6   music.1  61
7  1   music.2  48
8  2   music.2  70
9  3   music.2  53
10 4   music.2  80
11 5   music.2  66
12 6   music.2  68
13 1   music.3  50
14 2   music.3  69
15 3   music.3  57
16 4   music.3  83
17 5   music.3  70
18 6   music.3  71
```

以上结果显示两个新列:一列为背景音乐的类别,另一列为记忆测量。每位学生听每种背景音乐,因而每个 ID 被重复三次。

与函数 gather() 相反,函数 spread() 把数据长格式调整为宽格式,主要结构是 spread(data, key, value),其中 data 是数据框,key(键)列和 value(值)列是原有变量的名称。

我们来看一个例子。我们对一部小说分 6 章开展了情感分析。每章都有正面(positive)和负面(negative)情感值(n)。数据的 tibble 格式如下:

```
require(tibble)
data<-tibble(index = rep(1:6,each=2),
sentiment=rep(c("negative","positive"),6),
n=c(25,32,26,22,46,80,39,84,45,48,40,70)
)
```

加载数据包 tibble,将以上结果输入 R 界面,再输入对象名 data,按回车键得到以下结果:

```
# A tibble: 12 × 3
   index sentiment     n
   <int> <chr>     <dbl>
 1     1 negative     25
 2     1 positive     32
 3     2 negative     26
 4     2 positive     22
 5     3 negative     46
 6     3 positive     80
 7     4 negative     39
 8     4 positive     84
 9     5 negative     45
10     5 positive     48
11     6 negative     40
12     6 positive     70
```

以上结果显示三列:索引(index)、情感类别(sentiment)和情感值(n)。我们实际上希望把 sentiment 的两个类别展开,键变量是 sentiment,值变量是 n。本例 R 命令和统计分析结果如下:

```
> require(tidyr)
> data.wide<-spread(data,sentiment,n)
> data.wide
# A tibble: 6 × 3
  index negative positive
  <int>    <dbl>    <dbl>
1     1       25       32
2     2       26       22
3     3       46       80
4     4       39       84
5     5       45       48
6     6       40       70
```

以上结果显示三列：索引(index)、负面情感(negative)和正面情感(positive)。这样，我们就可以利用数据包 dplyr 中的函数 mutate( )和 pull( )计算每章的情感差异值，并且可以利用 R 自带函数 sum( )计算情感差异总值。关于函数 mutate( )和 pull( )的介绍，见下一节。本例计算小说各章情感差异总值的 R 命令和统计分析结果如下：

```
> require(dplyr)
> diff<-mutate(data.wide,score=positive-negative)
> diff
# A tibble: 6 × 4
   index  negative  positive  score
   <int>  <dbl>     <dbl>     <dbl>
1  1      25        32        7
2  2      26        22        -4
3  3      46        80        34
4  4      39        84        45
5  5      45        48        3
6  6      40        70        30
> sum(diff$score)
[1] 115
```

以上结果显示，本例正面情感值高于负面情感值，两者相差 115。输出中的⟨dbl⟩表示浮点数(小数)。

9.3 管道操作和数据包 dplyr

上一节我们利用前面介绍的重复 R 对象的方法逐步计算小说各章情感差异总值。这样做效率不高。如果中间步骤比较多，这样的操作就会创造出许多不必要的数据对象。我们可以通过加载数据包 dplyr 使用顺向性管道符"%>%"(pipe operator)来简化操作流程。例如，执行以下管道操作得到与上一节同样的结果：

```
> require(tibble)
> data<-tibble(index=rep(1:6,each=2),
    sentiment=rep(c("negative","positive"),6),
    n=c(25,32,26,22,46,80,39,84,45,48,40,70))
> require(tidyr)
> require(dplyr)
> data%>% spread(sentiment,n)%>%
mutate(score=positive-negative)%>%
pull(score)%>%
```

```
sum()
[1] 115
```

以上 R 命令中使用了多个管道符和两个 dplyr 包中以动词命名的函数 mutate()和 pull()。这种简化的操作是 R 编程的革命性变化。dplyr 函数或通常所说的 dplyr 动词(这些动词我们一看就懂,其作用一目了然)有助于使代码更简单,更具有可读性。但是,dplyr 的真正强大之处在于它能够将这些函数串联起来,体现直观的线性过程(Rhys,2020)[34]。把这些函数串联起来的符号就是管道符。管道符的功能就是通过管道把它左边函数的输出结果输送到它右边函数的第一个变元。管道符的意思就是"and then"(接下来),体现操作的序列性。

管道符"%>%"来源于斯蒂芬·米尔顿·贝琪(Stefan Milton Bache)创建的 R 数据包 magrittr(Wickham et al. ,2017)[152],但是在 R 4.1 版本之后可以用"|>"代替"%>%"。数据包 magrittr 是 tidyverse 集成包的一部分,tidyverse 和 dplyr 自动加载管道符。R 命令 x%>% f() 等同于 f(x),即求 x 的函数 f()值。例如:

```
> x<-1:10
> sum(x)
[1] 55
> require(dplyr) # or require(magrittr)
> x<-1:10
> x%>% sum()
[1] 55
x|>sum()
[1] 55
```

在上面的管道操作中,x%>% sum()等同于 x%>% sum。即函数的调用无须小括号。但是,大多数人喜欢用小括号,使函数的处理明确化。如果函数有多个变元,小括号是必须的(Mailund,2019)[74]。我们来看一个函数有两个变元的例子:

```
> require(tibble)
> data<-tibble(gender = rep(c("male", "female"), each = 4), weight = c(174, 140, 174, 140, 90, 102, 102, 100), height = c(175, 180, 165, 167, 155, 165, 172, 167))
> require(dplyr)
> data%>%
group_by(gender)%>%
summarise(m=mean(weight),n=n())%>%
ungroup()
# A tibble: 2 × 3
```

```
   gender      m       n
   <chr>     <dbl>   <int>
1  female    98.5     4
2  male      157      4
```

在以上命令中,我们调用数据包 dplyr 中的函数 group_by() 对数据框 data 中的数据按照 gender(性别)进行分组。函数 group_by() 的基本结构是 group_by(.data,...),其中 .data 是数据框或数据框扩展式(如 tibble),... 是分组变量。调用函数 group_by() 之后,管道符后面的函数计算分组进行。本例利用函数 summarise() 对数据开展描述性统计,计算的两个统计量是平均数(调用函数 mean())和观测值数或样本量(调用函数 n())。最后,调用函数 ungroup() 取消分组。函数 ungroup() 的主要变元 x 是 tibble 格式数据框。注意,在使用分组函数 group_by() 之后通常要使用取消分组函数 ungroup()。试比较以下结果:

```
# 第一个结果
> require(tibble)
> data<-tibble(gender = rep(c("male","female"),each = 4),weight = c(174,140,174,140,90,
  102,102,100),height = c(175,180,165,167,155,165,172,167))
> require(dplyr)
> result<-data%>%
    group_by(gender)%>%
    mutate(w=mean(weight),n=n())%>%
    mutate(h=mean(height))
> result
# A tibble: 8 × 6
# Groups:   gender [2]
   gender    weight   height     w       n      h
   <chr>     <dbl>    <dbl>    <dbl>   <int>  <dbl>
1  male       174     175     157      4     172.
2  male       140     180     157      4     172.
3  male       174     165     157      4     172.
4  male       140     167     157      4     172.
5  female      90     155     98.5     4     165.
6  female     102     165     98.5     4     165.
7  female     102     172     98.5     4     165.
8  female     100     167     98.5     4     165.
# 第二个结果
> result2<-data%>%
    group_by(gender)%>%
    mutate(w=mean(weight),n=n())%>%
    ungroup(gender)%>%
```

```
     mutate(h=mean(height))
> result2
# A tibble: 8 × 6
  gender  weight  height     w      n      h
  <chr>   <dbl>   <dbl>   <dbl>  <int>  <dbl>
1 male     174     175     157     4    168.
2 male     140     180     157     4    168.
3 male     174     165     157     4    168.
4 male     140     167     157     4    168.
5 female    90     155    98.5     4    168.
6 female   102     165    98.5     4    168.
7 female   102     172    98.5     4    168.
8 female   100     167    98.5     4    168.
```

第一组 R 命令调用函数 group_by() 对体重分组，并按性别分组计算平均体重，但是命令没有以 ungroup() 结束，因而对输出结果 result 的后续分析仍然延续着按性别分组。调用函数 mutate() 增加身高平均数一列时，平均数的计算仍然按性别分组进行，所以我们得到男性组和女性组不同的平均身高（172. 和 165.）。第二组 R 命令调用函数 group_by() 对体重分组，并以 ungroup() 结束，说明分组行为已经取消。调用函数 mutate() 增加身高平均数一列时，平均数的计算利用 height 列所有的值，所以我们得到所有参与者的平均身高（168.）。

我们下面回到计算情感值的例子。我们在本节开头 data 数据框后添加一个管道符，在管道符后添加函数 spread()，其意就是我们把 data 作为 spread(sentiment,n) 的第一个变元，然后把 data 中列变量 sentiment（函数 spread() 的第二个变元）的两个类别拓展为新的列变量，变量值为原列变量 n（函数 spread() 的第三个变元）的值。在得到宽式数据结构之后，我们使用另一个管道符%>%，在其后添加命令 mutate(score=positive-negative)，表示接下来要在前面得到的宽式数据中再增加一列，列变量名为 score，计算方法为正面情感值和负面情感值之差（positive-negative）。函数 mutate() 的主要结构是 mutate(.data,...,.keep=c("all","used","unused","none"))，其中 .data 是数据框或数据框扩展式（如 tibble），... 表示输出的列名与数值配对，本例为 score=positive-negative。变元 .keep="all" 是函数默认的形式，指保留 .data 中所有的列；.keep="used" 表示除 score 列之外保留在 ... 中用于创建新列的列，本例指除 score 列之外保留 negative 和 positive 列；.keep="unused" 表示除 score 列之外保留未在 ... 中用于创建新列的列；.keep="none" 表示不保留 .data 中任何额外的列。

接着看计算情感值的例子。在计算出 score 之后还要做什么呢？我们要把前面结果中的情感变量 score 包含的值提取出来。为此，我们调用 dplyr 包中的函数 pull()，其功能类似于我们前面使用的提取符 $。如果我们在前面的 R 命令 mutate(score=positive-negative) 中增加变元 .keep="none"，那么可以省去对函数 pull() 的调用，因为输出的数据框只有一列。函数 pull() 的主要结构是 pull(.data)，其中.data 是数据框或数据框扩展式（如 tibble）。最后，我们调用 R 基础函数 sum() 累计前一个命令得到的情感差异值。

　　除了 9.1 节介绍的函数 count() 和本节已经介绍的函数 mutate()、pull()、summarise()、group_by() 和 ungroup() 之外，数据包 dplyr 还提供了其他一些有用函数，包括 row_number()、inner_join()、anti_join()、filter()、select() 和 arrange()。

　　函数 row_number() 用于创建识别号，与转换函数 mutate() 连用增加一个连续数值序列。函数 inner_join() 的基本结构是 inner_join(x,y,by=NULL)，其中 x 和 y 是配对数据框或数据框扩展式（如 tibble）；by=NULL 表示默认使用 x 和 y 中的所有共同变量值进行匹配，研究者也可以自定义匹配变量。

　　我们调用数据包 tidytext 中的情感词库"bing"（见 9.6 节）对文档 hotel0.txt 开展情感分析。我们打算利用函数 row_number() 把每三句作为一个单位，计算其中的正面情感词数和负面情感词数。由于这个文本有 21 个句子，调用函数"%/%"，利用"%/%3"（行数或句子数除以 3，商取整数部分）得到 8 个单位，如故事开头两句作为第一个单位，索引为 0；第 3、4、5 句作为第二个单位，索引为 1，以此类推。为此，我们调用数据包 readtext 把文档 hotel0.txt 读入 R，然后调用数据包 tidytext 中的函数 unnest_tokens() 对文档开展句子划分，再调用 dplyr 包中的函数 mutate() 增加列变量 linenumber。由于情感分析依据文本词与情感词库中情感词的匹配数量，因此我们要把前面得到的句子进一步按词进行分割，然后调用函数 inner_join() 找出文本中的正面和负面情感词，词库的调用利用函数 get_sentiments("bing")。由于我们的目的是以每三个句子为单位计算文本中的情感词分布，因而需要再次利用函数 mutate() 增加索引列（即 index=linenumber%/%3）。最后，调用分组函数 group_by() 以索引 index 和情感 sentiment 列为分组变量，再调用函数 count() 计算各个单位中正面和负面情感值，然后通过函数 ungroup() 取消分组。本例 R 命令和统计分析结果如下：

```
> library(readtext)
> hotel0<-readtext("D:/Rpackage/hotel0.txt")
> require(tidytext)
> require(dplyr)
> require(tidyr)
> hotel0%>%
unnest_tokens(sentence,text,token="sentences")%>%
mutate(linenumber=row_number())%>%
unnest_tokens(word,sentence,token="words")%>%
inner_join(get_sentiments("bing"))%>%
mutate(index=linenumber%/%3)%>%
group_by(index,sentiment)%>%
count(sentiment)%>%
ungroup()
Joining, by = "word"
```

```
# A tibble: 13 × 3
# Groups:   index, sentiment [13]
   index sentiment     n
   <dbl> <chr>     <int>
 1     0 negative      2
 2     0 positive      4
 3     1 negative      1
 4     1 positive      2
 5     2 negative      1
 6     2 positive      1
 7     3 negative      1
 8     4 negative      3
 9     4 positive      1
10     5 negative      3
11     6 negative      1
12     6 positive      1
13     7 positive      2
```

以上结果显示,在有关史密斯先生入住酒店不愉快经历的故事中,情感的变化随故事的推进出现某种程度的小波动。例如,在故事开头的两个句子中,负面情感值为2(情感词为"cheap"和"spite",频次均为1),正面情感值为4(情感词为"clean""comfortable"和"grand",其中前两个词均出现1次,第三个词出现2次)。文本最后一句(第21句)使用了两个正面情感词("sharp"和"right"),得分为2。

我们最后简要介绍一下其他4个有用的函数：anti_join()、filter()、select()和arrange()。函数anti_join()与inner_join()相反,基本结构是anti_join(x,y,by=NULL),其中x和y是配对数据框或数据框扩展式(如tibble);by=NULL表示默认使用x和y中的所有共同变量值进行匹配,研究者也可以自定义匹配变量,返还x中与y变量值不匹配的所有排值。我们仍然以文档hotel0.txt为例计算文本中不包括停用词的词频分布。停用词表为tidytext包自带的英语词表(get_stopwords('en'))。具体做法是：先调用数据包readtext把文档hotel0.txt读入R,再调用tidytext包中的函数unnest_tokens()对文本开展形符化,然后调用dplyr包中的函数anti_join()排除文本中的停用词,最后调用函数count()计算词频分布。本例R命令和统计分析结果如下：

```
> library(readtext)
> hotel0<-readtext("D:/Rpackage/hotel0.txt")
> require(tidytext)
> require(dplyr)
> hotel0%>%
   unnest_tokens(word,text,token="words")%>%
   anti_join(get_stopwords('en'))%>%
   count(word,sort=TRUE)
```

```
Joining, by = "word"
readtext object consisting of 125 documents and 0 docvars.
# Description: df [125 × 3]
    word       n     text
    <chr>      <int> <chr>
1   mr         6     "\"\"..."
2   smith      6     "\"\"..."
3   manager    5     "\"\"..."
4   room       5     "\"\"..."
5   began      3     "\"\"..."
6   hammering  3     "\"\"..."
# ... with 119 more rows
```

以上结果显示，文档 hotel0.txt 中出现频次最高的词是"mr"和"smith"，其次是"manager"和"room"，其他词以此类推。

过滤函数 filter() 用于提取一个数据框的子集，只保留满足某个特定条件的所有排和列数据，基本结构是 filter(.data, ...)，其中.data 是数据框或数据框扩展式(如 tibble)，... 是条件表达式。在上面的例子中，有 125 个类符，有较多的词只出现 1 次和 2 次。如果我们要考察出现频次在 2 次以上的词的分布，那么可以调用函数 filter()，把条件表达式设为 n>2。本例 R 命令和统计分析结果如下：

```
> library(readtext)
> hotel0<-readtext("D:/Rpackage/hotel0.txt")
> require(tidytext)
> require(dplyr)
> hotel0%>%
   unnest_tokens(word,text,token="words")%>%
     anti_join(get_stopwords('en'))%>%
     count(word,sort=TRUE)%>%
       filter(n>2)
Joining, by = "word"
readtext object consisting of 7 documents and 0 docvars.
# Description: df [7 × 3]
    word       n     text
    <chr>      <int> <chr>
1   mr         6     "\"\"..."
2   smith      6     "\"\"..."
3   manager    5     "\"\"..."
4   room       5     "\"\"..."
5   began      3     "\"\"..."
6   hammering  3     "\"\"..."
# ... with 1 more row
```

函数 select( )的基本结构是 select(. data, ...),其中. data 是数据框或数据框扩展式(如 tibble), ... 是一个或多个用逗号隔开且不加引号的表达式。前面介绍的 R 对象 result2 储存了多个变量,如 gender、weight 和 height 等。如果我们只考察 gender 和 height 两个变量,那么实施选择的 R 命令和统计分析结果如下:

```
> require( tibble)
> data<-tibble( gender = rep( c( "male" , "female" ) , each = 4) , weight = c( 174, 140, 174, 140, 90,
102, 102, 100) , height = c( 175, 180, 165, 167, 155, 165, 172, 167) )
> require( dplyr)
> result2<-data% >%
  group_by( gender) % >%
  mutate( w = mean( weight) , n = n( ) ) % >%
  ungroup( gender)
> result2% >%
  select( gender, height)
# A tibble: 8 × 2
    gender    height
    <chr>     <dbl>
1   male      175
2   male      180
3   male      165
4   male      167
5   female    155
6   female    165
7   female    172
8   female    167
```

如果要选择除最后一列变量(n)之外的所有变量,可以使用-n,其中的"-"表示剔除的意思。例如:

```
> result2<-data% >%
  group_by( gender) % >%
  mutate( w = mean( weight) , n = n( ) ) % >%
  ungroup( gender)
> result2% >%
  select( -n)
# A tibble: 8 × 4
```

	gender ⟨chr⟩	weight ⟨dbl⟩	height ⟨dbl⟩	w ⟨dbl⟩
1	male	174	175	157
2	male	140	180	157
3	male	174	165	157
4	male	140	167	157
5	female	90	155	98.5
6	female	102	165	98.5
7	female	102	172	98.5
8	female	100	167	98.5

本节介绍的最后一个函数是arrange()。这个函数的基本结构是arrange(.data,...)，其中.data是数据框或数据框扩展式(如tibble)，...是排序依据的列变量。例如，在数据框results2的例子中，如果要对weight值按升序排列，那么使用R命令arrange(weight)；如果要对weight值按降序排列，那么使用R命令arrange(desc(weight))。试比较以下结果：

```
# 默认的升序
> result2% >%
   arrange(weight)
# A tibble: 8 × 5
```

	gender ⟨chr⟩	weight ⟨dbl⟩	height ⟨dbl⟩	w ⟨dbl⟩	n ⟨int⟩
1	male	90	155	98.5	4
2	male	100	167	98.5	4
3	male	102	165	98.5	4
4	male	102	172	98.5	4
5	female	140	180	157	4
6	female	140	167	157	4
7	female	174	175	157	4
8	female	174	165	157	4

```
# 降序使用命令desc(weight)
> result2% >%
   arrange(desc(weight))
# A tibble: 8 × 5
```

	gender ⟨chr⟩	weight ⟨dbl⟩	height ⟨dbl⟩	w ⟨dbl⟩	n ⟨int⟩
1	male	174	175	157	4
2	male	174	165	157	4
3	male	140	180	157	4
4	male	140	167	157	4
5	female	102	165	98.5	4

6	female	102	172	98.5	4
7	female	100	167	98.5	4
8	female	90	155	98.5	4

9.4 基于整洁数据的数据包 tidytext

上一节介绍了整洁数据包 tidytext 中对自然语言处理很重要的三个函数,即 get_sentiments()、get_stopwords()和 unnest_tokens()。本节介绍其他三个有用的函数,包括 unnest_ngrams()、unnest_regex()和 stop_words()。

提取 N 元组的函数 unnest_ngrams()是 unnest_tokens()的封装函数,主要结构是 unnest_ngrams(tbl,output,input,n=3L,n_min=n,ngram_delim=" "),其中变元 tbl 指 tibble 类数据框,变元 output 指被创建的输出列,变元 input 指被分割的输入列(指输入文本),n 是 N 元组中的词数,n=3L 是默认的三元组(trigrams),n_min=n 指函数默认的最小 n 元组包括的词数为 n,n_min 的取值范围在 1 和 n 之间。变元 ngram_delim=" "指函数默认 n 元组中词语之间的分隔符为空格,研究者也可以自定义其他符号,如 ngram_delim="_"或 ngram_delim=" * "。我们要考察文档 hotel0.txt 中频次不低于 2 的二元组(bigrams)。具体做法是,首先利用数据包 readtext 把文档 hotel0.txt 读入 R,其次调用 tidytext 包中的函数 unnest_ngrams()分割所有的二元组,然后调用函数 count()计算二元组频次并排序,最后调用函数 filter()过滤出所有频次不低于 2 的二元组。本例 R 命令和统计分析结果如下:

```
> library(readtext)
> hotel0<-readtext("D:/Rpackage/hotel0.txt")
> require(dplyr)
> require(tidytext)
> hotel0% >%
   unnest_ngrams(bigram,text,n=2L)% >%
   count(bigram,sort=TRUE)% >%
   filter(n>=2)
readtext object consisting of 14 documents and 0 docvars.
# Description: df [14 × 3]
  bigram         n      text
  <chr>        <int>   <chr>
1 mr smith       6    " \" \" ... "
2 the manager    5    " \" \" ... "
3 at the         3    " \" \" ... "
4 it is          3    " \" \" ... "
5 to the         3    " \" \" ... "
6 began to       2    " \" \" ... "
# ... with 8 more rows
```

以上结果显示，hotel0.txt 中出现频次不低于 2 的二元组有 14 个，显示的二元组有 6 个，其中出现频次最高的是"mr smith"，其次是"the manager"，"at the""it is"和"to the"出现的频次相同，其他二元组出现的频次均为 2。如果想要得到所有的二元组，那么在上述命令后添加 %>% pull(bigram)，再按回车键即可。

函数 unnest_regex( ) 是 unnest_tokens( ) 的封装函数，主要结构是 unnest_regex(tbl, output, input, pattern = "\\s+")，其中变元 tbl 指 tibble 类数据框，变元 output 指被创建的输出列，变元 input 指被分割的输入列，变元 pattern = "\\s+"是函数默认的正则表达式，表示以一个或多个连续的空格符切分词符。关于正则表达式的一般知识，详见第二章。如果使用函数默认的正则表达式 pattern = "\\s+" 对文档 hotel0.txt 开展形符化，那么原文的标点符号仍然保留。R 命令和统计分析结果如下：

```
> hotel0%>%
   unnest_regex(word,text, pattern = "\\s+")
readtext object consisting of 312 documents and 0 docvars.
# Description: df [312 × 3]
   doc_id       word        text
   <chr>        <chr>       <chr>
 1 hotel0.txt   whenever    "\"\"..."
 2 hotel0.txt   mr          "\"\"..."
 3 hotel0.txt   smith       "\"\"..."
 4 hotel0.txt   goes        "\"\"..."
 5 hotel0.txt   to          "\"\"..."
 6 hotel0.txt   westgate,   "\"\"..."
# … with 306 more rows
```

以上结果显示，文本形符化后标点符号和词连在一起，如"westgate,"。如果要去除标点符号，那么需要在变元 pattern 中增加"[[:punct:]]"，这样标点符号就会被作为划分词界限的字符而被去除。本例 R 命令和统计分析结果如下：

```
> hotel0%>%
   unnest_regex(word,text, pattern = "[\\s+[[:punct:]]]")
readtext object consisting of 312 documents and 0 docvars.
# Description: df [312 × 3]
   doc_id       word        text
   <chr>        <chr>       <chr>
 1 hotel0.txt   whenever    "\"\"..."
 2 hotel0.txt   mr          "\"\"..."
 3 hotel0.txt   smith       "\"\"..."
 4 hotel0.txt   goes        "\"\"..."
 5 hotel0.txt   to          "\"\"..."
 6 hotel0.txt   westgate    "\"\"..."
# … with 306 more rows
```

以上结果显示,"westgate"后不再有逗号(,)了。如果在以上命令后添加% >% pull(word),那么可以看到完整的形符化结果。

上一节介绍 dplyr 包中的过滤函数 filter()时,调用了去除函数 anti_join(),把在文档 hotel0.txt 中出现的包含在 get_stopwords('en')词表中的停用词从文档中排除出去。这个词表的词库来源是"snowball",包括 175 个词,以冠词、代词、助动词、连词和介词为主。安装 stopwords 包可以查看多个词库来源。例如,执行以下 R 命令得到 8 种词库:

```
> require(stopwords)
> stopwords_getsources()
[1] "snowball"    "stopwords-iso"   "misc"    "smart"
[5] "marimo"     "ancient"      "nltk"    "perseus"
```

我们在第三章了解到,R 数据包 tm 也自带停用词表,包括 174 个词,词表查看的 R 命令为:tm::stopwords("en")。这两个词表基本相同,只是 get_stopwords("en")词表中多出"will"一词。R 数据包 tidytext 还提供另外一个停用词表,名称为"stop_words"。词表 stop_words 来自三个词库:"onix""SMART"和"snowball",共计 1 149 个词。我们利用前面数据包中的函数计算各个词库中的词汇量:

```
> require(tidytext)
> require(dplyr)
> stop_words% >%
    group_by(lexicon)% >%
    count(word)% >%
    summarise(size=sum(n))% >%
    ungroup()
# A tibble: 3 × 2
  lexicon    size
  <chr>     <int>
1 onix       404
2 SMART      571
3 snowball   174
```

在以上代码中,从 dplyr 包中调用函数 group_by()对三个词库中的词进行分组统计,先调用函数 count()计算词频,然后调用函数 summarise()开展汇总统计。统计分析结果表明,"SMART""snowball"和"onix"库的词汇量依次为 571、174 和 404 个词。这里的"snowball"词表列出的词语与 tm 包中函数 stopwords("en")列出的词语相同。当然,这三个词库中的词有重复,如它们都包括"a""about"和"above"。要计算这三个词库出现的类符(types)数,首先需要调用 dplyr 包中的函数 pull()从 stop_words 的 tibble 中把词(word)提取出来,其次调用 R 基础函数 unique()计算类符,然后调用 R 基础函数 length

( )计算类符数。本例 R 命令和统计分析结果如下:

```
> require( tidytext)
> require( dplyr)
> stop_words% >%
   pull( word) % >%
   unique( ) % >%
   length( )
[1] 728
```

以上结果表明,来源于三个词库的 stop_words 包含的类符数为 728 个词。如果我们用三个词库中的词作为停用词,形符重复问题就不重要。当然,我们也可以根据研究需要,选择其中的一个词库制作停用词表,如"SMART"或"snowball"停用词表。

9.5 精美制图数据包 ggplot2

数据包 ggplot2 是威克姆的杰作,也是整洁数据集成包 tidyverse 的重要组成部分(Wickham,2016)。数据包 ggplot2 的制图原理依据利兰·威尔金森的《图形语法》(*The Grammar of Graphics*)(Wilkinson,2005)。该包命名中的 gg 代表 Grammar of Graphics 中的首字母缩写。《图形语法》认为,任何数据图形都可以通过数据与不同层级图形元素(如轴、刻度、网格线、点、线和条形)的有机结合进行创建。通过对图形元素的分层,我们可以使用 ggplot2 包很直观地制作出有交流价值的美图(Rhys,2020)[35]。这套语法告诉我们,统计图就是由数据向几何对象(Geometric Objects,GEOM,包括点、线和条形)美学属性(Aesthetic Attributes,AES,包括颜色、形状和大小)的映射(Wickham,2016)[4]。

本节以文档 hotel0.txt 为例简要介绍利用 ggplot2 绘制文本词频分布图和情感词分布图。

要绘制文本词频分布图,首先要计算文本词频分布。前面已经介绍,我们可以利用 R 数据包 readtext 把文档 hotel0.txt 读入 R,再加载 dplyr 包和 tidytext 包对文档开展形符化,得到各个词语,储存的 R 对象为 hotel_tks。本例具体的 R 代码操作如下:

```
> require( readtext)
> hotel0_df<-readtext( "D:/Rpackage/hotel0.txt" )
> require( dplyr)
> require( tidytext)
> hotel_tks<-hotel0_df% >% unnest_tokens( word, text)
```

接下来,调用 dplyr 包中的函数 count( )对 hotel_tks 中的词语开展词频统计。由于词频分布范围较大,通常只需考察高频词的分布,特别是有意义实词的分布。本例文本词汇量不是很大,我们调用过滤函数 filter( )筛选频次大于 2 的所有文本词。为了使词频分布图按照频次值顺序对文本词排列,我们调用 R 基础函数 reorder( )对筛选值排序,并调用函数

mutate()在数据框中增加新列 word。最后,调用 ggplot2 包中的函数 ggplot(),把美学映射函数 aes()(描述数据中的变量如何被映射到几何元素的视觉属性中)中的美学特征 x 轴设为 word,y 轴设为 n,使词语在横轴上显示,频次在纵轴上显示。注意,在函数中设置变量时,不要在变量上加引号,如 x="word"或 y="n"。函数 ggplot()的下一层是函数 geom_col(),其中 geom 指几何对象,图形元素为 col("col"为"column"的缩写),用于绘制条形图(bar chart)。分层函数命名为 geom_[graphical element](),如 geom_point()在图形中增加数据点。当调用函数 geom_col()时,为了把新层级添加到 ggplot()初始调用中,需要把用于连接的加号(+)置于每一行的结尾以增加代码的可读性,切不可把加号置于行首。设置 xlab(NULL)是为了在横轴上不显示变量名(word)。若词语过多,则横轴显示会过于拥挤,可以调用函数 coord_flip()把横轴和纵轴对调。当然,我们也可以直接设置 ggplot(aes(x=n,y=word)),省去调用 coord_flip()的麻烦。本例的 R 代码如下:

```
library(ggplot2)
hotel_tks%>%
count(word,sort=TRUE)%>%
filter(n>2)%>%
mutate(word=reorder(word, n))%>%
ggplot(aes(x=word,y=n))+geom_col()+xlab(NULL)+coord_flip()
```

执行以上 R 代码,得到如图 9.1 所示的词频分布条形图。

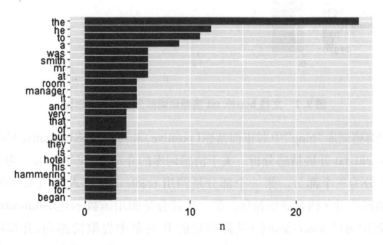

图 9.1 文档 hotel0.txt 高频词词频分布条形图

图 9.1 显示,定冠词"the"出现 26 次,出现频次最高。其他出现频次不低于 5 的词是"he""to""a""was""smith""mr""at""room""manager""it"和"and"。出现频次较低($n=3$)的词包括"they""is"和"hotel"等。

我们在制图过程中可以根据视觉表现或审美需要使用彩色图形。例如,我们希望用文本词频 n 作为填充色增加到图形中,可以设置 ggplot(aes(x=word,y=n,fill=n))。若不需要显示图例(legend),则可以在函数 geom_col()中增设 show.legend=FALSE。如果不喜欢图形自动填充的色彩,可以利用函数 scale_fill_gradient()重新设置连续性颜色梯度(gradient,即渐变色)。我们这里在函数 scale_fill_gradient()中把梯度低端和高端的颜色分别设为:low="blue",high="red"。执行以下 R 命令可以得到如图 9.2 所示的词频分布彩色条形图:

```
> hotel_tks%>%
  count(word,sort=TRUE)%>%
  filter(n>2)%>%
  mutate(word=reorder(word,n))%>%
  ggplot(aes(x=word,y=n,fill=n))+
  geom_col(show.legend=FALSE)+xlab(NULL)+coord_flip()+
  scale_fill_gradient(low="blue",high="red")
```

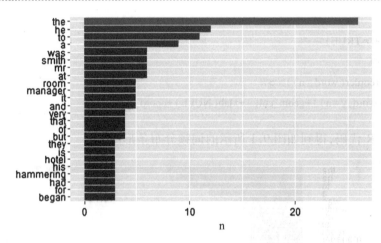

图 9.2　文档 hotel0.txt 高频词词频分布彩色条形图

下面利用情感词库"bing"中对正面情感(positive)和负面情感(negative)词的分类介绍如何对文档 hotel0.txt 开展情感分析。关于情感词库的介绍,详见下一节。为了便于理解,我们把 R 代码分为三个部分。第一个部分是调用 readtext 包读入文档 hotel0.txt,且调用 tidytext 包中的函数对文档开展形符化。第二个部分是调用函数 get_sentiments("bing")加载情感词库,调用函数 inner_join()从形符化的 R 对象中提取情感词,并调用函数 count()对情感词计数。然后,调用 R 基础函数 reorder()对词语排序(为的是使图形中的情感词按计数顺序排列),并调用函数 mutate()在数据框中增加列,列变量名仍为 word。第三个部分是调用 ggplot2 包绘制条形图。首先,在函数 ggplot()中的函数 aes()中设置 aes(word,n,fill=sentiment),即横坐标(x)和纵坐标(y)变量分别为 word 和 n,填充色依据变量 sentiment。注意,x 和 y 名称因为常用而常被忽略,但是其他美学属性(如变元 fill)必须被命名。如果不需要图例(legend),可以在条形图函数 geom_col()中设置 show.legend=FALSE。我们想要把文本情感词的分布按照正面和负面情感分开呈现,于是调用封装型分面函数 facet_wrap( )把一维面板条块封装到二维中,分面变量为 sentiment,输入时使用单侧公式:~sentiment。在正面和负面情感词分开放置时,我们允许转轴前 y 的标度可变,因而设置 scales="free_y"。函数变元 scales 的设置可以有四种形式:scales="fixed" 指 x 和 y 的标度在所有面板中都相同;scales="free" 指 x 和 y 的标度在每个面板都可以变化;scales="free_x" 指 x 的标度可变,y 的标度固定;scales="free_y" 指 y 的标度可变,x 的标度固定。最后,我们添加设置 labs(x=NULL,y="Contribution to sentiment")使 x 轴不显示变量名"word",y 轴显示标题"Contribution to sentiment"。为了使 x 轴上情感词显示不过度重合,调用函数 coord_flip( )把 x 轴和 y 轴对调。由于本例中的操作使用管道符,以上三个部分的代码被整合到了一起。R 命令和操作结果如下:

```
> require(readtext)
> require(dplyr)
> require(tidytext)
> require(ggplot2)
> hotel0<-readtext("D:/Rpackage/hotel0.txt")
> hotel0%>% unnest_tokens(word,text)%>%
  inner_join(get_sentiments("bing"))%>%
  count(word=word,sentiment)%>%
  mutate(word=reorder(word,n))%>%
  ggplot(aes(word,n,fill=sentiment))+
  geom_col(show.legend=FALSE)+
  facet_wrap(~sentiment,scales="free_y")+
  labs(x=NULL,y="Contribution to sentiment")+
  coord_flip()
Joining, by = "word"
```

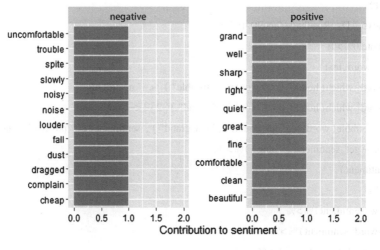

图 9.3　文档 hotel0.txt 情感词分布条形图

图 9.3 显示,负面情感词包括 12 个词,如"uncomfortable"和"trouble",每个词均出现 1 次;正面情感词包括 10 个词,如"grand"和"well",除"grand"一词出现 2 次外,其他词均出现 1 次。整体上看,文本中正面和负面情感词的数量和频次相近。如果各个情感词的情感权重相同,那么本例中的情感值为-1,即负面情感值略高于正面情感值。

以上只是对文本情感词的粗略分析。否定词的使用需要引起注意。如"trouble"用在否定句"he never has to go to the trouble of reserving a room",应该表示正面的情感。文中出现的第二个"grand"用于否定短语"not very 'grand'",应该表示负面的情感。一反一正,两者相互抵消,文本情感差异值不变。如果我们要对以上问题进行调整,把"trouble"改为"no trouble","grand"改为"not grand",需要在文本形符化之后增加以下 R 命令:

```
> hotel_tks $ word[which(hotel_tks $ word=="trouble")]<-"no trouble"
> hotel_tks $ word[which(hotel_tks $ word=="grand")][2]<-"not grand"
```

另外,由于"no trouble"和"not grand"分别表示正面情感和负面情感,我们需要在原词库中增加这两个短语,并重新对情感词库命名。R 代码如下:

```
add_sentiment<-tibble(word=c("no trouble","not grand"),
sentiment=c("positive","negative"))
new_sentiment<-get_sentiments("bing")%>%
rbind(add_sentiment)
```

针对本例,在调整情感词后开展情感分析使用的完整 R 代码如下:

```
require(readtext)
require(dplyr)
require(tidytext)
require(ggplot2)
hotel0<-readtext("D:/Rpackage/hotel0.txt")
hotel0_tks<-hotel0%>%
unnest_tokens(word,text)
hotel0_tks $ word[which(hotel0_tks $ word=="trouble")]<-"no trouble"
hotel0_tks $ word[which(hotel0_tks $ word=="grand")][2]<-"not grand"
add_sentiment<-tibble(word=c("no trouble","not grand"),sentiment=c("positive","negative"))
new_sentiment<-get_sentiments("bing")%>%
rbind(add_sentiment)
hotel0_tks%>%
inner_join(new_sentiment)%>%
count(word=word,sentiment)%>%
mutate(word=reorder(word,n))%>%
```

```
ggplot(aes(word,n,fill=sentiment))+
geom_col(show.legend=FALSE)+
facet_wrap(~sentiment,scales="free_y")+
labs(x=NULL,y="Contribution to sentiment")+
coord_flip()
```

执行以上 R 命令,得到如图 9.4 所示的文本词情感分析条形图。

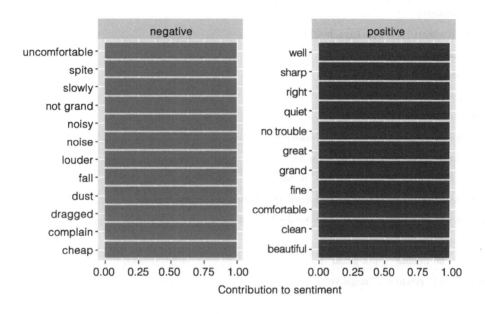

图 9.4 调整的文档 hotel0.txt 情感分析条形图

9.3 节调用数据包 tidytext 中的情感词库"bing"对文档 hotel0.txt 开展了情感分析。情感分析以每三个句子为单位计算其中的正面情感词数和负面情感词数。这个文档有 21 个句子,共有 8 个情感分析单位。我们下面据此把情感分析结果绘制成条形图,同时对情感词进行微调。

本例 R 代码的编写分为三个部分。第一个部分首先调用函数 unnest_tokens()对文本进行句子分割,并调用函数 mutate()增加句子索引,名称为 linenumber。然后,依据句子索引对分割的文本句子开展形符化,把得到的结果命名为 hotel0_tks。第二个部分首先对 hotel0_tks 中的个别情感词进行调整,同时把它们及其情感分类添加到原有情感词库"bing"中,并把得到的新情感库重新命名为 new_sentiment。其次,调用函数 inner_join()提取 hotel0_tks 中包含在情感库 new_sentiment 中的情感词,并调用函数 count()按每三个句子作为一段对情感词计数。然后,调用函数 spread()把长格式改为宽格式,即把 sentiment 的两个类别作为两个列变量,观测值为计数 n。注意,要在函数 spread()中设置变元 fill=0,使缺失值(一个片段中没有情感词,记作 NA)被"0"值所替代。最后,调用函数 mutate()在 tibble 中增加 sentiment 列,表示正面情感值与负面情感值的差异。第三个部分调用

ggplot2 包绘制情感分析条形图。在美学映射函数 aes()中设置横坐标为 index,纵坐标为 sentiment,颜色填充变量为 index。然后,调用函数 geom_col()添加几何图形。若不显示图例,则设置 show.legend=FALSE。如果对默认的条形图颜色不满意,可以调用函数 scale_fill_gradientn()进行自定义。该函数把多种颜色作为变元,构建一个标度,默认标度上的颜色是等距的。这里,我们调用 RColorBrewer 包中的调色函数 brewer.pal()设置多种颜色。本例完整的 R 代码如下:

```
library(readtext)
hotel0<-readtext("D:/Rpackage/hotel0.txt")
require(tidytext)
require(dplyr)
require(tidyr)
hotel0_tks<-hotel0%>%
unnest_tokens(sentence,text,token="sentences")%>%
mutate(linenumber=row_number())%>%
unnest_tokens(word,sentence,token="words")
hotel0_tks $ word[which(hotel0_tks $ word=="trouble")]<-"no trouble"
hotel0_tks $ word[which(hotel0_tks $ word=="grand")[2]]<-"not grand"
add_sentiment<-tibble(
word=c("no trouble","not grand"),
sentiment=c("positive","negative")
)
new_sentiment<-get_sentiments("bing")%>%
rbind(add_sentiment)
library(ggplot2)
hotel0_tks%>%
inner_join(new_sentiment)%>%
count(doc_id,index=linenumber%/%3,sentiment)%>%
spread(sentiment,n,fill=0)%>%
mutate(sentiment=positive-negative)%>%
ggplot(aes(index,sentiment,fill=index))+
geom_col(show.legend=FALSE)+
scale_fill_gradientn(colours=RColorBrewer::brewer.pal(8,"Dark2"))
```

执行以上 R 代码,得到如图 9.5 所示的文本词情感分析条形图。

图 9.5 显示,在文档 hotel0.txt 中,有三段情感呈中性,即索引号为 0、2 和 6 的三个句子片段使用的正面和负面情感词数和频次相同。例如,索引号为"0"的开头两句包含的正面情感词为"grand""clean"和"comfortable",负面情感词为"spite""not grand"和"cheap",在情感权重相同的情况下,正面和负面情感差异值为"0"。图 9.5 还显示,故事开头和结尾部

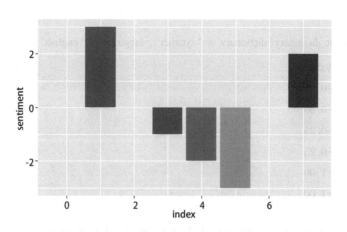

图 9.5 文档 hotel0.txt 情感词分布变化趋势

分显示正面情感,中间部分显示负面情感,整体上看,故事的情感表现不强烈。我们在下一章对同样故事的中文译文开展情感分析,有兴趣的读者可以比较中英文情感表现的差异。

9.6 利用整洁数据的文本情感分析

情感分析(sentiment analysis)用于探究文本中表现出的态度和观点。情感分析的前提条件是确定情感词库。上一节介绍 ggplot2 包时使用了情感词库"bing"对一则故事开展了情感分析。本节简要介绍 syuzhet 包中的函数 get_sentiment_dictionary( )包括的几个情感词库,然后利用情感词库对一个故事文本开展基础统计和情感分析。

9.6.1 情感词库

R 数据包 syuzhet 由乔克斯开发,当前版本号为 1.0.6,默认安装路径的 R 命令为:install.packages("syuzhet"),或者自定义安装路径(Jockers,2015)。根据乔克斯网站的描述(参见网址:https://github.com/mjockers/syuzhet),"syuzhet"这个名字来自俄罗斯形式主义者维克多·什克洛夫斯基(Victor Shklovsky)和弗拉基米尔·普罗普(Vladimir Propp)的叙事学理论。他们将叙事分为"fabula"和"syuzhet"两个部分。"fabula"指事件的时间顺序或故事元素,而"syuzhet"指叙事技巧,即故事元素的组织方式。乔克斯使用 syuzhet 作为情感分析数据包的名称意在通过情感分析揭示叙事的潜在结构,揭示冲突与化解之间叙事移动所表现出的情感变迁。

数据包 syuzhet 包括四个情感词库:"syuzhet""bing""afinn"和"nrc"。情感词库"syuzhet"在乔克斯指导下由内布拉斯加文学实验室(The Nebraska Literary Lab)开发。这个词库是一个 10 748×2 的数据框,包括 10 748 个情感词,有 word(词)和 value(情感值)2 列,其中负面情感词为 7 161 个,正面情感词为 3 587 个。syuzhet 库的调用使用以下 R 函数:get_sentiment_dictionary(dictionary="syuzhet", language="english"),前 6 行如下所示:

```
> require(syuzhet)
> head(get_sentiment_dictionary(dictionary = "syuzhet", language = "english"))
        word   value
1      abandon -0.75
2    abandoned -0.50
3    abandoner -0.25
4 abandonment -0.25
5     abandons -1.00
6     abducted -1.00
```

以上结果显示,这些词是负面情感词,最小值为-1,最大值为-0.25(整个词库中负面情感词的最大值为-0.10)。正面情感词的最大值为1,最小值为0.10。为了了解 syuzhet 库中情感词的分布,利用以下 R 代码得到如图 9.6 所示的条形图:

```
library(syuzhet)
library(ggplot2)
library(dplyr)
syuzhet<-get_sentiment_dictionary(dictionary="syuzhet",language="english")
as_tibble(syuzhet)%>%
count(value,sort=TRUE)%>%
mutate(word=reorder(value,n))%>%
ggplot(aes(word,n,fill=n))+geom_col(show.legend=FALSE)+
scale_fill_gradientn(colours=RColorBrewer::brewer.pal(8,"Set1"))+
labs(x=NULL)+
coord_flip()
```

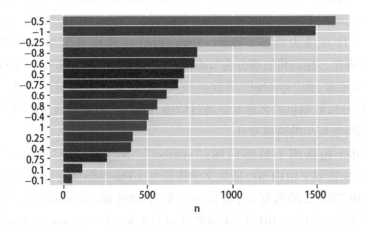

图 9.6　syuzhet 库情感词分布

图 9.6 显示,syuzhet 库的情感值有 16 类: -1.00,-0.80,-0.75,-0.60,-0.50,-0.40, -0.25,-0.10,0.10,0.25,0.40,0.50,0.60,0.75,0.80,1.00,因而 syuzhet 库的尺度范围很大。在各个情感值的分布中,包括情感词数最多的三个类别是负面情感类,情感值为-0.5、-1 和-0.25 的情感词数依次为 1 616 个、1 493 个和 1 229 个。包括正面情感词数最多的三个类别是 0.5(719 个)、0.6(614 个)和 0.8(561 个)。包括情感词数最少的三个类别是 0.75(260 个)、0.1(113 个)和-0.1(52 个)。

情感库 bing 由 Hu 等(2004)开发。该库的调用使用 R 函数 get_sentiment_dictionary (dictionary=" bing" ,language=" english"),前 6 行如下所示:

```
> library(syuzhet)
> bing<-get_sentiment_dictionary(dictionary=" bing" ,
language=" english" )
> head(bing)
       word     value
1        a+       1
2     abound      1
3    abounds     1
4   abundance    1
5    abundant    1
6   accessable   1
```

以上结果显示,这些词是正面情感词,赋值均为 1。syuzhet 库对情感词赋予不同的权重,而 bing 库只对情感词进行两极分类,即对正面情感词赋值 1,负面情感词赋值-1。bing 库是一个 6 789×2 的数据框,有 word(词)和 value(情感值)2 列,包括的情感词总数为 6 789 个,其中正面情感词为 2 006 个,负面情感词为 4 783 个。syuzhet 库和 bing 库共有情感词 5 910 个,包括正面情感词 1 732 个,如" accomplishments" 和" achievements" ,负面情感词 4 178 个,如" abominably" 和" abominate" 。

bing 库的调用还可以利用数据包 tidytext,调用函数是 get_sentiments(" bing")。该函数提供一个 6 786×2 的数据框,比 get_sentiment_dictionary(dictionary=" bing" ,language=" english")少 3 个词,即" a+" " 2-faced" 和" na\xefve" 。

第三个情感词库是 afinn,由芬恩·阿鲁普·尼尔森为微博(包括推特 Twitter)中的情感分析而开发(Nielsen,2011)。该库的调用利用 R 函数 get_sentiment_dictionary (dictionary=" afinn" ,language=" english")。afinn 库是一个 3 382×2 的数据框,有 word (词)和 value(情感值)2 列,包括的情感词总数为 3 382 个,其中正面情感词(情感值大于 0)为 1 176 个,负面情感词(情感值小于 0)为 2 204 个,中性情感词(情感值为 0)为 2 个。不同于 syuzhet 库,afinn 库利用等间隔的量表,情感值在-5(很负面)和+5(很正面)之间,有 11 个情感强度等级。图 9.7 显示 afinn 库情感词等级分布。

如图 9.7 所示,情感值为-2 的词数最多(1 411 个),其次是情感值为 2 的词数(627

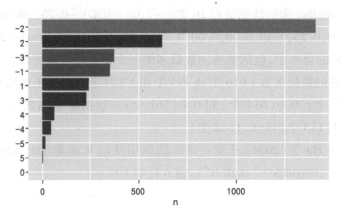

图9.7 afinn库情感词分布

个),情感值为-3和-1的词数相当(分别为376个和354个),情感值为1和3的词数也相当(分别为246个和233个),情感值为4和-4的词数较少,不到100个(分别为65个和47个),情感值为-5、5和0的词数最少(依次为16个、5个和2个)。

afinn库的调用还可以利用数据包tidytext,调用函数是get_sentiments("afinn")。该函数提供一个2477×2的数据框,比get_sentiment_dictionary(dictionary="bing",language="english")少905个词,如"aborted""aborts"和"abusing"等。

本节介绍的最后一个情感词库是nrc库,由穆罕默德和特尼开发,包括8种基本情绪和2类情感(正面和负面)(Mohammad et al.,2010)。这8种情绪是"anger"(愤怒)、"fear"(恐惧)、"anticipation"(期待)、"trust"(信任)、"surprise"(惊讶)、"sadness"(悲伤)、"joy"(喜悦)和"disgust"(厌恶)。nrc库的调用利用R函数get_sentiment_dictionary(dictionary="nrc",language="english")。该库是一个13 901×4的数据框,有language(语言)、word(词)、sentiment(情感)和value(情感值)4列,包括的情感词总数为13 901个。执行以下R命令得到各个情绪和情感类别上的情感词数:

```
> require(syuzhet)
> require(dplyr)
> nrc<-get_sentiment_dictionary(dictionary="nrc",
language="english")
> as_tibble(nrc)%>%
group_by(sentiment)%>%
count(sentiment,sort=TRUE)%>%
ungroup()
# A tibble: 10 × 2
  sentiment      n
  <chr>        <int>
1 negative     3 324
2 positive     2 312
```

```
3  fear             1 476
4  anger            1 247
5  trust            1 231
6  sadness          1 191
7  disgust          1 058
8  anticipation       839
9  joy                689
10 surprise           534
```

以上结果显示,nrc 库包括的负向情感词和正向情感词分别为 3 324 个和 2 312 个,在 8 种情绪中,表示恐惧、愤怒、信任、悲伤和厌恶的情绪词较多(依次为 1 476 个、1 247 个、1 231 个、1 191 个和 1 058 个),包含词数最少的 3 种情绪是期待(839 个)、喜悦(689 个)和惊讶(534 个)。

在以上四个词库中,nrc 库包括的情感词数量最多,其次是 syuzhet 库,bing 库位居第三,afinn 库的库容最小。

9.6.2 对故事文本的情感分析

本节以故事文档 hotel0.txt 为例,调用情感分析数据包 syuzhet 中不同的情感词库和整洁数据包(如 dplyr 和 ggplot2)计算文本中的情感值。

调用 syuzhet 库计算文本情感值比较方便。syuzhet 库提供以句子为分析单位和以词为分析单位的两种形符化函数。以句子为分析单位的 R 函数是 get_sentences(text_of_file),其中 text_of_file 是文本字符串。以词为分析单位的 R 函数是 get_tokens(text_of_file, pattern = "\\W", lowercase = TRUE),其中 text_of_file 是文本字符串,pattern = "\\W" 是函数默认的字符切分方式,lowercase = TRUE 指函数默认把所有的大写字母转化为小写字母。计算情感值的 R 函数是 get_sentiment(char_v, method = "syuzhet"),其中 char_v 是字符串向量,method = "syuzhet" 指函数默认调用 syuzhet 情感词库。利用 syuzhet 库调用函数 get_sentences()计算文档 hotel0.txt 情感值的 R 命令和统计分析结果如下:

```
> require(readtext)
> hotel0<-readtext("D:/Rpackage/hotel0.txt") $ text
> require(syuzhet)
> hotel0_sents<-get_sentences(hotel0)
> syuzhet_vector<-get_sentiment(hotel0_sents, method = "syuzhet")
> syuzhet_vector
 [1]   0.50   0.75   0.30   0.00   0.85   0.05   0.90
 [8]   0.00   0.00   0.60   0.00  -0.50   0.40  -1.30
[15]  -0.25  -0.15  -0.75   0.25  -0.90   0.00   0.30
> sum(syuzhet_vector)
[1] 1.05
```

以上结果显示,这个例子包括 21 个句子,每个句子的情感值包含在 syuzhet_vector 向量中,情感总值为 1.05,说明本文体现较弱的正面情感趋势。如果调用函数 get_tokens(),那么函数 get_sentiment()计算每个形符(词)的情感值(不在情感词库中的文本词情感值计为 0),情感分析结果相同。

使用 get_sentences()的一个好处是,如果文本不长,我们可以比较不同句段中的情感值。在本例中,为了探究故事中的情感随情节或时间的变化,我们以每三个句子为一段分析情感变化。利用分段(bins)计算情感均值的 R 函数是 get_percentage_values(raw_values,bins=100),其中 raw_values 是原始情感值,本例为 syuzhet_vector,bins=100 是函数默认的分段数,本例设为 bins=7。接着上面的 R 命令,利用函数 get_percentage_values()计算每三个句子的情感值,利用 ggplot2 包绘制情感值随句段变化的线图。R 代码如下:

```
percent_vals<-get_percentage_values(syuzhet_vector,bins=7)
require(dplyr)
require(ggplot2)
tibble(percent_vals)%>%
mutate(sents=row_number())%>%
ggplot(aes(x=sents,y= percent_vals))+
geom_line(color="red")+
geom_point(shape=23,fill="blue",color="darkred",size=2)+labs(x="sentences",y="sentiment value")
```

执行以上 R 命令,得到如图 9.8 所示的线图。

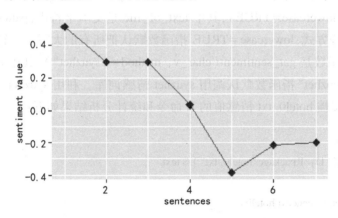

图 9.8 文档 hotel0.txt 情感值随句段变化的趋势

图 9.8 显示,以三句为一段,文档 hotel0.txt 中的情感值变化呈先降后升之势。整体上看,故事的前半部分主要叙述主人公史密斯先生对酒店的好感以及与酒店经理的交情,表现出较弱的正面情感,后半部分主要叙述主人公史密斯先生不愉快的酒店经历,因而表现出较弱的负面情感。

上一节利用数据包 tidytext 中的情感词库 bing 计算文档 hotel0.txt 中的综合情感值为

−1。syuzhet 包中的 bing 情感词库与 tidytext 中的情感词库 bing 几乎相同。调用 syuzhet 包中的 bing 情感词库对文档开展情感分析的 R 命令和统计分析结果如下：

```
> require(readtext)
> hotel0<-readtext("D:/Rpackage/hotel0.txt")$text
> require(syuzhet)
> bing<-get_sentiment_dictionary(dictionary="bing")
> hotel0_sents<-get_sentences(hotel0)
> bing_vector<-get_sentiment(hotel0_sents, method="bing")
> bing_vector
[1]  1  1  0  0  1 -1  1  0 -1  0  0 -1  0 -1 -1 -1 -1  1  1 -1  0  2
> sum(bing_vector)
[1] -1
```

以上结果显示文档 hotel0.txt 每个句子的情感值，如第一句的情感值为 1，综合情感值为 −1。

若要利用 syuzhet 包中的 afinn 情感词库计算文档 hotel0.txt 的情感值，使用与上面相似的命令，只是把 bing 改为 afinn，得到的综合情感值为 4，体现较弱的正面情感。同样，利用 syuzhet 包中的 nrc 情感词库计算文档 hotel0.txt 的情感值，得到的综合情感值为 1。除了利用 afinn 库计算出的情感值稍大以外，利用其他三个情感词库得到的情感值有很小的差异。造成以上差异的原因不仅是情感词库使用的测量尺度不同，而且情感词也不全相同。

情感词库 nrc 的一个优势是把情绪词分成 8 种，更有助于把握文本中情绪变化的具体特点。下面计算文档 hotel0.txt 中的 8 种情绪词分布和 2 类情感词分布。为了便于理解，我们分三个步骤编写 R 代码。首先，利用 readtext 包把文档 hotel0.txt 读入 R，利用 tidytext 包中的函数 unnest_tokens() 把文本字符串形符化，利用 dplyr 包中的函数 anti_join() 和过滤函数 filter() 把文本中的停用词（调用 tidytext 包中的 stop_words）和数字（利用正则表达式 grepl('[0-9]')）排除出去。然后，调用 syuzhet 包中的 nrc 情感词库，调用 dplyr 包中的函数 inner_join() 提取文本中的情感词，调用分组函数 group_by() 和 summarize() 对情绪和情感分类计数，取变量名 freq，再两次调用函数 mutate() 计算每类情绪词和情感词的占比，并对结果排序，以便按占比排序绘制条形图，利用函数 select() 选择绘图需要的两个变量 sentiment 和 percent。以上操作得到情绪词和情感词分类统计结果，取 R 对象名为 emotion_words。最后，调用整洁数据包 ggplot2 绘制条形图。函数 scale_fill_gradientn() 的调用是为了自定义图形的颜色。在本例中，R 代码的编写使用管道操作符 %>%，具体如下：

```
require(readtext)
hotel0<-readtext("D:/Rpackage/hotel0.txt")
require(syuzhet)
require(dplyr)
```

```
require(tidytext)
nrc<-get_sentiment_dictionary(dictionary="nrc",language="english")
emotion_words<-
tibble(hotel0)%>%
unnest_tokens(word,text)%>%
anti_join(stop_words,by="word")%>%
filter(!grepl('[0-9]',word))%>%
inner_join(nrc,by="word",relationship="many-to-many")%>%
group_by(sentiment)%>%
summarize(freq=n())%>%
mutate(percent=round(freq/sum(freq)*100))%>%
mutate(sentiment=reorder(sentiment,percent))%>%
select(-freq)%>%
ungroup()
library(ggplot2)
ggplot(emotion_words,aes(percent,sentiment,fill=percent))+
geom_col(show.legend=FALSE)+
labs(y=NULL)+
scale_fill_gradientn(colours=RColorBrewer::brewer.pal(9,"Set1"))
```

执行以上 R 代码得到如图 9.9 所示的情感分析条形图。

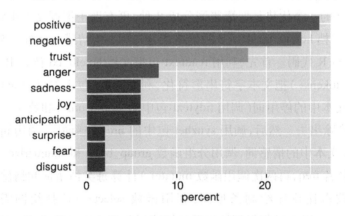

图 9.9 文档 hotel0.txt 中的情绪词与情感词分布

根据以上代码,得到如表 9.1 所示的各类情绪词、情感词以及情感值。

表 9.1 文档 hotel0.txt 中的情绪词与情感词

word	sentiment	value	word	sentiment	value
clean	positive	1	smith	trust	1
building	positive	1	clean	trust	1

（续表）

word	sentiment	value	word	sentiment	value
beautiful	positive	1	smith	trust	1
visit	positive	1	usual	trust	1
usual	positive	1	smith	trust	1
build	positive	1	smith	trust	1
library	positive	1	white	trust	1
attention	positive	1	smith	trust	1
white	positive	1	smith	trust	1
building	positive	1	spite	anger	1
immediately	positive	1	noisy	anger	1
quiet	positive	1	demand	anger	1
landed	positive	1	complain	anger	1
spite	negative	1	fall	sadness	1
cheap	negative	1	complain	sadness	1
noisy	negative	1	quiet	sadness	1
demand	negative	1	clean	joy	1
noise	negative	1	beautiful	joy	1
uncomfortable	negative	1	white	joy	1
dust	negative	1	white	anticipation	1
cough	negative	1	coming	anticipation	1
fall	negative	1	immediately	anticipation	1
immediately	negative	1	suddenly	surprise	1
complain	negative	1	forced	fear	1
forced	negative	1	cough	disgust	1

表9.1显示，依据nrc情感词库，文档hotel0.txt中出现包括重复计数的50个词，每个词的情感值均为1。同样的词可以体现不同的情绪或情感。例如，"noisy"既表示负面情感，又表示"愤怒"情绪；"white"既表示正面情感，又表示"信任""喜悦"和"期待"情绪。同样，"beautiful"既表示正面情感，又表示"喜悦"情绪。当然，有些词的情绪分类可能与语境有关。例如，"immediately"既体现正面情感又体现负面情感，还表现"期待"情绪。在故事中，酒店装修产生的噪声和灰尘使史密斯先生开始咳嗽，心情很糟糕。史密斯先生在忍无可忍的情况下只好即刻去找经理投诉。根据这一语境，文中出现的"immediately"一词应体现负面情感。

如图9.9所示，正面情感值占比（26%）和负面情感值占比（24%）相当。在所有情绪类别中，表示"信任"的情绪值占比最高（18%），"悲伤""喜悦"和"期待"情绪值占比相同

（6%），略低于"愤怒"情绪值占比（8%）。情绪值占比最低的三个类别是"惊讶""恐惧"和"厌恶"，占比均为 2%。

如果我们想要了解不同类情绪词或情感词出现在文本中的句子语境，可以调用 syuzhet 包中的函数 get_nrc_sentiment(char_v, language = "english")，其中 char_v 是字符串向量，language = "english" 指默认的语言为英语。例如，利用以下 R 命令检索文档 hotel0.txt 中表示"愤怒"的句子：

```
> require(readtext)
> hotel0<-readtext("D:/Rpackage/hotel0.txt")$text
> require(syuzhet)
> hotel0_sents<-get_sentences(hotel0)
> nrc_data <- get_nrc_sentiment(hotel0_sents,lowercase=TRUE)
> anger_items <- which(nrc_data$anger > 0)
> hotel0_sents[anger_items]
[1] "In spite of its name, it is really not very \"grand,\" but it is cheap, clean, and comfortable."
[2] "On his last visit, Mr Smith was told that he could have his usual room, but the manager added apologetically that it might be a little noisy."
[3] "So great was the demand for rooms, the manager said, that the hotel had decided to build a new wing."
[4] "Mr Smith went immediately to complain to the manager."
```

以上结果表明，表示"愤怒"情绪的词"spite""noisy""demand"和"complain"出现在 4 个句子中。

假如我们有以下以 R 向量形式储存的正面和负面情感词：

```
positive <-c("clean","comfortable","beautiful","visit","usual",
        "attention","white","quiet","landed")
negative <-c("spite","cheap","noisy","noise","uncomfortable",
        "dust","cough","fall","immediately","complain","forced")
```

自定义情感词库利用我们已经熟悉的 R 函数 get_sentiment()。函数默认形式是 get_sentiment(char_v, method = "syuzhet", lexicon = NULL)，其中 char_v 是字符串向量，method = "syuzhet" 指默认使用的情感词库是 syuzhet，lexicon 指至少包括"word"和"value"两列的数据框。在利用自定义情感词库开展情感分析时，我们要把 method = "syuzhet" 改为自定义形式 method = "custom"，把默认的词库 lexicon = NULL 改为自定义的 lexicon = custom_lexicon。利用以上自定义情感词库对文档 hotel0.txt 开展情感分析的 R 命令和统计分析结果如下：

```
> positive<-c("clean","comfortable","beautiful","visit","usual",
     "attention","white","quiet","landed")
```

```
> negative<-c("spite","cheap","noisy","noise","uncomfortable",
  "dust","cough","fall","immediately","complain","forced")
> require(readtext)
> hotel0<-readtext("D:/Rpackage/hotel0.txt")$text
> require(syuzhet)
> hotel0_v<-get_tokens(hotel0)
> custom_lexicon<-data.frame(word=c(positive,negative),value=
  c(rep(1,9),rep(-1,11)))
> custom_values<-get_sentiment(hotel0_v,method="custom",
  lexicon=custom_lexicon)
> sum(custom_values)
[1] -2
```

以上结果显示,把每个积极词的情感权重设为1,消极词的情感权重设为-1,利用自定义情感词库对文档开展情感分析,得到情感总值为-2,即文档体现较弱的负面情感。

9.7 对小说《傲慢与偏见》的文本特征分析

本节先介绍如何调用 R 数据包 janeaustenr 和 gutenbergr 加载小说《傲慢与偏见》,然后利用整洁数据包开展常见的文本词汇统计。

9.7.1 把小说《傲慢与偏见》读入 R

《傲慢与偏见》(*Pride and Prejudice*)是英国著名小说家简·奥斯汀(Jane Austen)的代表作之一,可以通过由塞尔吉开发的 R 数据包 janeaustenr 调用(Silge,2022),或者通过由约翰斯顿和鲁滨逊联合开发的 R 数据包 gutenbergr 中的函数 gutenberg_download(gutenberg_id,mirror=NULL,strip=TRUE,meta_fields=NULL,verbose=TRUE)在线下载《傲慢与偏见》英文版(Johnston et al.,2022)。在这个函数中,gutenberg_id 是古腾堡计划(Project Gutenberg)编号,mirror=NULL 是默认的从中检索图书或作品的网址或镜像 URL(Uniform Resource Locator,统一资源定位符)。在 R 界面输入 gutenberg_get_mirror(),按回车键就会得到图书下载镜像点:http://aleph.gutenberg.org。如果推荐的镜像点不能下载,可以更换镜像点。函数 gutenberg_download 中的变元 strip=TRUE 表示剥离页眉和页脚,meta_fields 指提供元信息,如标题(title)和作者(author)。变元 verbose=TRUE 指默认显示关于古腾堡计划所选镜像的信息。编号查询的 R 函数是 gutenberg_works(...,languages="en"),其中...指古腾堡元信息,包括作品编号(gutenberg_id)、标题(title)、作者(author)、作者编号(author_id)、语言(language)、古腾堡书架(gutenberg_bookshelf)、版权(rights)和文本(has_text),可以在 R 界面输入 gutenberg_metadata,再按回车键查询这些信息。例如,小说《傲慢与偏见》的编号查询可以利用以下两种 R 命令:

```
> require(gutenbergr)
> #设置 title
> gutenberg_works(title=="Pride and Prejudice")
# A tibble: 1 × 8
  gutenberg_id title              author guten…¹ langu…² guten…³ rights   has_t…⁴
         <int> <chr>              <chr>    <int> <chr>   <chr>   <chr>    <lgl>
1         1342 Pride and Prejudice Auste…      68 en      Best B… Publi…  TRUE
# … with abbreviated variable names ¹gutenberg_author_id, ²language,
#   ³gutenberg_bookshelf, ⁴has_text

> #设置 author
> gutenberg_works(author=="Austen, Jane")
# A tibble: 10 × 8
   gutenberg_id title                                                 author guten…¹ langu…² guten…³ rights   has_t…⁴
          <int> <chr>                                                 <chr>    <int> <chr>   <chr>   <chr>    <lgl>
 1          105 "Persuasion"                                          Auste…      68 en      <NA>    Publi…  TRUE
 2          121 "Northanger Abbey"                                    Auste…      68 en      Gothic… Publi…  TRUE
 3          141 "Mansfield Park"                                      Auste…      68 en      <NA>    Publi…  TRUE
 4          158 "Emma"                                                Auste…      68 en      <NA>    Publi…  TRUE
 5          161 "Sense and Sensibility"                               Auste…      68 en      <NA>    Publi…  TRUE
 6          946 "Lady Susan"                                          Auste…      68 en      <NA>    Publi…  TRUE
 7         1212 "Love and Freindship [sic]"                           Auste…      68 en      <NA>    Publi…  TRUE
 8         1342 "Pride and Prejudice"                                 Auste…      68 en      Best B… Publi…  TRUE
 9        31100 "The Complete Project Gutenberg Works of Jane Aust…"  Auste…      68 en      <NA>    Publi…  TRUE
10        42078 "The Letters of Jane Austen\r\nSelected from the c…"  Auste…      68 en      <NA>    Publi…  TRUE
# … with abbreviated variable names ¹gutenberg_author_id, ²language, ³gutenberg_bookshelf, ⁴has_text
```

以上两个命令显示，小说《傲慢与偏见》的编号是 1342。我们既可以利用编号，又可以使用管道符把小说读入 R：

```
# 第一种方法
> require(gutenbergr)
> austen_book<-gutenberg_download(gutenberg_id=1342,mirror=
"ftp:/mirrors.xmission.com/gutenberg/")
> austen_book
# A tibble:14,529×2
   gutenberg_id text
          <int> <chr>
 1         1342 "                              [Illustration:"
 2         1342 ""
 3         1342 "                              GEORGE ALLEN"
 4         1342 "                                PUBLISHER"
 5         1342 ""
 6         1342 "                       156 CHARING CROSS ROAD"
 7         1342 "                              LONDON"
 8         1342 ""
 9         1342 "                              RUSKIN HOUSE"
10         1342 "                                      ]"
# …with 14,519 more rows
# ℹ Use `print(n=…)` to see more rows
# 第二种方法
> require(gutenbergr)
> require(dplyr)
```

```
> austen_book<-gutenberg_works(title=="Pride and Prejudice")%>%
  gutenberg_download(mirror="ftp:/mirrors.xmission.com/gutenberg/")
> austen_book
# A tibble:14,529×2
    gutenberg_id text
           <int><chr>
  1         1342 "                              [Illustration:"
  2         1342 ""
  3         1342 "                             GEORGE ALLEN"
  4         1342 "                              PUBLISHER"
  5         1342 ""
  6         1342 "                        156 CHARING CROSS ROAD"
  7         1342 "                              LONDON"
  8         1342 ""
  9         1342 "                             RUSKIN HOUSE"
 10         1342 "                                 ]"
# …with 14,519 more rows
# ⅰ Use 'print(n=…)' to see more rows
```

R 数据包 janeaustenr 包括简·奥斯汀的 6 部小说,如以下 R 命令的执行结果所示:

```
> library(janeaustenr)
> library(dplyr)
> austen_books()%>%distinct(book)
# A tibble: 6 × 1
  book
  <fct>
1 Sense & Sensibility
2 Pride & Prejudice
3 Mansfield Park
4 Emma
5 Northanger Abbey
6 Persuasion
```

以上命令调用了 dplyr 包中的筛选函数 distinct(.data,...),其中.data 是数据框或数据框扩展式(如 tibble),...指用于确定唯一性的变量。

要从 janeaustenr 包中把小说《傲慢与偏见》文本读入 R,可以利用前面学到的 dplyr 包中的过滤函数 filter()。R 命令和操作结果如下:

```
> library(janeaustenr)
> library(dplyr)
> austen_books()%>%
  filter(book=="Pride & Prejudice")
```

```
# A tibble:13,030×2
   text                                                              book
   <chr>                                                             <fct>
 1 "PRIDE AND PREJUDICE"                                             Pride & Prejudice
 2 ""                                                                Pride & Prejudice
 3 "By Jane Austen"                                                  Pride & Prejudice
 4 ""                                                                Pride & Prejudice
 5 ""                                                                Pride & Prejudice
 6 ""                                                                Pride & Prejudice
 7 "Chapter 1"                                                       Pride & Prejudice
 8 ""                                                                Pride & Prejudice
 9 ""                                                                Pride & Prejudice
10 "It is a truth universally acknowledged, that a single man in possession"  Pride & Prejudice
# …with 13,020 more rows
# ⓘ Use 'print(n=…)' to see more rows
```

以上结果显示一个整洁数据框（tibble 格式），包括 text 和 book 两列，text 由最多包括约 70 个字符（〈chr〉）的各排组成，book 列包括作为因素（〈fct〉）的书名。从 janeaustenr 包和从 gutenbergr 包读入的小说文本都是 tibble 格式，但是在编排上有差异，如 janeaustenr 包读入的文本排数为 13 030，而 gutenbergr 包读入的文本排数为 14 529。

我们对小说《傲慢与偏见》的统计分析只针对正文部分，即从第 1 章到第 61 章，不含目录等部分。文本的提取需要用到 stringr 包中的函数 str_detect(text, regex(pattern, ignore_case=FALSE)，其中 text 是文本字符串，正则表达式函数 regex 中的 pattern 指匹配模式，ignore_case=FALSE 指函数默认匹配时不忽略大小写问题。设置 ignore_case=TRUE 会在匹配时忽略大小写问题。在本例中，我们编写以下 R 命令查找小说的每一章：

```
str_detect(text, regex("^chapter [\\divxlc]", ignore_case = TRUE))
```

在以上命令中，text 是文本字符串，在忽略字母大小写时（ignore_case=TRUE），正则表达式^chapter [\\divxlc]表示匹配以 chapter 开头、后面紧跟任意数字的短语，如 Chapter I 或 Chapter 1。字符串 divxlc 代表罗马数字组合，d、i 和 v 依次代表 500、1 和 5，x、l 和 c 依次代表 10、50 和 100。例如，4 写作 IV，9 写作 IX，11 写作 XI，28 写作 XXVIII，90 写作 XC，400 写作 CD。在正则表达式与文本串匹配时，str_detect()返还逻辑符 TRUE，否则返还 FALSE。要计算出各章顺序，如 Chapter 1 和 Chapter 2，则需要调用 R 累计函数 cumsum(x)，其中 x 是向量，这样所有显示为 TRUE 的逻辑符计为 1，所有显示为 FALSE 的逻辑符计为 0，累计后就会得到 1,2,3 等数字。例如：

```
> require(stringr)
> x<-c("Chapter I","Chapter II","Chapter III","Chapter IV","Chapter V")
> cumsum(str_detect(x,regex("^chapter [\\divxlc]", ignore_case = TRUE)))
[1] 1 2 3 4 5
```

本例对从 gutenbergr 包读入的小说文本数据进行预处理的 R 命令和部分统计结果(小说开头和结尾部分)如下：

```
> require(gutenbergr)
> require(dplyr)
> require(stringr)
> pride<-gutenberg_works(title=="Pride and Prejudice")%>%
gutenberg_download("ftp://mirrors.xmission.com/gutenberg/")%>%
mutate(linenumber=row_number(),chapter=cumsum(str_detect(text,
regex("^chapter [\\divxlc]",ignore_case=TRUE))))%>%
filter(chapter!=0)
> head(pride)

# A tibble:6×4
  gutenberg_id text                                                    linenumber chapter
         〈int〉 〈chr〉                                                        〈int〉   〈int〉
1         1342 "Chapter I.]"                                                  670       1
2         1342 ""                                                             671       1
3         1342 ""                                                             672       1
4         1342 "It is a truth universally acknowledged, that a single man in possession"  673  1
5         1342 "of a good fortune must be in want of a wife."                 674       1
6         1342 ""                                                             675       1

> tail(pride)

# A tibble:6×4
  gutenberg_id text                                                    linenumber chapter
         〈int〉 〈chr〉                                                        〈int〉   〈int〉
1         1342 ""                                                           14524      61
2         1342 ""                                                           14525      61
3         1342 ""                                                           14526      61
4         1342 ""                                                           14527      61
5         1342 "CHISWICK PRESS:——CHARLES WHITTINGHAM AND CO."              14528      61
6         1342 "     TOOKS COURT,CHANCERY LANE,LONDON."                     14529      61

> pride[13840:13848,]
# A tibble:9×4
  gutenberg_id text                                                    linenumber chapter
         〈int〉 〈chr〉                                                        〈int〉   〈int〉
1         1342 "to see how his wife conducted herself;and she condescended to wait on"  14509  61
2         1342 "them at Pemberley,in spite of that pollution which its woods had"       14510  61
3         1342 "received,not merely from the presence of such a mistress,but the"       14511  61
4         1342 "visits of her uncle and aunt from the city."                             14512  61
5         1342 ""                                                                       14513  61
6         1342 "With the Gardiners they were always on the most intimate terms. Darcy," 14514  61
7         1342 "as well as Elizabeth,really loved them;and they were both ever"         14515  61
8         1342 "sensible of the warmest gratitude towards the persons who,by bringing"  14516  61
9         1342 "her into Derbyshire,had been the means of uniting them."                 14517  61
```

如果我们要对从 janeaustenr 包读入的小说文本进行预处理，除了用命令 austen_books()%>%filter(book=="Pride & Prejudice")读入小说之外，对各章筛选的 R 命令与上面从

gutenbergr 包读入小说文本的 R 命令相同。本例 R 命令和部分统计结果（小说开头和结尾部分）如下：

```
> require(janeaustenr)
> require(dplyr)
> require(stringr)
> austen_pride<-austen_books()%>%
  filter(book=="Pride & Prejudice")%>%
  mutate(linenumber=row_number(),chapter=cumsum(str_detect(text,
  regex("^chapter [\\divxlc]",ignore_case=TRUE))))%>%
  filter(chapter!=0)
> head(austen_pride)
# A tibble:6×4
```

text	book	linenumber	chapter
⟨chr⟩	⟨fct⟩	⟨int⟩	⟨int⟩
1 "Chapter 1"	Pride & Prejudice	7	1
2 ""	Pride & Prejudice	8	1
3 ""	Pride & Prejudice	9	1
4 "It is a truth universally acknowledged,that a single man in possession"	Pride & Prejudice	10	1
5 "of a good fortune,must be in want of a wife."	Pride & Prejudice	11	1
6 ""	Pride & Prejudice	12	1

```
> tail(austen_pride)
# A tibble:6×4
```

text	book	linenumber	chapter
⟨chr⟩	⟨fct⟩	⟨int⟩	⟨int⟩
1 "visits of her uncle and aunt from the city"	Pride & Prejudice	13025	61
2 ""	Pride & Prejudice	13026	61
3 "With the Gardiners,they were always on the most intimate terms."	Pride & Prejudice	13027	61
4 "Darcy,as well as Elizabeth,really loved them;and they were both ever"	Pride & Prejudice	13028	61
5 "sensible of the warmest gratitude towards the persons who,by bringing"	Pride & Prejudice	13029	61
6 "her into Derbyshire,had been the means of uniting them."	Pride & Prejudice	13030	61

以上两种文档预处理结果都是 tibble 格式，主要差别在于分行的差异。

9.7.2 小说《傲慢与偏见》的基础描述性统计

本节以上一节调用 janeaustenr 包经过预处理得到的文本 R 对象 austen_pride 为例，介绍三种常见的统计分析方法，即计算小说文本中的词汇量和低频词数、绘制出现频次大于 20 的三元组（trigrams）条形图和词频分布词云图。

小说文本中的词汇量为小说文本中包括的类符（types）数。编写 R 代码时，我们首先要调用 tidytext 包中的函数 unnest_tokens()，把字符串对象 austen_pride 形符化；其次调用

dplyr 包中的函数 pull( )，把 word 从 tibble 数据框中提取出来；然后利用函数 gsub( )删除着重号"_"（如_you_和_may_）；最后调用函数 n_distinct(...)，其中...是数值或字符向量，目的是计算唯一字符总数。本例 R 命令和统计分析结果如下：

```
> require(janeaustenr)
> require(dplyr)
> require(stringr)
> austen_pride<-austen_books( )%>%
   filter(book=="Pride & Prejudice")%>%
   mutate(linenumber=row_number( ),chapter=cumsum(str_detect(text,
   regex("^chapter [\\divxlc]",ignore_case=TRUE))))%>%
   filter(chapter!=0)
> require(tidytext)
> austen_words<-austen_pride%>%
   unnest_tokens(word,text)%>%
   pull(word)
> pride_words<-gsub("_","",austen_words)
> pride_words%>%
   n_distinct( )
[1] 6408
```

以上结果显示，小说《傲慢与偏见》实际使用的类符数是 6 408 个，词汇量不大。这也是这部小说易读的主要原因之一。

下面编写 R 命令计算小说文本的低频词数（即复杂词数）。在这个例子中，低频词数被定义为常用 2 000 词以外的类符数。使用的常用 2 000 词表由保罗·内申开发（网址：https://www.wgtn.ac.nz/lals/resources/paul-nations-resources/vocabulary-analysis-programs），下载文件名为：BNC-14000-and-programs-and-instructions.zip。解压文件后，把 basewrd1.txt 和 basewrd2.txt 文件复制到 D:\Rpackage 文件夹下。这两个文件包括最常用的英语 2 000 词。调用 tidytext 等数据包对这两个文件进行预处理，包括形符化、删除数字和多余的占位符""。2 000 词常用词表制作的 R 代码如下：

```
require(readtext)
require(tidytext)
require(stringr)
require(dplyr)
base1<-readtext("D:/Rpackage/basewrd1.txt")%>%
 tibble( )%>%
 unnest_tokens(word,text)%>%
 pull(word)%>%
```

```
str_remove_all(regex(pattern="[[:digit:]]"))
base1<-base1[which(base1!="")]
base2<-readtext("D:/Rpackage/basewrd2.txt")%>%
tibble()%>%
unnest_tokens(word,text)%>%
pull(word)%>%
str_remove_all(regex(pattern="[[:digit:]]"))
base2<-base2[which(base2!="")]
base<-c(base1,base2)
```

执行以上 R 命令得到 R 对象 base。这是一个包括 11 940 个形符的词表。

针对前面得到的 R 对象 pride_words，编写 R 命令计算常用 2 000 词以外的类符数作为低频词数，但是数字并未包含在常用 2 000 词中，需要调用 R 函数 gsub( ) 和 which( ) 予以排除。然后，调用 dplyr 包中的函数 anti_join( ) 剔除 R 对象 base 中包含的常用词。最后，调用函数 dplyr 包中的函数 n_distinct( ) 计算低频类符数。本例 R 命令和统计分析结果如下：

```
> require(tidytext)
> require(dplyr)
> pride_words<-gsub("[[:digit:]]","",pride_words)
> pride_words<-pride_words[which(pride_words!="")]
> tibble(word=pride_words)%>%
   anti_join(tibble(word=base))%>%
   n_distinct()
Joining, by = "word"
[1] 2961
```

以上结果表明，小说《傲慢与偏见》包含的常用 2 000 词以外的低频词数为 2 961 个。尽管本例低频词的计算解决了数字和着重号问题，但是没有排除专有名词（如人名和地名），因而这个结果还不是很精确。不过，它大体反映了小说在词汇层面的可读性。

下面编写 R 命令计算频次大于 20 的三元组，并据此绘制条形图。我们首先使用与前面相同的方法从 janeaustenr 包中读入小说《傲慢与偏见》，并设置章和排数。其次，调用前面介绍过的函数 unnest_ngrams( )，设置变元 n=3，利用函数 pull( ) 调取三元组，并且调用数据包 stringr 中的函数 str_remove_all( ) 去除着重号"_"，再调用基础函数 table( ) 和 data.frame( )，把得到的结果保存为 R 对象 pride_tri。再次，把保存的 R 对象 pride_tri 转化为 tibble 格式，调用 dplyr 包中的函数 arrange(.data,desc( )) 把三元组频数按降序排列，其中.data 是数据框或数据框扩展式（如 tibble）。然后调用过滤函数 filter( ) 筛选频次大于 20 的三元组，接着调用函数 mutate( ) 对筛选结果排序，以便使三元组在图形中按顺序排列。最后，调用 ggplot2 包绘制条形图。由此，计算三元组频次和绘制频次条形图的完整 R 代码如下：

```
require(janeaustenr)
require(dplyr)
require(stringr)
require(tidytext)
pride_tri<-austen_books()%>%
filter(book=="Pride & Prejudice")%>%
mutate(linenumber=row_number(),chapter=cumsum(str_detect(text,regex("^chapter [\\divxlc]",
ignore_case=TRUE))))%>%
filter(chapter!=0)%>%
unnest_ngrams(trigram,text,n=3)%>%
pull(trigram)%>%
str_remove_all("_")%>%
table()%>%
data.frame()
require(ggplot2)
colnames(pride_tri)<-c("tri","n")
pride_tri%>%
tibble()%>%
arrange(desc(n))%>%
filter(n>20)%>%
mutate(word=reorder(tri,n))%>%
ggplot(aes(x=word,y=n,fill=n))+
geom_col(show.legend=FALSE)+
xlab(NULL)+coord_flip()+
scale_fill_gradientn(colours=RColorBrewer::brewer.pal(8,"Dark2"))
```

执行以上命令得到如图 9.10 所示的频次大于 20 的三元组频数分布。

图 9.10 显示 23 个三元组。出现频次最高的 5 个三元组（n≥35）是"i do not""i am sure""as soon as""she could not"和"that he had"；出现频次在 25 以上和 34 之间的 6 个三元组是"i dare say""i am not""in the world""that he was""it was not"和"could not be"。其他三元组，如"it would be"和"as well as"，出现频次在 25 和 21 之间。这些三元组或表示断言（如"i am sure"），或表示否定（如"i do not"），或使用句干（如"it would be"）和从句（如"that he had"）。三元组中也包括连词和介词短语（如"as soon as"和"by no means"），也有个别名词性短语（如"uncle and aunt"）。整体上看，这些高频三元组以主谓结构、介词短语和连词的使用为主，且口语化较强，较少体现小说的实质性内容。

本节的最后一项工作是绘制词频分布词云图。词云图的绘制可以分为两步。第一步是把小说《傲慢与偏见》文本转化为形符。第二步是利用 wordcloud2 把 200 个高频词绘制

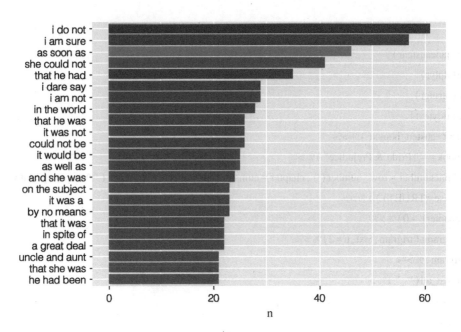

图 9.10　小说《傲慢与偏见》最高频次三元组分布

成词云图。在第一步,利用前面学习的函数,我们很容易得到小说中的所有形符。由于小说形符数较多(12 万多词),因此首先需要调用 dplyr 包中的函数 anti_join( )排除停用词,停用词表选择 tidytext 包自带的 stop_words(合计 1 149 个词)。其次,调用 stringr 包中的函数 str_remove_all( )去除所有的着重号,表达式为"_"。然后,调用 R 自带函数 table( )和 data.frame( ),同时把数据框的列变量改为"word"和"n"(词频)。最后,利用函数 tibble( )把数据框转化为 tibble 格式,调用 dplyr 包中的 arrange(. data, desc( ))把词频按降序排列,调用函数 top_n( )筛选 200 个高频词,将由此得到的结果保存为 R 对象 pride_words。这一部分完整的 R 代码如下:

```
require( janeaustenr)
require( dplyr)
require( stringr)
require( tidytext)
pride_tks<-austen_books( )%>%
filter( book = = " Pride & Prejudice" )%>%
mutate( linenumber = row_number( ) , chapter = cumsum( str_detect( text, regex( " ^chapter [ \\divxlc ]" ,
ignore_case = TRUE) ) ) )%>%
filter( chapter! = 0)%>%
unnest_tokens( word, text)%>%
anti_join( stop_words)%>%
```

```
pull(word)%>%
str_remove_all("_")%>%
table()%>%
data.frame()
colnames(pride_tks)<-c("word","n")
pride_words<-pride_tks%>%
tibble()%>%
arrange(desc(n))%>%
top_n(200)
```

第二步，根据 R 对象 pride_words 调用 wordcloud2 绘制词云图。词云图的背景图片是维多利亚时期时尚女士的装扮，如图 9.11 所示（下载网址： https://pixabay.com/illustrations/lady-victorian-crinoline-parasol-1382708/）。

在前面第三章和第六章我们学习了如何使用 wordcloud 包绘制词云图。数据包 wordcloud2 由 Lang 开发，当前版本号为 0.2.2（Lang, 2023）。调用的主要函数是 wordcloud2()，主要结构是 wordcloud2(data, size = 1, gridSize = 0, fontFamily = 'Segoe UI', color = 'random-dark', backgroundColor = "white", minRotation = −pi/4, maxRotation = pi/4, rotateRatio = 0.4, shape = 'circle', ellipticity = 0.65, figPath = NULL)，其中 data 是包括文本词和频数的数据框，size = 1 是默认的字体大小，大数值表示大号字。在这个例子中，我们把字号设置为 size = 0.9。变元 gridSize 显示词间隔大小，默认值为 0；fontFamily 用于设置字体，函数默认字体为 Segoe UI。我们把字体设置为 fontFamily = "Segoe Script"。如果要查看可供选择的字体，可以安装并调用 R 数据包 sysfonts，然后输入 R 命令 font_files()即可。

图 9.11 维多利亚时期时尚女士

函数 wordcloud2()中的变元 Color = 'random-dark' 是函数默认的深色色系，另外一个选择是 'random-light'（浅色色系）。我们也可以调用 R 数据包 RColorBrewer 中的函数 brewer.pal()设置自己想要的色系。变元 backgroundColor = "white" 指函数默认把图形背景色设置为"白色"；minRotation = −pi/4 和 maxRotation = pi/4 分别指字体旋转的最小值和最大值。变元 shape = 'circle' 指函数默认的词云形状为"圆圈"；rotateRatio = 0.4 指函数默认旋转词的比率，0.4 为默认值。如果所有的词均不旋转，可以设置 rotateRatio = 0。变元 ellipticity = 0.65 是函数默认绘制图形形状的扁平度。利用变元 figPath 可以设置背景图案的路径。我们在这个例子中用如图 9.11 所示的图片作为词云背景，名称为 woman.jpg，放在安装数据包 wordcloud2 的子文件夹 examples 中。数据包 wordcloud2 的安装可以采用默认方式，或者采用自定义路径的方式,如:devtools::install_github("lchiffon/wordcloud2")。查找 wordcloud2 包的路径,可以调用以下 R 命令:find.package("wordcloud2")。我们的电脑存放 wordcloud2 包的路径是:C:/Users/DELL/AppData/Local/R/win-library/4.2/wordcloud2。

本例绘制词云图完整的 R 代码如下:

```
require(wordcloud2)
figPath=system.file("examples/woman.jpg",package="wordcloud2")
wordcloud2(data=pride_words, figPath = figPath, size = 0.9, fontFamily = "Segoe Script")
```

执行以上 R 代码,出现一个 html 网页,点击重新加载此页图标(⟳)便可得到相应的图形。由于网页图形是随机生成的,每次刷新得到的图形都会有差异。如果我们对页面图形不满意,可以反复点击重新加载此页图标,直至得到我们想要的图形。图 9.12 显示小说《傲慢与偏见》中 200 个最高词频词的分布。

图 9.12 显示,最醒目的词是小说的女主人公"elizabeth"(在形符化时所有的文本词语均转化为小写字母),"elizabeth"一词在文中出现 597 次。其他出现频次很高的小说人物是"darcy"(出现 374 次)、"bennet"(出现 294 次)、"jane"(出现 263 次)、"bingley"(出现 257 次)、"wickham"(出现 162 次)、"collins"(出现 156 次)、"lydia"(出现 133 次)和"catherine"(出现 110 次)。词云图也显示了一些表示亲缘关系或称谓的

图 9.12 小说《傲慢与偏见》高频词词云图

词,如"miss"(出现 283 次)、"lady"(出现 183 次)、"sister"(出现 180 次)、"sisters"(出现 76 次)、"dear"(出现 158 次)、"father"(出现 116 次)、"mother"(出现 112 次)、"daughter"(出现 77 次)、"brother"(出现 66 次)、"friend"(出现 104 次)、"sir"(出现 78 次)、"aunt"(出现 78 次)和"uncle"(出现 60 次)。故事的发展围绕这些小说人物之间错综复杂的关系。主题关键词涉及家庭、爱情、婚姻和幸福,如"family"(出现 152 次)、"love"(出现 92 次)、"affection"(出现 58 次)、"pleasure"(出现 92 次)、"happy"(出现 83 次)、"happiness"(出现 72 次)和"marriage"(出现 66 次)。词云图也反映了故事发生的时间和地点,如"day"(出现 140 次)、"morning"(出现 77 次)、"evening"(出现 72 次)、"house"(出现 107 次)、"home"(出现 66 次)和"longbourn"(出现 88 次)。总之,图 9.12 较好地展示了小说的人物和主题,由此可以认为这是一部反映老百姓家庭与婚姻的小说。

9.7.3 对小说《傲慢与偏见》的情感分析

9.6.1 节介绍了四种情感词库,即 syuzhet、bing、afinn 和 nrc。本节利用 bing 情感词库对小说《傲慢与偏见》开展情感分析。这部小说有 61 章,根据每一章的文字计算正面情感值和负面情感值的变化有助于我们更好地了解小说故事情节的发展和主题的变化。因此,本节包括两个部分。第一个部分以章为依据绘制情感分析条形图和线图,第二个部分绘制高频情感词对比条形图和词云图。

9.7.3.1 调用 bing 库按章开展情感分析

为情感分析编写的 R 命令大致分为三个部分:把小说文本转化为以 word 为单位的整洁数据;提取整洁数据中的情感词,按章计算正面与负面情感值差异;调用 ggplot2 包绘制情感分析条形图和线图。

在第一个部分,我们首先要调用 janeaustenr 包中的小说文本,对之进行必要的处理,接下来调用 tidytext 包中的函数 unnest_tokens() 对文本开展形符化操作,并保存由此得到的 R 对象 pride_tks。在函数 unnest_tokens() 中,两个基本的变元是列名称。第一个变元是形符化后输出的列名,这里是 word;第二个变元是作为文本来源的输入列名,这里是 text。调用 stringr 包中的函数 str_remove_all(),并利用表达式去除所有的着重号"_"。这一部分的 R 代码如下:

```
require(janeaustenr)
require(dplyr)
require(stringr)
require(tidytext)
pride_tks<-austen_books()%>%
filter(book=="Pride & Prejudice")%>%
mutate(linenumber=row_number(),chapter=cumsum(str_detect(text,
regex("^chapter [\\divxlc]",ignore_case=TRUE))))%>%
filter(chapter!=0)%>%
unnest_tokens(word,text)
pride_tks$word<-str_remove_all(pride_tks$word,"_")
```

执行以上 R 代码得到的 R 对象 pride_tks 是一个 tibble 格式的数据框,每排有一个形符(即一个词)。

在阅读小说等文本时,我们会利用词的情感负荷推断文本的哪一个部分是表达正面情感的,哪一个部分是表达负面情感的,或者表达其他复杂的情绪的,如厌恶或快乐。一种常见的情感分析方法是把某类情感词相加起来,比较它们之间的计数差异。

通过文本处理得到的 R 对象 pride_tks 是整洁格式,即一排对应一个词,我们可以对之开展情感分析了。情感分析调用的 bing 情感词库来自 syuzhet 包。在第二个部分,我们在调用 bing 情感词库之后,再调用 dplyr 包中的函数 inner_join() 筛选 pride_tks 中包括在 bing

库中的所有情感词。然后,按章计算每章中的正面和负面情感词数,使用的 R 命令为 count(index=chapter, sentiment),其中函数 count()从数据包 dplyr 中调入。从 tidyr 包中调入函数 spread()把前面得到的情感类 sentiment 和计数 n 重新编排为列名 negative 和 positive,其观测值为计数 n。最后,调用 dplyr 包中的函数 mutate(),增加一个列名 sentiment,其数值等于每章正面情感值与负面情感值之差,把由此得到的 tibble 数据框命名为 R 对象 pride_sentiment。但是需要注意的是,"miss"一词在小说中表示称谓("小姐"),而在 bing 库中"miss"表示"错过",归为负面情感词。在大多数情况下,"object"一词在小说中表示"目标"或"对象"(42 例),只有少数用法表示"反对"(6 例),而 bing 库中的"object"表示"反对",归为负面情感词。比较简便的纠正方法是调用 dplyr 包中的函数 anti_join(),并将这两个词作为停用词从 bing 库中排除出去。为了便于使用管道操作,我们调整列名称,把情感值 1 和 -1 分别改为 positive 和 negative,把得到的数据框转化为 tibble 格式,命名为 bing。这一部分的 R 代码如下:

```
require(syuzhet)
stopwords<-tibble(word=c("miss","object"))
bing<-get_sentiment_dictionary(dictionary="bing",language="english")%>%
anti_join(stopwords)
colnames(bing)<-c("word","sentiment")
bing["sentiment"][bing["sentiment"]==1]<-"positive"
bing["sentiment"][bing["sentiment"]==-1]<-"negative"
bing<-tibble(bing)
library(tidyr)
pride_sentiment<-pride_tks%>%
inner_join(bing)%>%
count(index=chapter,sentiment)%>%
spread(sentiment,n)%>%
mutate(sentiment=positive-negative)
```

在第三部分,我们调用 ggplot2 包将各章的情感分析结果绘制成条形图。关于 ggplot2 函数及其各个变元的解释,见 9.5 节。这一部分的 R 代码如下:

```
library(ggplot2)
ggplot(pride_sentiment, aes(index, sentiment, fill=sentiment))+
geom_col(show.legend=FALSE)
```

执行以上 R 命令,得到如图 9.13 所示的情感变化条形图。

为了便于对图 9.13 的理解,我们根据 R 对象 pride_sentiment 列出各章的情感值,如表 9.2 所示。

表 9.2 小说《傲慢与偏见》各章情感值

章	负面	正面	情感值	章	负面	正面	情感值
1	22	33	11	32	22	56	34
2	20	22	2	33	55	64	9
3	37	82	45	34	111	72	−39
4	22	75	53	35	101	89	−12
5	18	50	32	36	93	73	−20
6	52	130	78	37	49	55	6
7	54	67	13	38	21	60	39
8	51	76	25	39	30	60	30
9	28	83	55	40	68	78	10
10	46	91	45	41	107	94	−13
11	59	71	12	42	63	86	23
12	19	34	15	43	89	222	133
13	48	77	29	44	59	111	52
14	23	59	36	45	48	54	6
15	33	76	43	46	136	88	−48
16	86	160	74	47	117	127	10
17	28	43	15	48	61	62	1
18	145	224	79	49	41	101	60
19	56	90	34	50	46	88	42
20	59	65	6	51	49	78	29
21	55	77	22	52	73	99	26
22	37	101	64	53	77	106	29
23	53	68	15	54	41	82	41
24	66	107	41	55	49	102	53
25	35	61	26	56	66	73	7
26	62	102	40	57	54	50	−4
27	35	64	29	58	90	96	6
28	29	63	34	59	70	118	48
29	34	97	63	60	41	91	50
30	18	43	25	61	40	67	27
31	24	65	41				

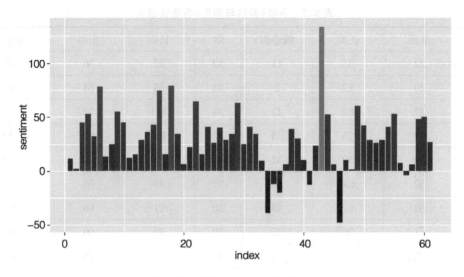

图 9.13　小说《傲慢与偏见》各章情感变化

图 9.13 显示，小说《傲慢与偏见》共 61 章，每章情感变化不一，正值表示正面情感，负值表示负面情感，绝对值越大，情感差异就越鲜明。

从图 9.13、表 9.2 来看，小说中的正面情感值远多于负面情感值。小说第 43 章的情感差异值最大，超过了 100。这一结果与小说中情节的起伏有着密切联系。在这一章中，伊丽莎白和达西分手后随舅父母来到彭伯里庄园。伊丽莎白开始了解到达西很受当地人欢迎，再次邂逅达西时，发现想象中高高在上的他竟然对自己的舅父母彬彬有礼，她的内心逐渐消除对达西的偏见。男女主人公所持有的傲慢与偏见从此处开始化解，使正面情感达到高潮。

随着故事情节的推进，第 46 章出现了全书情感的最低潮。此时的伊丽莎白在家书中得知小妹莉迪亚与身负赌债的威克汉姆私奔了。这无疑是一桩家丑，伊丽莎白感觉经此一事达西会更瞧不起自己和自己的家庭，于是情绪陷入谷底。

达西在知道"丑闻"之后，非但没有瞧不起伊丽莎白，还替威克汉姆还清赌债，并给了他一笔钱，让他与莉迪亚完婚。从此以后，伊丽莎白从前对达西的种种偏见化为真情实意。与图表对应，在第 46 章之后人物的情感出现回暖。在小说最后，一对因傲慢与偏见而好事多磨的有情人终成眷属，正面情感值也因此上升。

对照图 9.13、表 9.2 还发现，情感值差异比较小的地方并不意味着情感不发生波动或没有情感。例如，在第 43 章，负面情感值高达 89，而正面情感值更是高达 222，充分体现了主人公复杂的情感变化。

为了说明上面提及的这一点，我们继续调用 ggplot2 包中的函数绘制每章正面和负面情感值线图。由于我们前面得到的 R 对象 pride_sentiment 是宽式数据框，因此需要调用 tidyr 包中的函数 gather( ) 把它转化为窄式数据框，新增变量 sentiment 包括 negative 和 positive 两个类别，新增变量 value 代表每类情感值。本例 R 代码如下：

```
require( ggplot2 )
```

```
require(tidyr)
pride_sentiment<-gather(pride_sentiment,sentiment,value,negative:positive)
ggplot(data=pride_sentiment,
    aes(x=index,y=value,group=sentiment))+
geom_line(aes(linetype=sentiment))+
geom_point(aes(shape=sentiment))+
geom_line(aes(color=sentiment))+
geom_point(aes(color=sentiment))+
scale_colour_brewer(palette="Set1")
```

在以上代码中，函数 geom_line() 把各个观测值按顺序连成线，变元 linetype 用于设置不同线型。函数 geom_point() 绘制各个数据点，变元 shape 设置不同形状的点。要绘制不同的线型和点的颜色，可以利用变元 color。函数 scale_colour_brewer() 利用 RColorBrewer 包中的色系绘制彩色线条和点。在 R 界面输入 RColorBrewer::display.brewer.all()，按回车键后可以看到各个色系名称和颜色。我们可以通过 scale_colour_brewer() 中的主要变元 palette 设置色系。

执行以上 R 代码，得到如图 9.14 所示的线图。

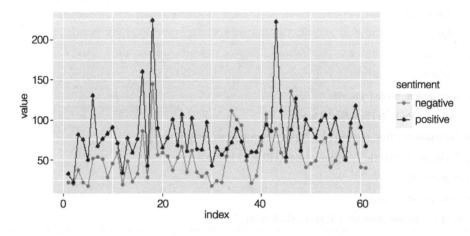

图 9.14　小说《傲慢与偏见》各章正面与负面情感对比

图 9.14 显示，最小的正面情感值 22 出现在第 2 章，与此章的负面情感值 20 很接近，体现出小说的情感预热阶段。除了第 1、第 2、第 12、第 17 和第 30 章之外，其他各章的正面情感值都在 50（含 50）以上，第 18 和第 43 章的正面情感值甚至超过 200。正面情感值的巨大变化体现出小说人物的情感变迁，烘托出小说的主题。相比之下，有 31 章的负面情感值小于 50，包括第 1 章、第 2 章、第 3 章、第 4 章、第 5 章、第 9 章、第 10 章、第 12 章、第 13 章、第 14 章、第 15 章、第 17 章、第 22 章、第 25 章、第 27 章、第 28 章、第 29 章、第 30 章、第 31 章、第 32 章、第 37 章、第 38 章、第 39 章、第 45 章、第 49 章、第 50 章、第 51 章、第 54 章、第 55 章、第 60 章和第 61 章。负面情感值在 101～145 之间的 6 章是第 18 章、第 34 章、第

35章、第41章、第46章和第47章。以第18章为例,负面情感值为145,而正面情感值高达224,说明本章的情感起伏比较大,情绪冲突明显。负面和正面情感值差异最小的3章是第2章、第48章和第57章。例如,第48章的负面情感值为61,正面情感值为62,说明两种对立情感都很丰富。这些情感值的变化充分体现小说人物情感的跌宕起伏,也更加使小说的情节扣人心弦。

9.7.3.2 高频情感词对比条形图和词云图

本节首先编写R命令绘制20个高频正面与负面情感词条形图,然后编写R命令绘制涵盖100个高频正面和负面情感词的词云对比图。

绘制20个高频正面与负面情感词条形图的R代码编写包括三个部分。第一个部分是按照上一节的方法得到R对象pride_tks。具体做法是:首先,从R数据包janeaustenr中调入奥斯汀小说集austen_books(),并调用dplyr包中的过滤函数filter()从austen_books()中筛选小说《傲慢与偏见》。然后,调用dplyr包中的函数mutate()和stringr包中的函数str_detect()把小说按章排列,再利用函数filter()选择小说第1章到第61章的文字。最后,调用tidytext包中的函数unnest_tokens()把文本字符串形符化,并调用stringr包中的函数str_remove_all()除去着重号"_",把由此得到的结果命名为R对象pride_tks。本部分的R代码重复如下:

```
require(janeaustenr)
require(dplyr)
require(stringr)
require(tidytext)
pride_tks<-austen_books()%>%
filter(book=="Pride & Prejudice")%>%
mutate(linenumber=row_number(),chapter=cumsum(str_detect(text,
regex("^chapter [\\divxlc]",ignore_case=TRUE))))%>%
filter(chapter!=0)%>%
unnest_tokens(word,text)
pride_tks$word<-str_remove_all(pride_tks$word,"_")
```

在第二部分,如上一节所示的那样,我们首先调入数据包syuzhet,把"miss"和"object"两个词从情感词库bing中排除,命名新的词库为"bing",并对之进行适当调整,以便能够更方便地使用整洁数据包中的函数。这个部分的R代码重复如下:

```
require(tidytext)
require(syuzhet)
stopwords<-tibble(word=c("miss","object"))
bing<-get_sentiment_dictionary(dictionary="bing",language="english")%>%anti_join(stopwords)
colnames(bing)<-c("word","sentiment")
bing["sentiment"][bing["sentiment"]==1]<-"positive"
```

```
bing["sentiment"][bing["sentiment"]==-1]<-"negative"
bing<-tibble(bing)
```

在第三部分,利用 dplyr 包中的函数 inner_join( )从小说文本形符化对象 pride_tks 中筛选出情感词,调用函数 count( )对情感词分类计数,再调用函数 group_by( )和 top_n( )从分类情感词中各选择 10 个高频词,并利用函数 mutate( )和 reorder( )新增变量和对情感词排序。接下来,调用 ggplot2 包中的函数 ggplot( )、geom_col( )和 facet_wrap( )等绘制正面和负面情感词比较条形图。这个部分的 R 代码如下:

```
require(ggplot2)
pride_tks%>%
inner_join(bing)%>%
count(word,sentiment,sort=TRUE)%>%
group_by(sentiment)%>%
top_n(10)%>%
ungroup( )%>%
mutate(word=reorder(word,n))%>%
ggplot(aes(word,n,fill=sentiment))+
geom_col(show.legend=FALSE)+
facet_wrap(~sentiment,scales="free_y")+
labs(y="Contribution to sentiment",x=NULL)+
coord_flip( )
```

执行以上命令,得到如图 9.15 所示的情感词对比条形图。

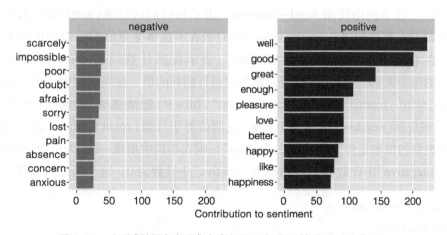

图 9.15　小说《傲慢与偏见》中高频正面与负面情感词对比条形图

图 9.15 显示负面情感词有 11 个,各个词频次分布在 26~45 之间,频次最高的两个词是"scarcely"(出现 45 次)和"impossible"(出现 44 次),频次最低的两个词是"concern"和"anxious"(均出现 26 次)。正面情感词有 10 个,频次最高的两个词是"well"(出现 224

次)和"good"(出现 201 次),频次最低的两个词是"like"(出现 77 次)和"happiness"(出现 72 次)。图 9.15 的最显著特点是,正面情感词出现的频次异常高于负面情感词出现的频次,说明小说情感的主调是正面情感。例如,正面情感词"well"出现的频次约为负面情感词"scarcely"出现频次的 5 倍,就连频次最低的正面情感词"happiness"也比频次最高的负面情感词"scarcely"较多地出现在小说文本中。

我们下面编写 R 命令绘制词云图对比小说《傲慢与偏见》中高频使用的正面和负面情感词,对比的总词数是 100。为了便于理解,我们把代码的编写分成四个部分。

第一个部分的主要目的是把小说文本形符化,并适当调整格式,保存得到的 R 对象 pride_tks。这一部分的代码与上面绘制条形图时在第一部分使用的 R 代码相同。

第二个部分的代码等同于上面绘制条形图时在第二部分使用的 R 代码。这部分的主要目标是调用 syuzhet 包中的情感词库 bing,并根据本例的特点对词库中的个别情感词进行调整,保存的 R 对象是 tibble 型数据框 bing。

在第三部分,我们首先调用函数 inner_join() 提取形符化对象 pride_tks 中的情感词,其次调用 dplyr 包中的函数 group_by() 和 summarize() 对情感词(列名为 word)进行分类计数,并在 summarize() 中设置 sentiment = first(sentiment),其中的 first(sentiment) 指提取向量 sentiment 的第一个值,目的是保留列名和正确的情感词计数。然后,调用 dplyr 包中的函数 arrange(.data,...) 给计数值排序。该函数中的.data 指数据框或数据框的扩展式(如 tibble),... 指定按照升序给数值排序的列名,把由此得到的处理结果命名为 pride_sentiment。最后,需要调用 reshape2 包中的数组(array)重组函数 acast() 把 tibble 数据框 pride_sentiment(列名为 word、count 和 sentiment)转化为绘图包 wordcloud 接受的矩阵格式。这个函数的主要结构是 acast(data, formula, value.var = guess_value(data), fill = NULL),其中变元 data 是数据框,formula 是类似于回归分析的公式,基本形式为:左变量(行名)~ 右变量(列名)。在本例中使用 word ~ sentiment,表示每行词的情感(sentiment)来源。变元 value.var 指储存数据的列名,guess_value(data) 指在不提供变量值时函数默认猜测 data 中的数值。在本例中设置 value.var = "count",表示按照计数列名 count 把数值分配给 sentiment 的每个类别。变元 fill 设置对缺失值进行插补的值。在当前的例子中,acast(pride_sentiment, word ~ sentiment, value.var = "count", fill = 0) 把情感类 sentiment 转化为正面(positive)和负面(negative)情感两个列名,观测值等于计数(count),fill = 0 表示把情感词不在某个类别的情形记作"0"。例如,"abrupt"一词在小说中出现一次,为负面情感词,因而在情感列"negative"中计数为 1,在情感列"positive"中计数为 0。这个部分的 R 代码如下:

```
pride_sentiment<-pride_tks%>%
inner_join(bing)%>%
group_by(word)%>%
summarize(count=n(), sentiment=first(sentiment))%>%
ungroup()%>%
arrange(count)
```

```
library(reshape2)
matrix<-acast(pride_sentiment, word~sentiment, value.var="count", fill=0)
```

在第四部分,我们调用数据包 wordcloud 中的函数 comparison.cloud() 绘制对比词云图。我们在第三章绘制关键词词云时调用了函数 wordcloud(),但是该函数不能用于比较不同文档的词频或同一文档不同类别的词频。函数 comparison.cloud() 不仅可以用于比较不同文档的词频,而且还可以用于比较不同类情感词。绘制对比词云图函数 comparison.cloud() 的主要结构是 comparison.cloud(term.matrix, scale=c(4,.5), max.words=300, random.order=FALSE, rot.per=.1, colors=brewer.pal(max(3, ncol(term.matrix)), "Dark2"), title.size=3, title.colors=NULL),其中变元 term.matrix 是词项矩阵,这里指的是正面和负面情感词矩阵 matrix。变元 scale 是长度为 2 的向量,表示词大小的范围。变量 scale=c(4,.5) 是函数默认的尺度范围。变元 colors=brewer.pal(max(3, ncol(term.matrix)), "Dark2") 指函数自定义词对比颜色,其中的 term.matrix 在这里指的是正面和负面情感词矩阵。变元 max.words 指图中绘制的最大词语数。变元 random.order 指图中词语的绘制按照随机顺序,random.order=FALSE 指按照词语频数由里向外降序排列。变元 rot.per 指旋转 90° 的词占比,rot.per=.1 指函数默认 10% 的词旋转 90°。变元 title.size 用于设置词云图标题的尺寸,title.colors 用于设置词云图标题的颜色。这个部分的 R 代码如下:

```
library(wordcloud)
comparison.cloud(matrix, scale=c(4.5,0.4), colors=brewer.pal(max(3, ncol(matrix)), "Dark2"),
max.words=100, title.size=2, title.colors= brewer.pal(max(3, ncol(matrix)), "Dark2"), random.
order=FALSE)
```

执行以上代码,得到如图 9.16 所示的情感词对比词云图。

图 9.16 显示的一个明显特点是,三个正面情感词 "well" "good" 和 "great" 字体很大,说明它们在小说《傲慢与偏见》中出现的频次非常高。出现频次较高的其他正面情感词包括 "pleasure" "love" "better" "happy" 和 "like" 等。从字体大小可以看出,负面情感词出现的频次明显少于正面情感词。出现频次相对高的负面情感词包括 "scarcely" "impossible" "poor" "afraid" 和 "doubt" 等。两类词分布的广度很不同,说明小说《傲慢与偏见》以正面情感为主调。

图 9.16 小说《傲慢与偏见》中
高频情感词对比词云图

第十章 中文自然语言处理

中文自然语言处理最基本也是最关键的一步是把字符串转化为词，即分词（word segmentation）。本章首先介绍处理分词的两个 R 数据包——quanteda 和 jiebaR，随后介绍如何利用 jiebaR 对中文文本开展基本的描述性统计、如何绘制词云图、如何开展关键词检索与提取以及如何开展中文情感分析。本章调用的 R 数据包包括：quanteda、quanteda.textplots、jiebaR、readtext、dplyr、tidytext、tidyr、purrr、stringr、wordcloud、wordcloud2、RColorBrewer、webshot、htmlwidgets、rJava、xlsx 和 ggplot2。

10.1 中文分词

10.1.1 利用数据包 quanteda 分词

本节和下一节介绍两种中文处理方法。一种方法是利用 R 数据包 quanteda，另一种方法是利用"结巴"（英文为"stutter"）分词 R 中文包 jiebaR。我们先介绍如何利用 quanteda 开展中文文字处理。本节举例使用的文本如下所示：

> 史密斯先生每次去韦斯特盖特市都住在格兰特大酒店。尽管酒店名有个"大"字，但它其实并不"大"，不过很便宜，干净，也很舒适。史密斯先生和酒店经理很熟络，所以他从来不用费工夫去预订房间。他总是住在同一间房，房间位于走廊的尽头，在那儿可以欣赏海湾美景。
> 史密斯先生上次来访时，经理告诉他，可以入住老房间，但是经理抱歉地补充说，客房需求量增大，酒店决定在侧翼扩建，所以房间可能有点吵。史密斯先生表示自己不介意。
> 入住的第一天，史密斯先生几乎没有听到噪音。第二天下午，他从酒店图书室借了本书，然后就上楼看书。他刚坐下来，就听到有人在大声砸墙。起初他没有在意，但是过了一会儿，他开始感到很不舒服。渐渐地，他的衣服上蒙上一层细小的白色粉末。不多久，房间里满是灰尘，他开始咳嗽起来。砸墙声越来越大，墙壁上的灰泥碎片也在噼里啪啦地掉落，整个楼好像都要倒塌了一样。史密斯先生立即去找经理投诉。他们两人一同回到房间，但是听不到一丝声响。两人站在那里面面相觑。史密斯先生觉得很尴尬，因为是他硬生生地把经理叫上了楼，白耽误了工夫。突然，砸墙声又出现了，一块大砖头落到了地板上。他们抬头一看，一把锋利的金属工具穿通了墙壁，在床的正上方捅出了一个大窟窿！

我们把文件以文本格式（.txt）储存，取文件名 hotel.txt，置于本地电脑 D:\Rpackage 文件夹中。我们先利用 R 数据包 quanteda 计算文本 hotel.txt 的词数、句子数和平均句长。R 命令如下：

```
> stringi::stri_locale_set("zh_hans_cn")
```

```
> require(quanteda)
> require(readtext)
> hotel<-readtext("D:/Rpackage/hotel.txt")
> hotel_toks<-tokens(hotel $ text, remove_punct = TRUE)
```

查看以上结果,我们发现两个专有名词"韦斯特盖特"和"格兰特大酒店"的切分结果很奇怪:

```
> hotel_toks[[1]][1:16]
 [1] "史密斯" "先生"  "每次"  "去"   "韦"   "斯"   "特"   "盖"   "特"   "市"
[11] "都"    "住在"  "格"   "兰"   "特大"  "酒店"
```

以上切分结果显示,"韦斯特盖特"被逐字切分,"格兰特大酒店"被切分为"格""兰""特大"和"酒店"。如果我们想把"韦斯特盖特"和"格兰特大酒店"作为一个专有名词,可以调用函数 tokens_compound()。R 命令和操作结果如下:

```
> hotel_toks<-tokens_compound(hotel_toks,list(c("韦","斯","特","盖","特"),c("格","兰","特大","酒店")))
> hotel_toks[[1]][1:9]
 [1] "史密斯"    "先生"    "每次"    "去"              "韦_斯_特_盖_特"
 [6] "市"        "都"      "住在"    "格_兰_特大_酒店"
```

10.1.2 利用数据包 jiebaR 分词

从前面的例子可以看出,数据包 quanteda 遇到异常词(如"韦斯特盖特"和"格兰特")时,处理不是很方便。R 数据包 jiebaR 则更好一些(Qin et al.,2019)。该数据包的当前版本号为 0.11。如果要把数据包安装在系统默认的路径,可以在 R 工作界面输入简单的命令:install.packages("jiebaR")。用户也可以自定义,输入带有指定路径的 R 命令,如:install.packages("jiebaR",lib="C:/Program Files/R/R-4.2.2/library"),按回车键对数据包进行自动在线安装。我们下面利用 R 数据包 jiebaR 通过自定义形符化方法处理异常词或切分不当的词。

首先,我们调用函数 worker() 对 jiebaR 的分词引擎初始化,并赋值给 cutter:

```
> require(jiebaR)
> cutter <-worker()
```

下面先以文本文件 hotel.txt 中的第一句为例。词切分可以调用函数 segment()。函数 segment() 的主要变元是 code(中文句子或文本文件路径)、jiebar(jiebaR 引擎,这里命名为 cutter)和 mod=NULL(模型)。根据 R 帮助文档(在 R 界面键入?segment,再按回车键即可获得帮助),至少有四种可选模型。最大概率分词模型(Maximum Probability Segmentation Model)使用 Trie 树结构构建一个有向无环图(Directed Acyclic Graph,DAG),并使用动态编程算法。它是核心分词算法。隐马尔可夫模型(Hidden Markov Model,HMM)利用 HMM 模型来确定状态集和观察到的词集。默认的 HMM 模型基于《人民日报》语言库。混合分词模型(MixSegment Model)同时使用最大概率分词模型和隐马尔科夫模型进行词切分,是函数默认的分词模型。索引分词模型(QuerySegment Model)使用混合分词模型进行词切分,然后在词典中枚举所有可能的长词。下面是利用四个模型的例子:

```
> text<-"史密斯先生每次去韦斯特盖特市都住在格兰特大酒店。"
> segment(text, cutter, "mp")
 [1] "史密斯"  "先生"    "每次"  "去"  "韦斯特"  "盖特"  "市"  "都"  "住"  "在"
[11] "格兰特"  "大酒店"
> segment(text, cutter, "hmm")
 [1] "史密斯"  "先生"  "每次"    "去"  "韦斯特盖"  "特市"  "都"  "住"
 [9] "在"      "格兰"  "特大酒"  "店"
> segment(text, cutter, "mix")
 [1] "史密斯"  "先生"    "每次"  "去"  "韦斯特"  "盖特"  "市"  "都"  "住"  "在"
[11] "格兰特"  "大酒店"
> segment(text,cutter,"query")
 [1] "密斯"  "史密斯"  "先生"  "每次"  "去"    "斯特"  "韦斯特"  "盖特"  "市"    "都"
[11] "住"    "在"      "兰特"  "格兰特"  "酒店"  "大酒店"
```

以上结果显示,利用最大概率分词模型("mp")和混合分词模型("mix")的分词结果相同,即把"韦斯特盖特"切分为"韦斯特"和"盖特","格兰特大酒店"切分为"格兰特"和"大酒店",不再像 quanteda 包中的函数 tokens()那样逐字切分。利用隐马尔可夫模型("hmm")时,"韦斯特盖特市"被错误地切分为"韦斯特盖"和"特市","格兰特大酒店"被错误地切分为"格兰""特大酒"和"店"。利用索引分词模型("query")时,出现包括重复的不同分词结果,如"韦斯特盖特"被错误地切分为"斯特""韦斯特"和"盖特"三个部分。我们通常使用函数 segment()默认的混合分词模型。在这个例子中,若使用函数默认的模型,我们可以简化 R 代码:

```
> segment(text,cutter)
 [1] "史密斯"  "先生"  "每次"  "去"  "韦斯特"  "盖特"  "市"  "都"  "住"  "在"
[11] "格兰特"  "大酒店"
```

另外一种词切分方法是使用操作符"<=",如:

```
> cutter<=text
```

```
[1] "史密斯"   "先生"     "每次"  "去"  "韦斯特" "盖特" "市" "都" "住" "在"
[11] "格兰特"   "大酒店"
```

第三种词切分方法是利用中括号"[]",直接把文本置于词切分器之后,如:

```
> cutter[text]
[1] "史密斯"   "先生"     "每次"  "去"  "韦斯特" "盖特" "市" "都" "住" "在"
[11] "格兰特"   "大酒店"
```

以上结果显示,三种方法得到的分词结果相同。

我们时常对有些分词结果不满意,这时需要利用自定义词典。例如,在上面的例子中,我们希望把"韦斯特盖特"和"格兰特大酒店"都作为一个词来处理。对于这种情况,我们可以利用自定义词典。自定义词典有三种方法。第一种方法是利用函数 new_user_word( ),主要结构是 new_user_word(worker, words),其中变元 worker 是"结巴"分词引擎,words 是新增词。针对上面的例子,我们可以利用以下命令得到希望的结果:

```
> require(jiebaR)
> cutter<-worker( )
> user_word<-c("韦斯特盖特","格兰特大酒店")
> new_user_word(cutter, user_word)
[1] TRUE
> segment(text, cutter)
[1] "史密斯"      "先生"       "每次"    "去"
[5] "韦斯特盖特"  "市"         "都"      "住"
[9] "在"          "格兰特大酒店"
```

如果自定义的词比较多,我们可以把这些词放在一个文本文件(后缀为.txt)里,置于 D:\Rpackage 文件夹中,然后利用函数 readLines( )进行调用。针对本例,我们把"韦斯特盖特"和"格兰特大酒店"放在名称为 user.txt 的文档中,文档放在 D:\Rpackage 文件夹中,每个词分行输入,即每个词各占一行。例如:

```
> cutter<-worker( )
> user<-readLines("D:/Rpackage/user.txt",warn=FALSE)
> new_user_word(cutter, user)
[1] TRUE
> segment(text, cutter)
[1] "史密斯"      "先生"       "每次"    "去"
[5] "韦斯特盖特"  "市"         "都"      "住"
[9] "在"          "格兰特大酒店"
```

另外一种自定义词典的方法是使用函数worker()中的参数user添加词典,在参数后设置自定义词典的路径。例如,我们把"韦斯特盖特"和"格兰特大酒店"放在文档名为dictionary.txt的文件中,文件编码格式为UTF-8,文档存放在D:\Rpackage文件夹中,从第二行开始每个词分行输入,即每个词各占一行。本例分词的R命令和结果如下:

```
> cutter_user<-worker(user="D:/Rpackage/dictionary.txt")
> segment(text,cutter_user)
[1] "史密斯"      "先生"           "每次"     "去"
[5] "韦斯特盖特"  "市"             "都"       "住"
[9] "在"          "格兰特大酒店"
```

10.2 文档基本描述性统计量

R数据包jiebaR的中文分词效果较好,加之自定义词典比较方便,本节以及后面各节的文字处理调用jiebaR包。本节以文本文档hotel.txt为例编写计算几种基本描述性统计量的R代码,基本统计量包括文本长度(总词数)、句子数、平均句长和出现频次至少为两次的词频分布。

我们首先利用上一节的自定义词典对文本hotel.txt分词。

```
> require(readtext)
> hotel<-readtext("D:/Rpackage/hotel.txt") $ text
> require(jiebaR)
> cutter_user<-worker(user="D:/Rpackage/dictionary.txt")
> segment(hotel,cutter_user)
[1]   "史密斯"   "先生"    "每次"    "去"      "韦斯特盖特" "市"    "都"     "住"
[9]   "在"       "格兰特大酒店" "尽管"  "酒店"   "名有"    "个"     "大"     "字"
[17]  "但"       "它"      "其实"    "并"      "不"      "大"     "不过"   "很"
[25]  "便宜"     "干净"    "也"      "很"      "舒适"    "史密斯" "先生"   "和"
[33]  "酒店"     "经理"    "很"      "熟络"    "所以"    "他"     "从来不" "用"
[41]  "费工夫"   "去"      "预订"    "房间"    "他"      "总是"   "住"     "在"
[49]  "同一"     "间房"    "房间"    "位于"    "走廊"    "的"     "尽头"   "在"
[57]  "那儿"     "可以"    "欣赏"    "海湾"    "美景"    "史密斯" "先生"   "上次"
[65]  "来访"     "时"      "经理"    "告诉"    "他"      "可以"   "入住"   "老"
[73]  "房间"     "但是"    "经理"    "抱歉"    "地"      "补充"   "说"     "由于"
[81]  "客房"     "需求量"  "增大"    "酒店"    "决定"    "在"     "侧翼"   "扩建"
[89]  "所以"     "房间"    "可能"    "有点"    "吵"      "史密斯" "先生"   "表示"
[97]  "自己"     "不介意"  "入住"    "的"      "第一天"  "史密斯" "先生"   "几乎"
[105] "没有"     "听到"    "噪音"    "第二天"  "下午"    "他"     "从"     "酒店"
[113] "图书室"   "借"      "了"      "本书"    "然后"    "就"     "上楼"   "看书"
[121] "他"       "刚"      "坐下"    "来"      "就"      "听到"   "有人"   "在"
```

[129]	"大声"	"砸"	"墙"	"起初"	"他"	"没有"	"在意"	"但是"
[137]	"过"	"了"	"一会儿"	"他"	"开始"	"感到"	"很"	"不"
[145]	"舒服"	"渐渐"	"地"	"他"	"的"	"衣服"	"上蒙上"	"一层"
[153]	"细小"	"的"	"白色"	"粉末"	"不"	"多久"	"房间"	"里"
[161]	"满"	"是"	"灰尘"	"他"	"开始"	"咳嗽"	"起来"	"砸"
[169]	"墙声"	"越来越"	"大"	"墙壁"	"上"	"的"	"灰泥"	"碎片"
[177]	"也"	"在"	"噼里啪啦"	"地"	"掉落"	"整个"	"楼"	"好像"
[185]	"都"	"要"	"倒塌"	"了"	"一样"	"史密斯"	"先生"	"立即"
[193]	"去"	"找"	"经理"	"投诉"	"他们"	"两人"	"一同"	"回到"
[201]	"房间"	"但是"	"听"	"不到"	"一丝"	"声响"	"两人"	"站"
[209]	"在"	"那里"	"面面相觑"	"史密斯"	"先生"	"觉得"	"很"	"尴尬"
[217]	"因为"	"是"	"他"	"硬生生"	"地"	"把"	"经理"	"叫"
[225]	"上"	"了"	"楼"	"白"	"耽误"	"了"	"工夫"	"突然"
[233]	"砸"	"墙声"	"又"	"出现"	"了"	"一块"	"大"	"砖头"
[241]	"落到"	"了"	"地板"	"上"	"他们"	"抬头"	"一看"	"一把"
[249]	"锋利"	"的"	"金属"	"工具"	"穿通"	"了"	"墙壁"	"在"
[257]	"床"	"的"	"正上方"	"捅出"	"了"	"一个"	"大"	"窟窿"

以上分词结果总体上令人较满意，但是一些词切分效果似乎不太理想，如"名有个"应切分为"名"和"有个"，"同一间房"可切分为"同""一间"和"房"。我们按照以下方式自定义词典，取文件名dictionary2.txt，文件编码格式为UTF-8，文档存放在D:\Rpackage文件夹中：

```
韦斯特盖特
格兰特大酒店
名
有个
并不
从来
同
一间
房
介意
本
书
坐下来
蒙上
满是
墙壁
一同
墙
声
```

本例利用以上自定义词典,编写的 R 命令和分词结果如下:

```
> cutter_user<-worker(user="D:/Rpackage/dictionary2.txt")
> hotel_words<-segment(hotel,cutter_user)
> hotel_words
```

[1]	"史密斯"	"先生"	"每次"	"去"	"韦斯特盖特"	"市"	"都"	"住"
[9]	"在"	"格兰特大酒店"	"尽管"	"酒店"	"名"	"有个"	"大"	"字"
[17]	"但"	"它"	"其实"	"并不"	"大"	"不过"	"很"	"便宜"
[25]	"干净"	"也"	"很"	"舒适"	"史密斯"	"先生"	"和"	"酒店"
[33]	"经理"	"很"	"熟络"	"所以"	"他"	"从来"	"不用"	"费工夫"
[41]	"去"	"预订"	"房间"	"他"	"总是"	"住"	"在"	"同"
[49]	"一间"	"房"	"房间"	"位于"	"走廊"	"的"	"尽头"	"在"
[57]	"那儿"	"可以"	"欣赏"	"海湾"	"美景"	"史密斯"	"先生"	"上次"
[65]	"来访"	"时"	"经理"	"告诉"	"他"	"可以"	"入住"	"老"
[73]	"房间"	"但是"	"经理"	"抱歉"	"地"	"补充"	"说"	"由于"
[81]	"客房"	"需求量"	"增大"	"酒店"	"决定"	"在"	"侧翼"	"扩建"
[89]	"所以"	"房间"	"可能"	"有点"	"吵"	"史密斯"	"先生"	"表示"
[97]	"自己"	"不"	"介意"	"入住"	"的"	"第一天"	"史密斯"	"先生"
[105]	"几乎"	"没有"	"听到"	"噪音"	"第二天"	"下午"	"他"	"就"
[113]	"从"	"酒店"	"图书室"	"借"	"了"	"本"	"书"	"然后"
[121]	"上楼"	"看"	"书"	"他"	"刚"	"坐下来"	"就"	"听到"
[129]	"有人"	"在"	"大声"	"砸"	"墙"	"起初"	"他"	"没有"
[137]	"在意"	"但是"	"过"	"了"	"一会儿"	"他"	"开始"	"感到"
[145]	"很"	"不"	"舒服"	"渐渐"	"地"	"他"	"的"	"衣服"
[153]	"上"	"蒙上"	"一层"	"细小"	"的"	"白色"	"粉末"	"不"
[161]	"多久"	"房间"	"里"	"满是"	"灰尘"	"他"	"开始"	"咳嗽"
[169]	"起来"	"砸"	"墙"	"声"	"越来越"	"大"	"墙壁"	"上"
[177]	"的"	"灰泥"	"碎片"	"也"	"在"	"噼里啪啦"	"地"	"掉落"
[185]	"整个"	"楼"	"好像"	"都"	"要"	"倒塌"	"了"	"一样"
[193]	"史密斯"	"先生"	"立即"	"去"	"找"	"经理"	"投诉"	"他们"
[201]	"两人"	"一同"	"回到"	"房间"	"但是"	"听"	"不到"	"一丝"
[209]	"声响"	"两人"	"站"	"在"	"那里"	"面面相觑"	"史密斯"	"先生"
[217]	"觉得"	"很"	"尴尬"	"因为"	"是"	"他"	"硬生生"	"地"
[225]	"把"	"经理"	"叫"	"上"	"了"	"楼"	"白"	"耽误"
[233]	"了"	"工夫"	"突然"	"砸"	"墙"	"声"	"又"	"出现"
[241]	"了"	"一块"	"大"	"砖头"	"落到"	"了"	"地板"	"上"
[249]	"他们"	"抬头"	"一看"	"一把"	"锋利"	"的"	"金属"	"工具"
[257]	"穿通"	"了"	"墙壁"	"在"	"床"	"的"	"正上方"	"捅出"
[265]	"了"	"一个"	"大"	"窟窿"				

以上结果显示,此次调整后文本词语数为 268 个,较上一次调整后的词语数 264 多了 4 个,因而变化不大。当然,我们可以利用 R 函数 length( ) 计算文本长度:

```
> hotel_leng<-length(hotel_words)
> hotel_leng
[1] 268
```

另外一种分词方法是执行以下 R 命令:

```
> require(jiebaR)
> cutter_user<-worker(user = "D:/Rpackage/dictionary2.txt", symbol = FALSE, bylines = TRUE)
> require(readtext)
> hotel<-readtext("D:/Rpackage/hotel.txt")
> require(dplyr)
> require(tidytext)
> hotel_toks<-hotel%>%
   unnest_tokens(word, text, token
   = function(x) segment(x, jiebar = cutter_user))
> length(hotel_toks $ word)
[1] 268
```

在以上命令中,调用数据包 readtext 是为了把文本文档 hotel.txt 以数据框的格式读入 R。自定义词典(用变元 user 设置路径)利用函数 worker( ),symbol=TRUE 表示保留文本中的符号(如标点符号和换行符" \n"),symbol=FALSE 则表示不保留文本中的符号。本例设置 symbol=FALSE。变元 bylines=TRUE 表示根据输入文档的行返还结果。调用数据包 dplyr 和 tidytext 是为了进行管道操作和调用函数 unnest_tokens( ) 开展形符化。在函数 unnest_tokens( ) 中定义函数 token=function(x) 以便按照自定义分词。

要计算文本中包括的句子,可以利用 R 数据包 tidytext 中的函数 unnest_sentences(tbl, output, input) 得到句子。该函数的主要变元包括 tibble 数据框(在管道操作中数据框置于管道符"%>%"之前),第二个变元 output 指输出的字符串列(这里指 sentence),第三个变元 input 指输入的字符串列(这里指 text)。本例计算文本句子数和平均句长的 R 命令和统计分析结果如下:

```
> require(readtext)
> require(dplyr)
> hotel_tbl<-tibble(readtext("D:/Rpackage/hotel.txt"))
> require(tidytext)
> hotel_sent<-hotel_tbl%>%
   unnest_sentences(sentence, text)
> hotel_sent[,2]
```

```
[1] "史密斯先生每次去韦斯特盖特市都住在格兰特大酒店。"
[2] "尽管酒店名有个"大"字,但它其实并不"大",不过很便宜,干净,也很舒适。"
[3] "史密斯先生和酒店经理很熟络,所以他从来不用费工夫去预订房间。"
[4] "他总是住在同一间房,房间位于走廊的尽头,在那儿可以欣赏海湾美景。"
[5] "史密斯先生上次来访时,经理告诉他,可以入住老房间,但是经理抱歉地补充说,由于客房需求
    量增大,酒店决定在侧翼扩建,所以房间可能有点吵。"
[6] "史密斯先生表示自己不介意。"
[7] "入住的第一天,史密斯先生几乎没有听到噪音。"
[8] "第二天下午,他从酒店图书室借了本书,然后就上楼去看书。"
[9] "他刚坐下来,就听到有人在大声砸墙。"
[10] "起初他没有在意,但是过了一会儿,他开始感到很不舒服。"
[11] "渐渐地,他的衣服上蒙上一层细小的白色粉末。"
[12] "不多久,房间里满是灰尘,他开始咳嗽起来。"
[13] "砸墙声越来越大,墙壁上的灰泥碎片也在噼里啪啦地掉落,整个楼好像都要倒塌了一样。"
[14] "史密斯先生立即去找经理投诉。"
[15] "他们两人一同回到房间,但是听不到一丝声响。"
[16] "两人站在那里面面相觑。"
[17] "史密斯先生觉得很尴尬,因为是他硬生生地把经理叫上了楼,白耽误了工夫。"
[18] "突然,砸墙声又出现了,一块大砖头落到了地板上。"
[19] "他们抬头一看,一把锋利的金属工具穿通了墙壁,在床的正上方捅出了一个大窟窿!"
> sents<- nrow(hotel_sent[,2])
> sents
[1] 19
> sents_leng<-length(hotel_toks $ word)/sents
> sents_leng
[1] 14.10526
```

以上结果显示,这篇短文包括 19 个句子。查看文档发现,函数 unnest_sentences() 对句子的划分正确。本例中,短文的平均句长约为每句 14 个词。

如果我们在前面计算文本长度的 R 命令中,设置 symbol = TRUE,那么得到的形符结果(hotel_toks)包括词、标点符号和换行符" \n"。这部分的 R 代码如下:

```
require(jiebaR)
cutter_user<-worker(user = "D:/Rpackage/dictionary2.txt", symbol = TRUE, bylines = TRUE)
require(readtext)
hotel_tbl<-readtext("D:/Rpackage/hotel.txt")
require(dplyr)
require(tidytext)
```

```
hotel_toks<-hotel_tbl%>%unnest_tokens(word, text, token
= function(x) segment(x, jiebar = cutter_user))
```

利用函数 grep() 查找所有包括句末停顿的符号位置。再利用函数 length() 计算位置数,即为句子数。本例 R 命令和计算结果如下:

```
> length(grep("[。?!……]",hotel_toks $ word))
[1] 19
> length(grep("[。|?!|!|……]",hotel_toks $ word))
[1] 19
```

利用以上两个正则表达式得到的结果相同,即文档 hotel.txt 包括 19 个句子。

我们也可以调用字符处理 R 数据包 stringr 中的函数 str_split() 计算文本句子数。第一步,调用 R 数据包 readtext 以数据框格式读入文本文档 hotel.txt。第二步,调用 R 数据包 quanteda,把上一步得到的数据框转化为语料库格式。第三步,调用 R 数据包 jiebaR,按照与前面相同的方式自定义词典,利用函数 worker() 分词,保留原文本的标点符号。第四步,确定划分句子边界的标点符号,常用的符号是:"。""!""?""……",调用 R 数据包 stringr 中的函数 str_split() 按照设定的句子边界划分方法进行字符串切分,设定函数变元 simplify = TRUE 使返还结果为字符矩阵。调用 R 数据包 require(purrr) 中的函数 map() 是为了把函数 str_split() 用于列表或原子向量中的每个元素,返还一个列表。第五步,利用 R 函数 dim() 返还矩阵的排数和列数,列数即为句子数。本例 R 命令和统计分析结果如下:

```
> require(readtext)
> hotel<-readtext("D:/Rpackage/hotel.txt")
> require(stringr)
> stringi::stri_locale_set("zh_hans_cn")
> require(quanteda)
> hotel_corpus <- corpus(hotel)
> require(jiebaR)
> cutter_user<-worker(user ="D:/Rpackage/dictionary2.txt",
    symbol = TRUE, bylines = TRUE)
> boundary<-"[。|!|?|……]+"
> require(purrr)
> hotel_sents<-hotel_corpus%>%
map(str_split, pattern = boundary, simplify = TRUE)
> dim(hotel_sents[[1]])[2]
[1] 19
```

接下来绘制词频至少为两次的文本词词频分布条形图。我们在前面的章节学过,在词频分析时通常排除一些没有太大意义的词,包括代词、连词和虚词等。我们这里使用 stopwords 包调用中文停用词表或 quanteda 包自带的停用词表(研究者实际研究中可以根据需要添加若干停用词),调用命令为:stopwords("zh",source="misc")。我们把这个停用词表以文本格式(文件名为 stop_zh.txt)保存在 D:\Rpackage 文件夹中,文件编码格式为 UTF-8。保存停用词表的 R 命令如下:

```
stringi::stri_locale_set("zh_hans_cn")
require(quanteda)
stop_zh<-stopwords("zh", source = "misc")
write.table(stop_zh,
"D:/Rpackage/stop_zh.txt",quote=FALSE,row.names=FALSE,sep="\t")
```

接下来,调用中文分词包 jiebaR 和自定义词典,并且使用停用词表排除无太大意义的词:

```
> require(jiebaR)
> cutter_user<-worker(user = "D:/Rpackage/dictionary2.txt", symbol = FALSE, bylines = TRUE,
stop_word = "D:/Rpackage/stop_zh.txt")
```

下面利用数据包 readtext 把文本文档 hotel.txt 调入 R,调用 dplyr 和 tidytext 开展管道操作,得到中文词符:

```
> require(readtext)
> require(dplyr)
> hotel_tbl<-tibble(readtext("D:/Rpackage/hotel.txt"))
> require(tidytext)
> hotel_toks<-hotel_tbl%>%unnest_tokens(word, text, token
  = function(x) segment(x, jiebar = cutter_user))
```

最后,利用管道操作符提取出现频次不小于 2 的词语,并利用函数 reorder() 排序,再调用数据包 ggplot2 中的函数绘制词频分布条形图。R 命令和图形绘制结果如下:

```
> require(ggplot2)
> hotel_toks%>%count(word, sort = TRUE) %>% filter(n>1)%>% mutate(word = reorder(word,
n))%>%
ggplot(aes(word, n)) + geom_col() + xlab(NULL) + coord_flip()
```

图 10.1 显示,"史密斯"和"先生"共同出现的频次最高(均为 7 次),"房间""经理"和"酒店"依次出现 6 次、5 次和 4 次,"砸"和"墙"均出现 3 次,其他词,如"住"和"听到"等,出现 2 次,出现频次较低。

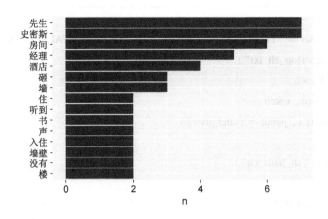

图 10.1　文本高频词语分布条形图

10.3　文本词频分布词云图

我们可以使用与英文文本处理相似的方法处理中文文本和绘制词云图。首先,利用上一节的 R 代码对文本 hotel.txt 分词,得到以数据框格式呈现的词符,不包括停用词和标点符号,文件名为 hotel_toks。然后,调用数据包 quanteda 中的函数 tokens( ) 和 dfm( ) 依次得到形符化结果和文档-特征矩阵(document-feature matrix)。最后,加载数据包 quanteda.textplots,调用其中的函数 textplot_wordcloud( ) 绘制词云图。该函数的主要变元是 quanteda.textplots(x, min_size=0.5, max_size=4, min_count=3, max_words=500, color="darkblue", rotation=0.1, random_order=FALSE),其中变元 x 是文档-特征矩阵。变元 min_size=0.5 是词的最小尺寸;max_size=4 是词的最大尺寸。变元 min_count 设置图中显示词的最小频数,默认值是 3,max_words 指图中包含的最大词数,默认值为 500。变元 color="darkblue" 表示函数默认的文字颜色为"深蓝色"。变元 rotation 指 90°旋转的词占比,默认值是 0.1。变元 random_order=FALSE 表示函数默认由里向外按词频降序排列词。由于文本 hotel.txt 包含的词语数量较少,我们决定把所有的词都显示出来,因而设置 min_count=1。颜色设置调用数据包 RColorBrewer 中的函数 brewer.pal( ),其他变元采用函数默认的设置。如果要查看各种调色板,可以在 R 界面输入:display.brewer.all( ),并按回车键。本例选择的调色板是"Dark2",编写的 R 代码和执行代码后绘制的词云图如下所示(图 10.2):

```
# read in the text in the tibble format
require(jiebaR)
require(readtext)
require(dplyr)
hotel_tbl<-tibble(readtext("D:/Rpackage/hotel.txt"))
# get tokens
```

```
require(tidytext)
cutter_user<-worker(user = "D:/Rpackage/dictionary2.txt", symbol =FALSE, bylines = TRUE, stop_
word = "D:/Rpackage/stop_zh.txt")
hotel_toks<-hotel_tbl%>%
unnest_tokens(word, text, token
= function(x) segment(x, jiebar = cutter_user))
# construct a dfm
stringi::stri_locale_set("zh_hans_cn")
require(quanteda)
hotel_dfm<-hotel_toks $ word%>%
dfm()
# plot a wordcloud
library(quanteda.textplots)
require(RColorBrewer)
pal<-brewer.pal(8, "Dark2")
textplot_wordcloud(hotel_dfm, min_count=1, rotation = 0, color = pal)
```

图 10.2 文本词频分布词云图(一)

我们在第三章介绍了调用 R 数据包 wordcloud 绘制词云图的方法,与调用 textplot_wordcloud 绘制词云图的方法大致相同。主要区别在于,函数 wordcloud()的处理对象是词和频数,而函数 textplot_wordcloud()的处理对象是文档-特征矩阵。本例调用数据包 wordcloud 绘制词云图的 R 命令和操作结果如下(图 10.3):

```
> require(dplyr)
> words<-hotel_toks%>%
count(word, sort=TRUE)
```

```
> require(wordcloud)
> require(RColorBrewer)
> pal<-brewer.pal(8,"Dark2")
> wordcloud(words = words $ word, freq = words $ n, random.order = FALSE, min.freq = 1,
  colors = pal)
```

图 10.3 文本词频分布词云图(二)

图 10.2、10.3 表明,"史密斯"和"先生"作为故事的主人公出现的频次最高,故事中的另外一个主要人物是"经理",出现的频次也很高。出现频次较高的词"酒店"和"房间"凸显故事发生的地点。"砸""墙"和"墙壁"等词体现故事中的主要事件。图中绝大部分词的大小相同,说明它们出现的频次较低。

如果我们对上面的词云图不是很满意,想增加一些创意,那么可以调用数据包 wordcloud2 制作更精美的图形。关于 wordcloud2 的介绍,详见第九章。我们先用以下代码把文档 hotel.txt 读入 R,利用管道操作,得到排除停用词后的词符。这一部分的命令等同于上一节绘制条形图前的 R 命令:

```
> require(readtext)
> require(dplyr)
> hotel_tbl<-tibble(readtext("D:/Rpackage/hotel.txt"))
> require(jiebaR)
> cutter_user<-worker(user = "D:/Rpackage/dictionary2.txt", symbol = FALSE, bylines = TRUE,
  stop_word = "D:/Rpackage/stop_zh.txt")
> require(tidytext)
> hotel_toks<-hotel_tbl%>%unnest_tokens(word, text, token
 = function(x) segment(x, jiebar = cutter_user))
```

然后，调用数据包 dplyr 中的函数 count( ) 计算词符频数，词语按照词频降序排列：

```
> hotel_freq<-hotel_toks%>%
count(word, sort = TRUE)
```

最后，调用数据包 wordcloud2，利用函数 wordcloud2( ) 绘制词云图。这个例子设置字体大小 size=0.56，字体设置为方正姚体(fontFamily = 'FZYaoTi')。利用变元 figPath 设置背景图案的路径。我们使用一张男性形体的图片作为词云背景，名称为 man.png，放在安装数据包 wordcloud2 的子文件夹 examples 中。如果要查找 wordcloud2 包的路径，可以调用以下 R 命令：find.package("wordcloud2")。

本例的 R 命令和绘制的图形如下所示(图 10.4)：

```
> require(wordcloud2)
> figPath=system.file("examples/man.png", package = "wordcloud2")
> wordcloud2(data=hotel_freq, figPath = figPath, fontFamily = 'FZYaoTi', size = 0.56)
```

图 10.4　文本词频分布词云图(三)

图 10.4 以人物轮廓为背景，凸显主人公史密斯先生对砸墙带来的噪声的不满，生动地体现史密斯先生向酒店经理抱怨的情形。同前面两幅词云图一样，图 10.4 显示故事主人公"史密斯"和"先生"出现的频次最高，"房间""经理"和"酒店"出现的频次依次降低，但是都是高频词。频次相对较低的"砸""墙"和"墙壁"等词体现故事中发生的主要事件。

如果我们要保存包括词云的 HTML 文件，可以使用以下命令把它保存在当前工作簿中：

```
> require(webshot)
> require(htmlwidgets)
> cloud <- wordcloud2(data=hotel_freq, figPath = figPath, fontFamily = 'FZYaoTi', size = 0.56)
> saveWidget(cloud,"body.html",selfcontained=FALSE)
```

如果我们要打开保存的文件 body.html，那么可以在以下路径中查找：C:\Users\DELL\Documents，可以在 R 界面输入命令 getwd() 查看。我们也可以在 R 界面输入以下命令打开文件：

```
> browseURL(paste(getwd(),'body.html',sep='/'))
```

我们也可以在生成的图片上右击鼠标，把图片另存为.png 格式的文件。由于图片文字排放的随机性，每次得到的结果都会有些许差异。建议多试几次，选择最佳图形显示效果。

10.4 关键词检索

在第二章我们学习了利用 for 循环编写 R 代码开展关键词检索，在第六章学习 R 数据包 quanteda 时了解到利用函数 kwic() 能够更方便地开展检索。该函数的最基本形式为 kwic(x,pattern,window=5,valuetype=c("glob","regex","fixed"))，其中 x 是字符或形符对象，pattern 指关键词模式或短语。如果一个模式中有空白，最好把它放在短语函数 phrase() 中。变元 window=5 是函数默认的关键词前后语境长度。变元 valuetype 提供三种匹配类型，"glob"指"glob"式通配符，"regex"指正则表达式，"fixed"指精确匹配。下面先看一个英文的例子：

```
> Text1<-" Whenever Mr Smith goes to Westgate, he stays at the Grand Hotel. Since he knows the manager well, he never has to go to the trouble of reserving a room."
> require(quanteda)
> Text1<-tokens(Text1,remove_punct=TRUE)
> kwic(Text1,pattern=" go",valuetype='glob',window=3)
Keyword-in-context with 1 match.
[text1, 23] never has to | go | to the trouble
> kwic(Text1,pattern=" go * ",valuetype='glob',window=3)
Keyword-in-context with 2 matches.
   [text1, 4]  Whenever Mr Smith | goes | to Westgate he
   [text1, 23]          never has to | go | to the trouble
> kwic(Text1,pattern=" go",valuetype='fixed',window=3)
Keyword-in-context with 1 match.
[text1, 23] never has to | go | to the trouble
> kwic(Text1,pattern=" go * ",valuetype='fixed',window=3)
```

```
Keyword-in-context with 0 matches.
> kwic(Text1,pattern=" go",valuetype='regex',window=3)
Keyword-in-context with 2 matches.
   [text1, 4]  Whenever Mr Smith | goes | to Westgate he
   [text1, 23]             never has to | go | to the trouble
> kwic(Text1,pattern= phrase(" go * to"),valuetype='regex',window=3)
Keyword-in-context with 2 matches.
   [text1, 4:5] Whenever Mr Smith | goes to | Westgate he stays
   [text1, 23:24]          never has to | go to | the trouble of
```

以上结果表明,'glob' 用于匹配整个字符,R 命令 kwic(Text1,pattern=" go",valuetype='glob')只用于检索文本中出现"go"的词,不包括"goes"。如果在"go"后面增加元字符" * "(通配符)从而选择模糊匹配,那么"goes"及其语境也一并被检索出来。如果使用固定匹配,kwic(Text1,pattern=" go",valuetype='fixed',window=3)只检索与"go"完全匹配的词及其语境。由于文本词符中没有" go * ",因而检索结果显示无匹配。使用正则表达式 pattern=" go",R 命令 kwic(Text1,pattern=" go",valuetype='regex',window=3)则返还精确匹配和模糊匹配的结果。上面最后一行代码的操作结果显示,如果检索短语,且在正则表达式中使用元字符" * ",那么检索结果包括短语的精确匹配和模糊匹配。

我们可以用相似的方法检索文本 hotel.txt 中的中文词语,如"房间"。代码的编写包括以下三个步骤。第一步,利用 R 数据包 readtext 把文本读入 R 界面。第二步,加载 R 数据包 jiebaR,对中文分词,不保留文本中的标点符号和换行符。第三步,先调用 R 数据包 purrr 中的函数 map(),把分词结果按列表排列,再使用函数 unlist()取消列表格式,最后利用 R 数据包 quanteda 中的函数 as.tokens()把分词结果转化为函数 kwic()可接受的词符格式。函数 map()把应用函数(这里是 segment())用于列表或原子向量中的每一个元素,返还一个列表。最后,调用函数 kwic()进行文本检索。本例 R 命令和执行命令对"房间"开展检索的结果如下:

```
> require(readtext)
> hotel_df <-readtext(" D:/Rpackage/hotel.txt")
> require(jiebaR)
> require(quanteda)
> require(purrr)
> cutter_user<-worker(user = " D:/Rpackage/dictionary2.txt", symbol = FALSE, bylines = TRUE)
> hotel_tokens<-hotel_df $ text% >%
  map(segment, cutter_user)% >%
  map(unlist)% >%
  as.tokens()
```

```
> kwic(hotel_tokens, pattern = "房间", valuetype = "glob")
Keyword-in-context with 6 matches.

 [text1, 43]   从来 不用 费工夫 去 预订 | 房间 | 他 总是 住 在 同
 [text1, 51]            住 在 同 一间 房 | 房间 | 位于 走廊 的 尽头 在
 [text1, 73]       告诉 他 可以 入住 老 | 房间 | 但是 经理 抱歉 地 补充
 [text1, 90]       决定 在 侧翼 扩建 所以 | 房间 | 可能 有点 吵 史密斯 先生
 [text1, 162]         的 白色 粉末 不 多久 | 房间 | 里 满是 灰尘 他 开始
 [text1, 204]     投诉 他们 两人 一同 回到 | 房间 | 但是 听 不到 一丝 声响

> kwic(hotel_tokens, pattern = "房间", valuetype = "regex")
Keyword-in-context with 6 matches.

 [text1, 43]   从来 不用 费工夫 去 预订 | 房间 | 他 总是 住 在 同
 [text1, 51]            住 在 同 一间 房 | 房间 | 位于 走廊 的 尽头 在
 [text1, 73]       告诉 他 可以 入住 老 | 房间 | 但是 经理 抱歉 地 补充
 [text1, 90]       决定 在 侧翼 扩建 所以 | 房间 | 可能 有点 吵 史密斯 先生
 [text1, 162]         的 白色 粉末 不 多久 | 房间 | 里 满是 灰尘 他 开始
 [text1, 204]     投诉 他们 两人 一同 回到 | 房间 | 但是 听 不到 一丝 声响

> kwic(hotel_tokens, pattern = "房间", valuetype = "fixed")
Keyword-in-context with 6 matches.

 [text1, 43]   从来 不用 费工夫 去 预订 | 房间 | 他 总是 住 在 同
 [text1, 51]            住 在 同 一间 房 | 房间 | 位于 走廊 的 尽头 在
 [text1, 73]       告诉 他 可以 入住 老 | 房间 | 但是 经理 抱歉 地 补充
 [text1, 90]       决定 在 侧翼 扩建 所以 | 房间 | 可能 有点 吵 史密斯 先生
 [text1, 162]         的 白色 粉末 不 多久 | 房间 | 里 满是 灰尘 他 开始
 [text1, 204]     投诉 他们 两人 一同 回到 | 房间 | 但是 听 不到 一丝 声响
```

以上结果表明,三种匹配类型得到同样的结果。具体而言,文档 hotel.txt 中"房间"一共出现 6 次,包括"预订房间""老房间""回到房间"和"房间里"等 4 个短语和"房间"的 2 次使用。

10.5 关键词提取

关键词提取(keyword extraction)的一种经典算法是 TF-IDF(Term Frequency-Inverse Document Frequency,词项频率-逆文档频率)算法。第八章简要介绍了词项频率-逆文档频率。词项频率是一个词在某个文档中出现的次数。除了常用词和没有多少实际意义的停用词(如"这""的""是"和"但是")之外,一个词在文档中出现的频率越高,这个词对文档就

越重要。逆文档频率是文档比较时一个词在不同文档里被赋予的权重。一个词在不同文档中出现的频率越高,被赋予的权重就越小。相反,一个词只在个别文档中出现,则被赋予的权重就越大。换言之,逆文档频率赋予不同文本中较常见的词以较小的权重,却赋予不常见词以更大的权重。因此,逆文档频率与一个词的常见程度成反比。

jiebaR 包包含一个 idf.utf8 文件,该文件包括逆文档频率,当前版本(版本号 0.1)的文件涉及 258 826 个词项。安装 jiebaRD 的 R 代码为:install.packages('jiebaRD')。解压 idf 文件夹,idf.utf8 文件的路径为:C:\users\DELL\AppData\Local\R\win-library\4.2\jiebaRD\dict\idf\idf.utf8。

利用以下 R 命令可以查看 10 个词项的逆文档频率:

```
> require(jiebaR)
> scan(file = "C:/users/DELL/AppData/Local/R/win-library/4.2/jiebaRD/dict/idf/idf.utf8", what = character(), nlines = 10, sep = "\n", encoding = "utf-8")
Read 10 items
 [1] "劳动防护 13.900677652"     "生化学 13.900677652"
 [3] "奥萨贝尔 13.900677652"     "考察队员 13.900677652"
 [5] "岗上 11.5027823792"        "倒车挡 12.2912397395"
 [7] "编译 9.21854642485"        "蝶泳 11.1926274509"
 [9] "外委 11.8212361103"        "故作高深 11.9547675029"
```

以上结果表明,"劳动防护"的逆文档频率为 13.900 677 652,"生化学"的逆文档频率也为 13.900 677 652,以此类推。一个词项在某个文档中出现的频率与该词的逆文档频率的乘积即为词项频率-逆文档频率。一个词的词项频率-逆文档频率值越大,它对文档的重要性就越大。我们先看个句子:生化学有意思,但是有人不喜欢生化学。在这个句子中,"但是"和"不"是常用词,因而作为停用词。"生化学"在这个句子中出现 2 次,它的逆文档频率为 13.900 677 652,则这个词的词项频率-逆文档频率值为 13.900 677 652×2 ≈ 27.801 4。我们利用以下命令在 idf.utf8 文件中查找其他三个词"有意思""有人"和"喜欢"的逆文档频率:

```
> require(jiebaR)
> words<-c("有意思","有人","喜欢")
> IDF_ZH<-
scan(file = "C:/users/DELL/AppData/Local/R/win-library/4.2/jiebaRD/dict/idf/idf.utf8", what = character(), sep = '\n', quiet = TRUE)
> for (i in 1:length(words)) { id<-grep(words[i],IDF_ZH)
    idf<-IDF_ZH[id]
    print(idf)
}
```

```
[1] "有意思 8.02574692112"    "蛮有意思 13.900677652"
[1] "大有人在 10.0505300503"  "人外有人 13.2075304714"
[3] "所有人 7.6701962044"     "持有人 8.0229418702"
[5] "素有人望 13.900677652"   "后继有人 11.3357282945"
[7] "有人 5.01870246773"      "未见有人 13.900677652"
[9] "现有人口 11.598092559"
[1] "喜不喜欢 11.8212361103"  "喜欢 5.70258840302"
[3] "逗人喜欢 13.2075304714"  "鹿喜欢 13.2075304714"
[5] "讨人喜欢 10.6818018271"
```

以上结果显示,"有意思""有人"和"喜欢"的逆文档频率依次约为8.025 7、5.018 7和5.702 6。这些词在句子中都出现1次,因而它们的词项频率-逆文档频率也依次为8.025 7、5.018 7和5.702 6。

下面以文档hotel.txt为例,从文中提取10个关键词。R代码和统计分析结果如下:

```
> require(readtext)
> hotel<-readtext("D:/Rpackage/hotel.txt")$text
> require(jiebaR)
> keys = worker("keywords", user = "D:/Rpackage/dictionary2.txt", symbol = FALSE, bylines = TRUE, topn = 10)
> keywords(hotel, keys)
 66.289    40.686 3  38.948 3  35.217 6  35.217 6
 "史密斯"  "房间"   "先生"    "墙"      "砸"
 29.256 3  23.478 4  23.478 4  23.478 4  21.592 5
 "经理"    "声"     "楼"      "住"      "酒店"
```

在以上R命令中,R数据包readtext的调入是为了把文档hotel.txt读入R操作界面。调用数据包jiebaR中的函数worker(),在变元设置中,函数默认排除数据包自带的停用词(stop_word),指定结巴引擎的类型"keywords",利用变元user调用自定义词典。变元symbol=FALSE表示排除标点符号和换行符;bylines=TRUE表示按输入文档的行返还结果。变元topn=10设置输出的10个关键词。最后,利用关键词函数keywords()得到关键词。该函数的第一个变元是字符串,在本例中是hotel;第二个变元是结巴引擎类型,在本例中是keys。输出结果中的10个关键词基本勾勒出故事的轮廓:"史密斯""先生""住"在一家"酒店"的一个"房间"。"楼"上有"砸""墙""声","史密斯""先生"找"经理"投诉。从关键性值(即词项频率-逆文档频率)来看,"史密斯"和"房间"最重要,但是从10.2节的词频分布来看,"史密斯"和"先生"出现的频次最高。尽管"砸"和"墙"在文中均出现3次,频次低于出现5次的"经理"一词,但是"砸"和"墙"的关键性值比"经理"一词的关键性值高。虽然"酒店"一词在文中出现4次,高于出现3次的"砸"和"墙"两个词,也高于出现2次的

"住"和"声"等词,但是根据关键性值,"酒店"一词对文本内容区分和识别的重要性却低于其他词。这意味着,"砸"和"墙"等词比"酒店"更能体现文本的主要内容。造成以上差异的主要原因是,词项频率-逆文档频率利用参照文本的加权值确定词项在某个文本中的重要性,而词频统计只考虑一个词项在某个文本中出现的次数,不使用加权值。

10.6 中文情感分析

第九章介绍如何利用整洁方法对英文文本开展情感分析。情感分析的质量很大程度上取决于情感词库的确定。目前,质量较高的中文情感词库是徐琳宏等编制的情感词汇本体库(网址:http://ir.dlut.edu.cn/)(徐琳宏 等,2008)。该库包括以下类别:词语、词性种类、词义数、词义序号、情感分类、强度和极性,情感分为 7 大类 21 小类,详见网页上的"中文情感词汇本体说明文档"。

本节利用以上情感词汇本体库对文本 hotel.txt 开展情感分析。为了不使问题复杂化,我们对原文本"情感词汇本体"(.xlsx 格式)的列进行调整,只保留以下五列:词语、词性种类、情感分类、强度和极性,重点考察文中表示正面情感和负面情感的词语数。在原文本的每类情感词极性标注中,0 代表中性,1 代表褒义,2 代表贬义,3 代表兼有褒贬两性。为了减少主观性,我们不考虑情感的强度,对标注的褒义词和贬义词赋予同样的权重。即是说,褒义词(正面情感词)赋值为 1,贬义词(负面情感词)赋值为-1,把两类情感词赋值加在一起得到情感分值,正值代表正面情感,负值代表负面情感。本节先计算文档 hotel.txt 中的情感词频数和文档体现出的整体情感倾向,然后绘制正面和负面情感词分布条形图。

编写计算文档 hotel.txt 情感值的 R 代码分以下几个步骤。第一步,可根据需要,先在 Windows 系统里安装 Java 软件,如 jdk-8u271-windows-x64.exe。在 R 中安装 rJava 和 xlsx 数据包,目的是使调用的函数 read.xlsx2()以正确的格式把"情感词汇本体.xlsx"(保存在 D:\package 文件夹下)读入 R 工作界面。R 命令如下所示:

```
# install.packages("rJava")
# install.packages("xlsx")
> require(rJava)
> require(xlsx)
> senti_zh<-read.xlsx2("D:/Rpackage/情感词汇本体.xlsx",1)
```

以上命令中,read.xlsx2()中的第一个变元是读入文档的路径,第二个变元是表单索引(sheet index)。通过 R 命令 nrow(senti_zh)发现"情感词汇本体.xlsx"包括 27 466 个情感词。

第二步,首先,从词库中筛选正面情感词和负面情感词,变量名分别为"pos"和"neg",并把第五列数值(值分别为"1"和"2")分别改为"positive"和"negative"。然后,利用函数 rbind()(按排合并,r 是 row 的缩写)把得到的正面情感词和负面情感词合并在一起,筛选变量名为"词语"和"极性"的两列,并将列名称分别改为"word"和"sentiment"。改用变量名"word"是为了使之与下面数据框(即 hotel_toks)中列变量名一致。改用变量名"sentiment"是为了对"positive"和"negative"进行恰当的归类。本例的 R 命令如下所示:

```
> pos<-senti_zh[senti_zh[,5]==1,]
> pos[,5]<-'positive'
> neg<-senti_zh[senti_zh[,5]==2,]
> neg[,5]<-'negative'
> senti<-rbind(pos,neg)
> senti<-senti[,c(1,5)]
> colnames(senti)<-c("word","sentiment")
```

第三步,按照 10.3 节介绍的方法得到文档 hotel.txt 的中文分词结果,格式为数据框。R 命令如下:

```
> require(readtext)
> hotel<-readtext("D:/Rpackage/hotel.txt")
> require(jiebaR)
> cutter_user<-worker(user = "D:/Rpackage/dictionary2.txt", symbol = FALSE, bylines = TRUE,
   stop_word = "D:/Rpackage/stop_zh.txt")
> require(tidytext)
> require(dplyr)
> hotel_toks<-hotel%>%unnest_tokens(word, text, token
   = function(x) segment(x, jiebar = cutter_user))
```

第四步,利用数据包 dplyr 中的函数 inner_join() 提取情感词及其类别标注。在默认状态下,该函数按照相同的变量把两个数据框合并在一起,生成一个新数据框。具体到本例,该函数从数据框 hotel_toks 中提取出词符包含在数据框 senti 中的所有排。例如,执行 R 命令 inner_join(hotel_toks, senti) 或 hotel_toks%>%inner_join(senti) 得到以下数据框:

```
> inner_join(hotel_toks, senti)
Joining, by = "word"
readtext object consisting of 15 documents and 1 docvar.
# Description: df [15 × 4]
    doc_id      word    sentiment   text
    <chr>       <chr>   <chr>       <chr>
1   hotel.txt   先生     positive    "\"\"..."
2   hotel.txt   干净     positive    "\"\"..."
3   hotel.txt   舒适     positive    "\"\"..."
4   hotel.txt   先生     positive    "\"\"..."
```

```
5 hotel.txt  费工夫   negative   "\"\"..."
6 hotel.txt  欣赏     positive   "\"\"..."
# … with 9 more rows
```

以上结果表明，本例按照"word"变量把两个数据框合并在一起，得到15个情感词。作为情感词库词的"先生"一词在数据框 hotel_toks 中出现，有关"先生"所在排的所有信息，如文档编号 doc_id 和情感类别 positive，全部被提取出来。接下来，我们利用数据包 dplyr 中的函数 count() 把出现在 hotel_toks 中所有情感词的词频计算出来。利用以下管道命令得到文档 hotel.txt 中包括的所有正面情感词和所有负面情感词的计数：

```
> hotel_toks %>% inner_join(senti) %>%
    count(index=doc_id, sentiment)
Joining, by = "word"
readtext object consisting of 2 documents and 1 docvar.
# Description: df [2 × 4]
  index      sentiment   n       text
  <chr>      <chr>       <int>   <chr>
1 hotel.txt  negative    4       "\"\"..."
2 hotel.txt  positive    11      "\"\"..."
```

以上结果显示，本例文档使用负面情感词的频次是4，正面情感词的频次是11。如果我们赋予所有正面和负面情感词以相同的权重，那么两类情感词频次之差整体上体现了文档的情感倾向。

最后，调用数据包 tidyr 中的函数 spread() 把排变量转化为列变量。即是说，把前面按照情感类 sentiment 和频数 n 分类得到的结果重新按照 negative 和 positive 分类，并记录对应的频次。函数 spread() 把数据框由长格式转化为宽格式，主要结构是 spread(data, key, value, fill=0)，其中 data 是数据框，key（键）指要转化为多个列的列变量水平或类别，value（值）指要转化为多个列的列变量水平的对应值，fill=0 表示缺失值由"0"替代，通常可以省略。调用数据包 dplyr 中的函数 mutate()，增加一个新变量 sentiment，数值等于正面情感累计值减去负面情感累计值。本例 R 命令如下：

```
> require(tidyr)
> hotel_toks %>%
    inner_join(senti) %>%
    count(sentiment) %>%
    spread(key=sentiment, value=n, fill=0) %>%
    mutate(sentiment = positive - negative)
Joining, by = "word"
```

```
  negative positive sentiment
1    4      11       7
```

以上结果显示,文档 hotel.txt 包含的正面情感值为 11,负面情感值为 4,整体上的文档情感值为 7,即文档整体上体现正面情感。

我们再利用函数 inner_join() 提取文档 hotel.txt 中的情感词,得到以下结果:

```
> senti_words<-hotel_toks%>%
   inner_join(senti)
   Joining, by = "word"
> data.frame(senti_words $ word,senti_words $ sentiment)
   senti_words.word senti_words.sentiment
1       先生            positive
2       干净            positive
3       舒适            positive
4       先生            positive
5       费工夫           negative
6       欣赏            positive
7       先生            positive
8       先生            positive
9       先生            positive
10      噪音            negative
11      起来            positive
12      先生            positive
13      先生            positive
14      硬生生           negative
15      耽误            negative
```

以上结果显示,"费工夫"被当作了消极情感词,但是原文档用的是否定表达"不用费工夫",体现了正面情感。另外,"起来"出现的原文语境是"他开始咳嗽起来",应视作中性词。我们利用以下 R 命令对原情感词库进行了微调:

```
> require(tidyr)
> hotel_senti<-hotel_toks%>%
   inner_join(senti)
   Joining, by = "word"
> hotel_senti $ word[which(hotel_senti $ word=="费工夫")]<-"不用费工夫"
> hotel_senti $ sentiment[which(hotel_senti $ word==
   "不用费工夫")]<-"positive"
> hotel_senti%>%filter(!row_number()%in% which(hotel_senti $ word=="起来"))%>%
```

```
count(index=doc_id,sentiment)%>%
spread(sentiment, n, fill = 0)%>%
mutate(sentiment = positive - negative)
readtext object consisting of 1 document and 2 docvars.
# Description: df [1 × 5]
    index      negative   positive   sentiment   text
    <chr>      <dbl>      <dbl>      <dbl>       <chr>
1   hotel.txt  3          11         8           "\"\"..."
```

以上命令显示,在原词库中增加了表示正面情感的"不用费工夫"一词,删除了表示中性情感的"起来"一词,由此得到整体上的正面情感值为8。虽然微调使结果没有发生明显变化,但是在情感分析中我们需要结合实际问题确定情感词,使情感分析趋于准确,因而这个增加的步骤通常是必不可少的。

至此,我们对文档 hotel.txt 开展情感分析的代码编写已经结束。R 完整代码如下:

```
# Read in 情感词汇本体.xlsx
require(rJava)
require(xlsx)
senti_zh<-read.xlsx2("D:/Rpackage/情感词汇本体.xlsx",1)
# Get sentiment lexicon needed
pos<-senti_zh[senti_zh[,5]==1,]
pos[,5]<-'positive'
neg<-senti_zh[senti_zh[,5]==2,]
neg[,5]<-'negative'
senti<-rbind(pos,neg)
senti<-senti[,c(1,5)]
colnames(senti)<-c("word",'sentiment')
# Read in hotel.txt and get a data frame involving words
require(readtext)
hotel_tbl<-readtext("D:/Rpackage/hotel.txt")
require(jiebaR)
cutter_user<-worker(user="D:/Rpackage/dictionary2.txt", symbol=FALSE, bylines=TRUE, stop_
word="D:/Rpackage/stop_zh.txt")
require(tidytext)
require(dplyr)
hotel_toks<-hotel_tbl%>%unnest_tokens(word, text, token
= function(x) segment(x,jiebar=cutter_user))
# Sentiment analysis
```

```
require(tidyr)
hotel_senti<-hotel_toks%>%inner_join(senti)
hotel_senti$word[which(hotel_senti$word=="费工夫")]<-"不用费工夫"
hotel_senti$sentiment[which(hotel_senti$word=="不用费工夫")]<-"positive"
hotel_senti%>%filter(!row_number()%in% which(hotel_senti$word=="起来"))%>%
count(index=doc_id,sentiment)%>%
spread(sentiment,n,fill=0)%>%
mutate(sentiment=positive-negative)
```

执行以上 R 代码,得到负面情感值 3 和正面情感值 11,正面与负面情感值之差为 8。9.6.2 节调用 syuzhet 包中的 nrc 情感词库计算了英语文本 hotel0.txt 中的正面和负面情感词和 8 种情绪词。我们这里只考虑正面和负面情感词。表 9.1 显示文本 hotel0.txt 中使用了 13 个正面情感词,12 个负面情感词,正面与负面情感值之差为 1。造成以上差异的主要原因是"情感词汇本体"词库把"先生"(在 hotel0.txt 中出现 7 次)视作正面情感词,而 nrc 情感词库把"Mr"视作中性情感词,把"sir"视作正面情感词。这两个词都可以译作"先生",中文的翻译没有体现出这两个英文词的差异。

在本节的最后,我们利用 R 数据包 ggplot2 中的函数 ggplot() 绘制文档情感词分布条形图。关于函数 ggplot() 的基本介绍,参见第九章。本例绘制文档 hotel.txt 中情感词分布条形图的 R 命令和绘图结果(图 10.5)如下:

```
> require(ggplot2)
> hotel_senti%>%filter(!row_number()%in% which(hotel_senti$word=="起来"))%>%
  count(word=word,sentiment) %>%
  ggplot(aes(word, n, fill = sentiment)) +
  geom_col(show.legend=FALSE)+
  facet_wrap(~sentiment, scales="free_y") +
  labs(y="Contribution to sentiment",
  x = NULL)+coord_flip()
```

图 10.5 显示,文档 hotel.txt 中的负面情感词包括"噪音""硬生生"和"耽误",每个词均出现 1 次;正面情感词包括"欣赏""先生""舒适""干净"和"不用费工夫",其中"先生"一词出现 7 次,其他词均出现 1 次。

我们想要用另外一种方法检验以上结果的正确性。首先,我们提取数据框 senti 中的情感词,并利用 R 基础函数 c() 把"不用费工夫"纳入情感词表。其次,调用函数 which() 删除 senti 中的虚假情感词"起来",调用函数 gsub() 把 hotel_w 中的"费工夫"替换为"不用费工夫"。然后,利用查找符 %in% 提取 hotel_w 中的所有情感词。最后,调用数据包 jiebaR 中的函数 freq() 计算情感词词频。R 命令和处理结果如下:

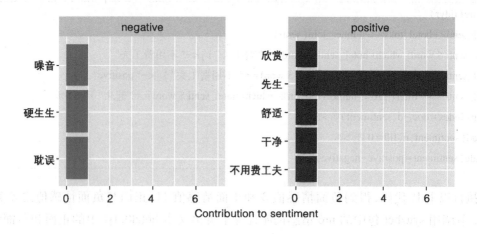

图 10.5　文档 hotel.txt 情感词分布条形图

```
> require(jiebaR)
> senti_w<-c(senti[,1],"不用费工夫")
> hotel_w<-hotel_toks $ word
> hotel_w<-gsub("费工夫","不用费工夫",hotel_w)
> hotel_w<- hotel_w[which(hotel_w!="起来")]
> sentiment<-hotel_w[hotel_w%in%senti_w]
> freq(sentiment)
       char freq
1      硬生生  1
2      欣赏    1
3   不用费工夫 1
4      舒适    1
5      耽误    1
6      噪音    1
7      干净    1
8      先生    7
```

以上结果表明,除"先生"一词出现 7 次之外,其他词,如"硬生生"和"欣赏",均出现 1 次,与图 10.5 显示的结果一致。

第十一章 词性与句法分析

第三章在介绍 koRpus 包的主要功能时提到了词性标注,但是没有展开讨论。本章主要介绍如何对文本词汇开展词性标注、句法成分依存分析和短语结构分析,分五节内容,包括:数据包 udpipe 的安装与初试、文本中的短语提取、句法分析、词语共现和快速自动关键词提取。本章调用的数据包包括:udpipe、textplot、rsyntax、stanford-parser-4.2.0、data.table、L2SCA_R、stanford-tregex-4.2.0、readtext、ggraph、lattice、wordcloud2 和 quanteda。

11.1 数据包 udpipe 的安装与初试

R 数据包 udpipe 由扬·维菲尔斯开发,当前版本号为 0.8.11(Wijffels,2023)。数据包名称中的 ud 是 universal dependencies(通用依存标注体系)的首字母缩写。通用依存标注研究的理念是使用统一的分类和原则对不同语言相似的结构开展一致性标注,同时允许有语言特有的扩展方式。数据包 udpipe 的安装方式与 quanteda 等数据包的安装方式相同。若要把数据包安装在系统默认的路径,可以在 R 工作界面输入以下简单的命令:install.packages("udpipe")。用户也可以自定义路径进行安装,如在以上命令中增加路径:lib="C:/Program Files/R/R-4.2.2/library"),按回车键对数据包进行自动在线安装。要调用 udpipe 数据包开展自然语言处理,需要根据语言需求加载对应的模型。该包提供 65 种以上的语言,包括英语和汉语。要加载语言模型,首先去以下网站下载语言模型数据 zip 压缩文件:https://github.com/jwijffels/udpipe.models.ud.2.5。下载完成后,解压文件,取文件夹名称为 udpipe.models.ud.2.5。本章处理英语文本,我们需要在文本处理之前先调用数据包 udpipe 中的函数 udpipe_load_model(file),其中 file 包括语言模型文件名以及存放的完整路径。本章使用的英语模型名称包括词串 english-ewt,该词串是 the English Web Treebank 的简称。我们把文件夹 udpipe.models.ud.2.5 置于 D:\Rpackage 文件夹中,调用英语模型的 R 代码如下(存储的 R 对象为 ud_en):

```
require(udpipe)
ud_en<-udpipe_load_model(file = "D:/Rpackage/udpipe.models.ud.2.5/english-ewt-ud-2.5-191206.udpipe")
```

下面以文档 story.txt 中的开头句为例初步了解 udpipe 数据包的使用。开头句保存的 R 对象是 text:

```
text<-"Once there was a little boy who was hungry for success in sports."
```

我们要调用的函数是 udpipe_annotate()，用于对字符串向量开展形符化、词条化、词性（Part of Speech, POS）标注和依存分析（dependency parsing）。词性标注（POS tagging）体现对句法结构的最基本理解。依存分析是句法分析，通过节点或形符之间的依存关系分析句子结构。函数 udpipe_annotate() 标注句子中各个形符之间的句法关系，主要结构是 udpipe_annotate(object, x, doc_id = paste("doc", seq_along(x), sep = ""), tokenizer = "tokenizer", tagger = c("default", "none"), parser = c("default", "none"))，其中 object 是加载的 udpipe 语言模型，x 是按照 UTF-8 编码的字符向量。变元 doc_id 是与 x 有同样长度的文档编号，tokenizer = "tokenizer" 是函数默认的 udpipe 形符化方法。变元 tagger = c("default", "none") 中有两个选项，tagger = "default" 指函数默认使用 udpipe 中的词性标注和形符化方法，tagger = "none" 则指不开展词性标注和形符化。变元 parser = "default" 指函数默认开展 udpipe 依存分析，parser = "none" 则指不开展依存分析。函数 udpipe_annotate() 对字符串形符化和标注的结果以 CoNLL-U 格式呈现（CoNLL 是 Conference on Computational Natural Language Learning（计算自然语言学习会议）的简称，CONLL-U 格式是数据标注的标准格式），可以利用函数 as.data.frame() 把结果转化为数据框格式。数据框列名包括：文档编号（doc_id）、段落编号（paragraph_id）、句子编号（sentence_id）、句子（sentence）、形符编号（token_id）、形符（token）、词目或词条（lemma）、通用词性标注（upos）、语言特有词性标注（xpos）、形态学特征（feats）、中心形符编号（head_token_id）、依存关系（dep_rel）、依存关系图（deps）和其他标注（misc）。我们也可以调用函数 udpipe(x, object, ...)，其中 x 是字符向量，object 是加载的 udpipe 语言模型，... 指输送给 udpipe_annotate() 的其他变元。

如果我们只对文本 text 进行词性标注和形符化，可以执行以下 R 命令：

```
> text<-"Once there was a little boy who was hungry for success in sports."
> require(udpipe)
> ud_en<-udpipe_load_model(file = "D:/Rpackage/udpipe.models.ud.2.5/english-ewt-ud-2.5-191206.udpipe")
> text_anno<-udpipe_annotate(ud_en,text,parser="none")
> text_anno<-as.data.frame(text_anno)
> # 或者用函数 udpipe()
> text_anno<-udpipe(text,ud_en,parser="none")
> text_anno
   doc_id paragraph_id sentence_id                                                          sentence token_id    token   lemma  upos xpos
1    doc1            1           1  Once there was a little boy who was hungry for success in sports.        1     Once    once SCONJ   IN
2    doc1            1           1  Once there was a little boy who was hungry for success in sports.        2    there   there  PRON   EX
3    doc1            1           1  Once there was a little boy who was hungry for success in sports.        3      was      be  VERB  VBD
4    doc1            1           1  Once there was a little boy who was hungry for success in sports.        4        a       a   DET   DT
5    doc1            1           1  Once there was a little boy who was hungry for success in sports.        5   little  little   ADJ   JJ
6    doc1            1           1  Once there was a little boy who was hungry for success in sports.        6      boy     boy  NOUN   NN
7    doc1            1           1  Once there was a little boy who was hungry for success in sports.        7      who     who  PRON   WP
8    doc1            1           1  Once there was a little boy who was hungry for success in sports.        8      was      be   AUX  VBD
9    doc1            1           1  Once there was a little boy who was hungry for success in sports.        9   hungry  hungry   ADJ   JJ
10   doc1            1           1  Once there was a little boy who was hungry for success in sports.       10      for     for   ADP   IN
11   doc1            1           1  Once there was a little boy who was hungry for success in sports.       11  success success  NOUN   NN
12   doc1            1           1  Once there was a little boy who was hungry for success in sports.       12       in      in   ADP   IN
13   doc1            1           1  Once there was a little boy who was hungry for success in sports.       13   sports   sport  NOUN  NNS
14   doc1            1           1  Once there was a little boy who was hungry for success in sports.       14        .       . PUNCT    .
```

以上结果显示，函数 udpipe_annotate() 把文本 text 形符化，每行一个形符，也对形符进行了词条化处理,如"was"和"sports"被依次转化为"be"和"sport"。以上结果还报告通用词性标注(universal part-of-speech tag,upos)和语言特有词性标注(language-specific part-of-speech tag,xpos)。表 11.1 列出了 17 种通用词性标注及其描述(参考网址:https://universaldependencies.org/u/pos/index.html)。

表 11.1　主要通用词性标注一览表

编号	开放词类	编号	封闭词类	编号	其他
1	ADJ:形容词	7	ADP:介词	15	PUNCT:标点
2	ADV:副词	8	AUX:助动词	16	SYM:符号,如 $ 、% 和 §
3	INTJ:感叹词	9	CCONJ:并列连词	17	X:其他,指对词性不能赋码的词
4	NOUN:名词	10	DET:限定词		
5	PROPN:专有名词	11	NUM:数词		
6	VERB:动词	12	PART:小品词,如所有格标记 's		
		13	PRON:代词		
		14	SCONJ:从属连词		

我们看一下对文本 text 中的句子进行通用词性标注的结果:

```
> text_anno[,c("token","upos")]
        token    upos
1        Once   SCONJ
2       there    PRON
3         was    VERB
4           a     DET
5      little     ADJ
6         boy    NOUN
7         who    PRON
8         was     AUX
9      hungry     ADJ
10        for     ADP
11    success    NOUN
12         in     ADP
13     sports    NOUN
14          .   PUNCT
```

以上结果显示,本例中的词性标注并不完全准确,如"Once"应为副词,却被标注成从属连词,尽管它有这样的用法。不过,词性标注器(POS tagger)是序列统计模型,即一个词的词性标注不仅取决于词本身,而且取决于邻近词和邻近词标注(Altinok,2021)[71],因而总体上看,标注器的词性标注还是很准确的。我们这里再举个例子"We will fish for a fish tomorrow."。这个例子包括两个"fish",第一个"fish"是动词,第二个"fish"是名词。对本例词性标注的 R 命令和主要结果如下:

```
> text1='We will fish for a fish tomorrow.'
> text1_anno<-udpipe(text1,ud_en)
> text1_anno
  doc_id paragraph_id sentence_id                       sentence start end term_id token_id    token    lemma  upos xpos
1   doc1            1           1 We will fish for a fish tomorrow.     1   2       1        1       We       we  PRON  PRP
2   doc1            1           1 We will fish for a fish tomorrow.     4   7       2        2     will     will   AUX   MD
3   doc1            1           1 We will fish for a fish tomorrow.     9  12       3        3     fish     fish  VERB   VB
4   doc1            1           1 We will fish for a fish tomorrow.    14  16       4        4      for      for   ADP   IN
5   doc1            1           1 We will fish for a fish tomorrow.    18  18       5        5        a        a   DET   DT
6   doc1            1           1 We will fish for a fish tomorrow.    20  23       6        6     fish     fish  NOUN   NN
7   doc1            1           1 We will fish for a fish tomorrow.    25  32       7        7 tomorrow tomorrow  NOUN   NN
8   doc1            1           1 We will fish for a fish tomorrow.    33  33       8        8        .        . PUNCT    .
```

以上结果显示,如我们所期望的那样,第一个"fish"出现在助动词(AUX)之后,标注器把它正确地标注为动词(VERB),第二个"fish"与限定词("a")连用,标注器则把它正确地标注为名词(NOUN)。

表 11.2 列出了 udpipe 数据包中宾夕法尼亚大学树库(the Penn Treebank)使用的 36 种英语词性标注及其描述。

表 11.2 udpipe 英语词性标注一览表

	标注	描述
1	CC	coordinating conjunction,并列连词
2	CD	cardinal number,基数
3	DT	determiner,限定词
4	EX	existential there,表示存在的 there
5	FW	foreign word,外来词
6	IN	preposition or subordinating conjunction,介词或从属连词
7	JJ	adjective,形容词
8	JJR	adjective, comparative,形容词,比较级
9	JJS	adjective, superlative,形容词,最高级
10	LS	list item marker,计量词
11	MD	modal,情态动词
12	NN	noun, singular or mass,单数名词或物质名词
13	NNS	noun, plural,复数名词
14	NNP	proper noun, singular,专有名词,单数
15	NNPS	proper noun, plural,专有名词,复数
16	PDT	predeterminer,前限定词
17	POS	possessive ending,所有格结尾
18	PRP	personal pronoun,人称代词
19	PRP$	possessive pronoun,代词所有格
20	RB	adverb,副词

(续表)

	标注	描述
21	RBR	adverb, comparative, 副词, 比较级
22	RBS	adverb, superlative, 副词, 最高级
23	RP	particle, 小品词
24	SYM	symbol, 符号
25	TO	TO, 动词不定式
26	UH	interjection, 感叹词
27	VB	verb, base form, 动词, 原形
28	VBD	verb, past tense, 动词, 过去时
29	VBG	verb, gerund or present participle, 动词, 动名词或现在分词
30	VBN	verb, past participle, 动词, 过去分词
31	VBP	verb, non-3rd person singular present, 动词, 非第三人称单数现在时
32	VBZ	verb, 3rd person singular present, 动词, 第三人称单数现在时
33	WDT	wh-determiner, wh－限定词
34	WP	wh-pronoun, wh－代词
35	WP$	possessive wh-pronoun, wh－代词所有格
36	WRB	wh-adverb, wh－副词

对文本 text 中句子的英语词性标注的结果如下：

```
> text_anno[ ,c( "token" ,"xpos" )]
      token    xpos
1      Once      IN
2     there      EX
3       was     VBD
4         a      DT
5    little      JJ
6       boy      NN
7       who      WP
8       was     VBD
9    hungry      JJ
10      for      IN
11  success      NN
12       in      IN
13   sports     NNS
14        .
```

以上结果中，"Once"在本句中应为副词, 被标注成了 IN(介词或从属连词), 未能解决一词多义问题。不过,词性标注整体上看很准确。

表 11.3 列出了 65 种依存关系标注及其描述(参考网址:https://universaldependencies.org/u/dep/index.html)。

表 11.3　udpipe 依存关系标注一览表

	标注	描述
1	acl	clausal modifier of noun（adjectival clause）,名词从句修饰语(形容词从句)
2	acl:relcl	relative clause modifier,关系从句修饰语
3	advcl	adverbial clause modifier,状语从句修饰语
4	advmod	adverbial modifier,副词修饰语
5	advmod:emph	emphasizing word, intensifier,强调词,加强语
6	advmod:lmod	locative adverbial modifier,方位状语修饰语
7	amod	adjectival modifier,形容词修饰语
8	appos	appositional modifier,同位语修饰语
9	aux	auxiliary,助词
10	aux:pass	passive auxiliary,被动助词
11	case	case marking,格标记,如所有格 's
12	cc	coordinating conjunction,并列连词
13	cc:preconj	preconjunct,前连词,如 neither... nor 中的 neither
14	ccomp	clausal complement,从句补足语
15	clf	classifier,分类词,如"两本书"中的"本"
16	compound	compound,复合词
17	compound:lvc	light verb construction,轻动词结构
18	compound:prt	phrasal verb particle,动词短语小品词
19	compound:redup	reduplicated compounds,重叠复合词
20	compound:svc	serial verb compounds,连动复合词
21	conj	conjunct,连词,如 and 和 or
22	cop	copula,系动词
23	csubj	clausal subject,子句主语
24	csubj:outer	outer clausal subject,外围子句主语
25	csubj:pass	clausal passive subject,子句被动主语
26	dep	unspecified dependency,不确定依存
27	det	determiner,限定词
28	det:numgov	pronominal quantifier governing the case of the noun,支配名词格的代词量词
29	det:nummod	pronominal quantifier agreeing in case with the noun,与名词格一致的代词量词

(续表)

	标注	描述
30	det:poss	possessive determiner,物主限定词
31	discourse	discourse element,语篇元素,如 oh 和 um
32	dislocated	dislocated elements,错位元素
33	expl	expletive,填充词,如表示存在的 there
34	expl:impers	impersonal expletive,非人称填充词
35	expl:pass	reflexive pronoun used in reflexive passive,反身被动语态中的反身代词
36	expl:pv	reflexive clitic with an inherently reflexive verb,带有自反动词的附着语
37	fixed	fixed multiword expression,固定短语
38	flat	flat multiword expression,扁平短语,如 Hilary Clinton
39	flat:foreign	foreign words,外来语
40	flat:name	names,名称短语
41	goeswith	goes with 连用词,如 never the less 各部分应连在一起
42	iobj	indirect object,间接宾语
43	list	list,列举关系
44	mark	marker,从句标记词,如 that 和 although
45	nmod	nominal modifier,名词性修饰语
46	nmod:poss	possessive nominal modifier,所有格名词修饰语
47	nmod:tmod	temporal modifier,时间性名词修饰语
48	nsubj	nominal subject,名词性主语
49	nsubj:outer	outer clause nominal subject,子句外的名词性主语
50	nsubj:pass	passive nominal subject,被动态名词性主语
51	nummod	numeric modifier,数字修饰语
52	nummod:gov	numeric modifier governing the case of the noun,支配名词格的数字修饰语
53	obj	object,宾语
54	obl	oblique nominal,间接名词
55	obl:agent	agent modifier,被动结构施动者
56	obl:arg	oblique argument,间接论元
57	obl:lmod	locative modifier,方位修饰语
58	obl:tmod	temporal modifier,时间性修饰语
59	orphan	orphan,孤立词
60	parataxis	parataxis,无连词合并
61	punct	punctuation,标点

(续表)

	标注	描述
62	reparandum	overridden disfluency,待修正语
63	root	root,根节点;根词
64	vocative	vocative,称呼语
65	xcomp	open clausal complement,(动词或形容词)开放性子句补足语

把 R 命令 udpipe_annotate(ud_en,text,parser=" none")中的变元设置 parser=" none"改为 parser=" default",或省略该变元设置,采用默认形式,得到文本 text 中句子依存关系标注的结果:

```
> text_anno<-udpipe_annotate(ud_en,text,parser=" default")
> text_anno<-as.data.frame(text_anno)
> # 或者用函数 udpipe()
> text_anno<-udpipe(text,ud_en,parser=" default")
> text_anno[,c(" token"," dep_rel")]
        token    dep_rel
1        Once       mark
2       there       expl
3         was       root
4           a        det
5      little       amod
6         boy      nsubj
7         who      nsubj
8         was        cop
9      hungry   acl:relcl
10        for       case
11    success        obl
12         in       case
13     sports       nmod
14          .      punct
```

以上结果显示,"Once"应为副词修饰语(advmod),却被错误地标注为从句标记词(mark),其他词的依存关系标注正确,如第一个"was"被标注为根节点,第二个"was"被标注为系动词(cop),"hungry"被标注为关系从句修饰语(acl:relcl)。

要对词性和依存关系的标注结果采用线性图形显示,可以调用 textplot 数据包中的函数 textplot_dependencyparser()(Wijffels,2022)。该函数默认的主要形式为 textplot_dependencyparser(x,title=" Dependency Parser",subtitle=" tokenisation,parts of speech tagging & dependency relations",vertex_color=" darkgreen",edge_color=" red",size=3,layout=" linear"),其中 x 为函数 udpipe()返还的数据框,变元 title 和 subtitle 用于设置主标

题和副标题,vertex_color="darkgreen"指函数默认每个节点的颜色为深绿色,edge_color="red"指函数默认标注颜色为红色,size=3指函数默认字体的大小为3,layout="linear"指函数默认图形布局为线性。若对"Once"的标注进行调整,对文本text开展形符化和标注的R命令和图形绘制结果如下:

```
> text_anno<-udpipe(text,ud_en)
> text_anno[1,"dep_rel"]<-"advmod"
> text_anno[1,"upos"]<-"ADV"
> require(textplot)
> textplot_dependencyparser(text_anno,title="",subtitle="")
```

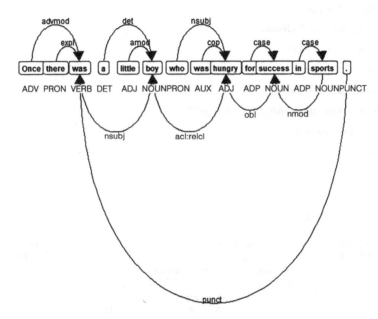

图11.1 字符串形符化、词性标注和依存关系图

如图11.1所示,这个句子主句的主语是名词"boy"(nsubj);"who"是代词,充当关系从句的主语(nsubj),关系从句的修饰语是"hungry";句子中有两个介词短语"for success"和"in sports"(结构是"ADP"+"NOUN")。

我们也可以调用rsyntax数据包中的函数plot_tree()绘制依存关系树形图(tree diagram)(Welbers et al.,2022)。该函数的主要结构是plot_tree(tokens,…,sentence_i=1, doc_id=NULL,pdf_file=NULL,all_lower=FALSE,textsize=1,use_color=TRUE,max_curve=0.3,palette=grDevices::terrain.colors,rel_on_edge=FALSE,pdf_viewer=FALSE, viewer_mode=TRUE,viewer_size=c(100,100)),其中tokens是调用函数as_tokenindex()得到的数据框,…指定图形中包括的数据框列名标签,如plot_tree(tokens,token)指图形中呈现数据框tokens中包括的形符($token)。变元sentence_i=1表示函数默认绘制第一个句子中各个句法成分依存关系的树形图,doc_id用于设置文档名。变元pdf_file用于指

定直接保存 PDF 格式文件的名称。变元 all_lower=FALSE 指函数默认不改变形符的大小写。若 all_lower=TURE,则所有形符转化为小写。变元 textsize=1 是函数默认的文本尺寸,use_color=TRUE 指函数默认彩色绘图。变元 max_curve=0.3 是函数允许的曲线最大弧度。变元 palette=grDevices::terrain.colors 指函数默认调用 grDevices 数据包中的色系。用户可以自定义颜色,如 grDevices::rainbow。变元 rel_on_edge=FALSE 指函数默认把表示关系的标签置于节点上方。变元 pdf_viewer=FALSE 表示函数默认不阅览 PDF 格式的树形图,viewer_mode=TRUE 表示函数默认以.html 格式呈现图形,viewer_size=c(100,100)则指函数默认的视图宽度和高度。绘制文本 text 中句子各个成分之间依存关系树形图的 R 命令和执行结果如下:

```
> text_anno<-udpipe(text,ud_en)
> text_anno[1,"dep_rel"]<-"advmod"
> text_anno[1,"upos"]<-"ADV"
> library(rsyntax)
> tokens<-as_tokenindex(text_anno)
> plot_tree(tokens, token, upos, pdf_file="tree.pdf",
    palette = grDevices::rainbow)
Document: doc1
Sentence: 1
```

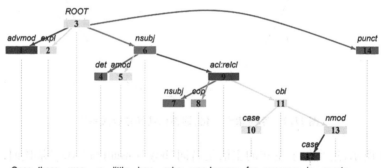

图 11.2　句子成分依存关系树形图

图 11.2 保存在 R 当前工作簿中。图 11.2 显示句子中各个成分的词性和依存关系。在这个句子中,"Once there was..."中的"was"是一个根节点(ROOT),编号为 3;"there"是填充词(expl),编号为 2。在句子中,根节点是主要动词(main verb)。如果对短语而非完整的句子开展依存分析,那么根节点是短语的中心词(head),如名词短语"a little boy"的根节点是"boy"。在与根节点的关系中,"boy"是名词,做主语(nsubj),编号为 6。在主干线上,编号为 9 的"hungry"充当关系从句修饰语,编号为 11 和 13 的"success"和"sports"分别作间接名词(obl)和名词性修饰语(nmod)。

11.2 文本中的短语提取

11.2.1 短语提取函数和词性标注缩略符

数据包 udpipe 的主要优势之一是词性标注。根据词性标注,我们能够方便地从文本中提取所需要的句法结构或短语,如名词短语和动词短语,把第七章讨论的搭配提取向前推进了一步,不再限于两个词的搭配提取。短语结构提取调用 udpipe 包中的函数 phrases(x, term = x, pattern, is_regex = FALSE, sep = " ", ngram_max = 8, detailed = TRUE),其中 x 是词性标注字符串向量,term = x 是与 x 中的标注向量对应的形符向量,pattern 是匹配模式,或为字符向量,或为正则表达式。变元 is_regex = FALSE 表示函数默认匹配模式不是正则表达式。若匹配模式是正则表达式,则应设置 is_regex = TRUE。变元 sep 用于设置短语之间的分隔符,默认设置为空格(" ")。变元 ngram_max 用于设置短语的最大长度,即短语包含的最大词语数,函数默认最大长度为 8 个词。变元 detailed 为逻辑符,若为 TRUE,函数返还短语的确切位置;若为 FALSE,函数返还短语出现的频次。

为了方便使用正则表达式提取文本中的结构模式,通用词性标注和英语词性标注均改为只有一个字母的缩略形式,如表 11.4 所示。

表 11.4 词性标注缩略符

编号	通用词性标注	缩写	编号	英语词性标注	缩写	编号	英语词性标注	缩写
1	ADJ	A	1	CC	C	19	PRP$	O
2	ADP	P	2	CD	A	20	RB	M
3	ADV	M	3	DT	D	21	RBR	M
4	AUX	V	4	EX	O	22	RBS	M
5	CCONJ	C	5	FW	N	23	RP	M
6	DET	D	6	IN	P	24	SYM	O
7	INTJ	O	7	JJ	A	25	TO	P
8	NOUN	N	8	JJR	A	26	UH	O
9	NUM	A	9	JJS	A	27	VB	V
10	PART	M	10	LS	O	28	VBD	V
11	PRON	N	11	MD	M	29	VBG	V
12	PROPN	N	12	NN	N	30	VBN	V
13	PUNCT	O	13	NNP	N	31	VBP	V
14	SCONJ	C	14	NNPS	N	32	VBZ	V
15	SYM	O	15	NNS	N	33	WDT	O
16	VERB	V	16	PDT	D	34	WP	O
17	X	O	17	POS	O	35	WP$	O
			18	PRP	O	36	WRB	O

11.2.2 名词短语的提取

本节以第二章2.6节提到的听后复述故事为例,介绍如何从故事文本中提取名词短语。这个故事以纯文本的形式(.txt)保存在 D:\Rpackage 文件夹中,文件名为 story.txt。

从文本中提取名词短语包括三个步骤。首先,调用 readtext 包中的函数 readtext 把文本读入 R。然后,加载 udpipe 英语模型,并调用函数 udpipe()对文本字符串开展形符化、词性标注和依存关系标注或赋码。最后,调用函数 as_phrasemachine()把文本中的词性标注转化为缩略符,利用函数 tolower()把形符转换为小写,再调用函数 phrases()从形符向量中提取名词短语。函数 as_phrasemachine()的结构是 as_phrasemachine(x, type = c("upos", "penn-treebank")),其中 x 是词性标注字符向量,type 用于设置词性标注类型,两个选项是通用词性标注("upos")和宾夕法尼亚大学树库("penn-treebank")。匹配名词短语的一个正则表达式为"(A|N)+N(P+D*(A|N)*N)*"。在函数 phrases()的变元设置中,重新设置 is_regex = TRUE, ngram_max = 4 和 detailed = FALSE。在这个例子中,我们分别利用通用词性标注和英语词性标注提取名词短语,R 命令和执行结果如下:

```
> #利用通用词性标注
> require(readtext)
> story<-readtext("D:/Rpackage/story.txt") $ text
> require(udpipe)
> ud_en<-udpipe_load_model(file =
"D:/Rpackage/udpipe.models.ud.2.5/english-ewt-ud-2.5-191206.udpipe")
> story_ud<-udpipe(story,ud_en)
> story_ud $ phrase_tag<-as_phrasemachine(story_ud $ upos,
   type = "upos")
> np_upos<-phrases(story_ud $ phrase_tag,
   term = tolower(story_ud $ token), pattern = "(A|N)+N(P+D*(A|N)*N)*", is_regex =
TRUE, ngram_max = 4, detailed = FALSE)
> np_upos
         keyword    ngram   freq
1       little boy      2      9
2    wise old man      3      4
3         old man      2      4
4        wise man      2      4
5     frail granny      2      2
6        blind man      2      2
7         race you      2      2
8    little boy who      3      1
9        boy who      2      1
10        one day      2      1
```

	keyword	ngram	freq
11	his village	2	1
12	his competitors	2	1
13	two other young boys	4	1
14	other young boys	3	1
15	young boys	2	1
16	large crowd	2	1
17	sporting spectacle	2	1
18	two new competitors	3	1
19	new competitors	2	1
20	only one	2	1
21	only one who	3	1
22	one who	2	1
23	his arms	2	1
24	his arms in delight	4	1
25	my success	2	1
26	finishing line	2	1
27	same time	2	1
28	his head	2	1
29	other race	2	1
30	other race you	3	1

```
> # 利用英语词性标注
> story_ud $ phrase_tag2<-as_phrasemachine( story_ud $ xpos,
type = "penn-treebank" )
> np_xpos<-phrases( story_ud $ phrase_tag2,
term = tolower( story_ud $ token), pattern = "(A|N)+N(P+D*(A|N)*N)*",is_regex = TRUE,
ngram_max = 4, detailed = FALSE)
> np_xpos
```

	keyword	ngram	freq
1	little boy	2	9
2	wise old man	3	4
3	old man	2	4
4	wise man	2	4
5	frail granny	2	2
6	blind man	2	2
7	one day	2	1
8	two other young boys	4	1
9	other young boys	3	1
10	young boys	2	1
11	large crowd	2	1
12	sporting spectacle	2	1
13	two new competitors	3	1
14	new competitors	2	1
15	only one	2	1
16	finishing line	2	1
17	same time	2	1
18	other race	2	1

以上结果显示,使用 upos 标注比使用 xpos 标注多提取了 12 个名词短语。究其原因,这主要是词性赋码的不同缩略形式造成的。例如,在对 upos 标注使用缩略形式时,同名词(NOUN)一样,人称代词或物主代词(PRON)、疑问代词或连接代词(如 who)使用同一个缩略符 N,因而"race you""boy who"和"his village"均被作为名词短语提取出来。

11.2.3 动词短语的提取

从调用函数 udpipe() 得到的词性标注中提取形符中的动词短语与从词性标注中提取名词短语在方法上相同,只是提取动词短语使用的正则表达式要复杂一些。动词短语不仅包括主谓结构和系表结构,还包括动宾结构、动词+副词结构、副词+动词结构、动词不定式和动名词短语等。针对文本 story.txt 这个例子,动词短语的匹配模式为:

```
"(((A|N)*N(P+D*(A|N)*N)*P*(M|V)*V(M|V)*I(M|V)*V(M|V)*D*(A|N)*N
(P+D*(A|N)*N)*I(M|V)*V(M|V)*(P+D*(A|N)*N)+I(A|N)*N(P+D*(A|N)*
N)*P*((M|V)*V(M|V)*D*(A|N)*N(P+D*(A|N)*N)*I(M|V)*V(M|V)*(P+D*
(A|N)*N)+))"
```

本例提取动词短语的 R 命令和执行结果如下:

```
> require(readtext)
> story<-readtext("D:/Rpackage/story.txt") $ text
> require(udpipe)
> ud_en<-udpipe_load_model(file=
"D:/Rpackage/udpipe.models.ud.2.5/english-ewt-ud-2.5-191206.udpipe")
> story_ud<-udpipe(story,ud_en)
> pattern<-
"(((A|N)*N(P+D*(A|N)*N)*P*(M|V)*V(M|V)*I(M|V)*V(M|V)*D*(A|N)*N
(P+D*(A|N)*N)*I(M|V)*V(M|V)*(P+D*(A|N)*N)+I(A|N)*N(P+D*(A|N)*
N)*P*((M|V)*V(M|V)*D*(A|N)*N(P+D*(A|N)*N)*I(M|V)*V(M|V)*(P+D*
(A|N)*N)+))"
> story_ud $ phrase_tag<-as_phrasemachine(story_ud $ upos,
type="upos")
> vp_upos<-phrases(story_ud $ phrase_tag,
term=tolower(story_ud $ token),pattern=pattern,is_regex=TRUE,ngram_max=4,detailed=FALSE)
> vp_upos[1:30,]
            keyword     ngram   freq
1           boy was      2       2
2       crowd cheered    2       2
3      little boy felt   3       2
4           boy felt     2       2
```

5	there was	2	1
6	was a little boy	4	1
7	little boy who was	4	1
8	boy who was	3	1
9	who was	2	1
10	winning was everything	3	1
11	was everything	2	1
12	success was	2	1
13	success was measured	3	1
14	was measured by results	4	1
15	measured by results	3	1
16	boy took	2	1
17	boy took part	3	1
18	took part	2	1
19	his competitors were	3	1
20	competitors were	2	1
21	large crowd gathered	3	1
22	large crowd gathered to	4	1
23	crowd gathered	2	1
24	crowd gathered to	3	1
25	crowd gathered to watch	4	1
26	to watch this sporting	4	1
27	watch this sporting	3	1
28	watch this sporting spectacle	4	1
29	hearing about the boy	4	1
30	also came from afar	4	1

文本 story.txt 中包括四词和四词以下的动词短语为 111 个,以上结果只显示前 30 个动词短语。由于文本较短,由两个词、三个词和四个词构成的动词短语重复频次低,出现频次最高的四个短语"boy was""crowd cheered""boy felt"和"little boy felt"实际在文本中只出现两次,其他短语均出现一次。

11.3 句法分析

前面两节介绍了 R 数据包 udpipe 对句子成分词性和依存关系的赋码或标注,并且举例说明如何从文本的形符对象中提取名词短语和动词短语。本节介绍如何使用斯坦福自然语言处理小组(The Stanford Natural Language Processing Group)开发的句法分析软件 Stanford Parser version 4.2.0(网址:https://nlp.stanford.edu/software/lex-parser.html)和树形关系模式匹配软件 Tregex version 4.2.0(网址:https://nlp.stanford.edu/software/tregex.html)开展句法分析,使用的例子为以下短句:"Once there was a little boy who was hungry for success in sports. Winning was everything and success was measured by results."我们的目的是利用上述两个软件对例子开展句法分析,并计算子句数。

我们先简要介绍树形关系的节点和模式。表 11.5 概括了树形结构关系中的主要节点。

表 11.5 斯坦福树形结构的节点

编号	标注	描述
1	ADJP	adjective phrase,形容词短语
2	ADVP	adverb phrase,副词短语
3	CC	coordinating conjunction,并列连词
4	CD	cardinal number,基数
5	CONJP	conjunction phrase,连词短语,如 as well as
6	DT	determiner,限定词
7	EX	existential there,表示存在的 there
8	FRAG	fragment,句子片段,如 Stop, uh.
9	FW	foreign word,外来语
10	IN	preposition or subordinating conjunction,介词或从属连词,如连词 because
11	INTJ	interjection,感叹词
12	JJ	adjective,形容词
13	JJR	adjective, comparative,形容词,比较级
14	JJS	adjective, superlative,形容词,最高级
15	LS	list item marker,计量词
16	MD	modal,情态动词
17	NN	noun, singular or mass,单数名词或物质名词
18	NNS	noun, plural,名词,复数
19	NNP	proper noun, singular,专有名词,单数
20	NNPS	proper noun, plural,专有名词,复数,如 Americans
21	NP	noun phrase,名词短语
22	PDT	predeterminer,前置限定词
23	POS	possessive ending,所有格结尾
24	PP	prepositional phrase,介词短语
25	PRP	personal pronoun,人称代词
26	PRP$	possessive pronoun,代词所有格

(续表)

编号	标注	描述
27	PRT	particle,小品词,如 took out 中的 out
28	RB	adverb,副词
29	RBR	adverb, comparative,副词,比较级
30	RBS	adverb, superlative,副词,最高级
31	ROOT	root,根节点
32	S	simple declarative clause,简单陈述句
33	SINV	inverted declarative sentence,倒装陈述句
34	SBAR	clause introduced by a subordinating conjunction,从属连词引导的子句
35	SBARQ	direct question introduced by a wh-word or a wh-phrase,由 wh-词或短语引导的直接问句
36	SQ	inverted yes/no question, or main clause of a wh-question, following the wh-phrase in SBARQ,一般疑问句,或 wh-问句(特殊问句)的主句,在 SBARQ 的 wh-短语之后
37	TO	to,动词不定式标记
38	UH	interjection,感叹词
39	VB	verb, base form,动词,原形
40	VBD	verb, past tense,动词,过去时
41	VBG	verb, gerund or present participle,动词,动名词或现在分词
42	VBP	verb, non-3rd person singular present,动词,非第三人称单数现在时
43	VBZ	verb, 3rd person singular present,动词,第三人称单数现在时
44	VBN	verb, past participle,动词,过去分词
45	VP	verb phrase,动词短语
46	WDT	wh-determiner,wh-限定词
47	WHADJP	wh-adjective phrase,wh-形容词短语,如 how hot
48	WHADVP	wh-adverb phrase,wh-副词短语,如 why 短语
49	WHNP	wh-noun phrase,wh-名词短语,如 which book
50	WP	wh-pronoun,wh-代词,如 what, who 和 whom
51	WP$	possessive wh-pronoun,wh-代词所有格
52	WRB	wh-adverb,wh-副词,如 how
53	X	unknown or uncertain,未知或不确定

表达式 Tregex 用于匹配树节点结构模式,节点之间的关系及符号表达如表 11.6 所示(详见 https://nlp.stanford.edu/~manning/courses/ling289/Tregex.html)。

表 11.6 斯坦福树形结构中的节点关系

编号	符号	说明	编号	符号	说明
1	A<<B	A 支配 B	20	A<iB	B 是 A 的第 i 个子女
2	A>>B	A 由 B 支配	21	A>iB	A 是 B 的第 i 个子女
3	A<B	A 直接支配 B	22	A<:B	B 是 A 的独生子女
4	A>B	A 直接由 B 支配	23	A>:B	A 是 B 的独生子女
5	A $ B	A 是 B 的姊妹	24	A<<:B	A 通过一个不间断的单元局部树链支配 B
6	A..B	A 在 B 之前	25	A>>:B	A 通过一个不间断的单元局部树链由 B 支配
7	A.B	A 直接在 B 之前	26	A $ ++B	A 是 B 左边的姊妹
8	A,,B	A 在 B 之后	27	A $ --B	A 是 B 右边的姊妹
9	A,B	A 直接在 B 之后	28	A $ +B	A 是 B 左边直接的姊妹
10	A<<,B	B 是 A 最左边的子嗣	29	A $ -B	A 是 B 右边直接的姊妹
11	A<<-B	B 是 A 最右边的子嗣	30	A $ .. B	A 是 B 的姊妹,且位于 B 之前
12	A>>,B	A 是 B 最左边的子嗣	31	A $,,B	A 是 B 的姊妹,且跟随 B
13	A>>-B	A 是 B 最右边的子嗣	32	A $. B	A 是 B 的姊妹,且直接位于 B 之前
14	A<,B	B 是 A 的第一个子女	33	A $,B	A 是 B 的姊妹,且直接位于 B 之后
15	A>,B	A 是 B 的第一个子女	34	A<<#B	B 是短语 A 的中心词
16	A<-B	B 是 A 的最后一个子女	35	A>>#B	A 是短语 B 的中心词
17	A>-B	A 是 B 的最后一个子女	36	A<#B	B 是短语 A 的直接中心词
18	A<'B	B 是 A 的最后一个子女	37	A>#B	A 是短语 B 的直接中心词
19	A>'B	A 是 B 的最后一个子女	38	!	否

在本节的例子中,子句匹配的树形正则表达式为:"S|SINV|SQ<(VP<#MD|VBD|VBP|VBZ)"。也可以使用以下表达式:"S|SINV|SQ[>ROOT<,(VP<#VB)|<#MD|VBZ|VBP|VBD|<(VP[<#MD|VBP|VBZ|VBD]<CC<(VP<#MD|VBP|VBZ|VBD)])]"。在关系链中,所有的关系都与链条的第一个节点相连。例如,"S<VP<NP"表示一个简单陈述句既直接支配一个动词短语,又直接支配一个名词短语。如果要表示一个简单陈述句直接支配一个动词短语,动词短语又直接支配一个名词短语,那么使用模式"S<(VP<NP)"。作为常用操作符,|和&分别表示或者与和之意,中括号[ ]表示对节点关系分组。例如,"NP[<NN|<NNS]&>S"匹配一个名词短语,名词短语支配一个单数(或物质)名词或支配一个复数名词,同时又受一个简单陈述句支配。子句匹配的第一个树形正则表达式的意思是,一个子句(简单陈述句、倒装陈述句、一般疑问句或特殊疑问句的主句)支配一个动词短语,而动词短语的直接中心词是情态动词、动词过去时、动词第三人称或非第三人称单数现在时。第二个树形正则表达式的意思是,一个子句(简单陈述句、倒装陈述句、一般疑问句或特殊疑问句的主句)受根节点支配;或者它的第一个子女是动词短语,而动词短语的直接中心词是

动词原形;或者它的直接中心词是情态动词、动词第三人称或非第三人称单数现在时或过去时;或者它直接支配一个动词短语,而动词短语的直接中心词是情态动词、动词第三人称或非第三人称单数现在时或过去时;或者动词短语直接支配并列连词;或者动词短语直接支配另一个动词短语,而另一个动词短语的中心词是情态动词、动词第三人称或非第三人称单数现在时或过去时。

掌握了树形结构和节点关系之后,我们要开展句法分析,下载软件包 stanford-parser-4.2.0.zip 和 stanford-tregex-4.2.0.zip,下载后解压,把解压后的文件分别放在以下两个文件夹中:"D:\Rpackage\stanford-parser-full-2020-11-17" 和 "D:\Rpackage\stanford-tregex-2020-11-17"。

要对例句开展句法分析,就要把例句放在文本文件中,取文件名为 example.txt,并把它放在 stanford-parser-full-2020-11-17 自带的文件夹 data 中,文件查找路径为:D:\Rpackage\stanford-parser-full-2020-11-17\data。在文件路径 D:\Rpackage\stanford-parser-full-2020-11-17 中查找文件 stanford-parser.jar。打开该文件,进入句法分析界面,先点击"Load File"("加载文件"),再点击"Browse"("浏览")查找上述文件 example.txt,最后点击"Open"("打开")结束。在界面菜单栏中点击"Load Parser"("加载分析器"),找到文件 stanford-parser-4.2.0-models.jar,点击"打开",在表格中选择文件 edu/stanford/nlp/models/lexparser/englishPCFG.caseless.ser.gz,点击"Okay"结束。以上操作结果如图 11.3 所示。

图 11.3 句法分析界面

接下来,先点击句法分析界面中的菜单"Parse"("句法分析"),再点击"Save Output"("保存输出"),把分析结果保存在 D:\Rpackage\stanfordParser2020\stanford-tregex-2020-11-17\examples\ 文件夹中,文件名仍为 example,关闭句法分析窗口。

最后,利用树形标注匹配软件 Tregex 提取文本中的子句数。在路径 D:\Rpackage\

stanford-tregex-2020-11-17\中查找文件 stanford-tregex.jar,点击打开该文件,进入 Tregex 匹配窗口。点击窗口菜单"File"("文件"),找到被树形标注过的文件 example。加载成功后可以在左上方"Tree Files"("树形标注文件")面板里看到树形文件名 example。在中间"Pattern"("模式")栏中输入子句匹配模式：S|SINV|SQ<(VP<#MD|VBD|VBP|VBZ)。点击"Search"("搜索")按钮,再点击"Statistics"("统计结果")按钮,可以看到例句中的模式匹配结果。操作结果如图 11.4 和图 11.5 所示。匹配结果显示,文本 text.txt 包括两个句子,每个句子均包括两个子句,第一个句子由一个主句和一个关系从句组成,第二个句子由两个并列句组成。

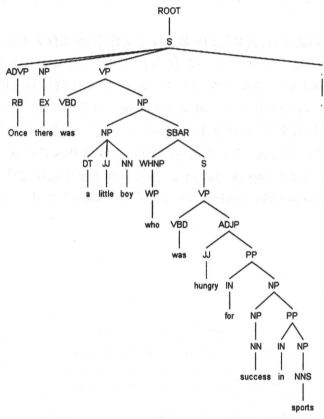

图 11.4　斯坦福句法分析树形图(一)

我们在本章第一节对句子"Once there was a little boy who was hungry for success in sports."绘制了 udpipe 依存关系树形图,与本节斯坦福句法分析树形图形成对照。斯坦福句法分析树是短语结构树,既体现单词和短语(如 NP,名词短语)之间的关系,又体现短语之间的关系。udpipe 依存关系树只体现单词之间的依存关系。同 udpipe 树形图相比,斯坦福树形图也使用根节点,但是根节点位置不同;斯坦福树形图分析包括多个层次,如使用 NP(名词短语)和 VP(动词短语)。udpipe 用 acl:relcl 显示关系从句,而斯坦福树形图用 SBAR(从属连词引导的子句),且分层细致。句法成分关系标注比较如表 11.7 所示。

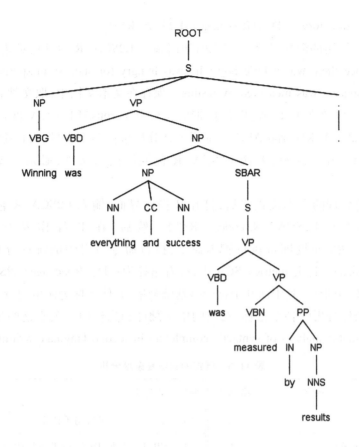

图 11.5 斯坦福句法分析树形图(二)

表 11.7 udpipe 与斯坦福句法关系标注比较

	Once	there	was	a	little	boy	who	was	hungry	for	success	in	sports	.
udpipe	mark	expl	root	det	amod	nsubj	nsubj	cop	acl:relcl	case	obl	case	nmod	punct
斯坦福	RB	EX	VBD	DT	JJ	NN	WP	VBD	JJ	IN	NN	IN	NNS	

表 11.7 显示,udpipe 与斯坦福句法关系标注差异甚大,分属不同的标注体系。对照表 11.2 可以发现,斯坦福句法关系标注系统使用了宾夕法尼亚大学树库的句法成分标注体系(xpos),但是用树形结构显示句法成分之间的关系。

利用斯坦福树形关系分析与模式匹配软件开发出的二语句法复杂度分析 R 数据包是 L2SCA_R。该数据包由凯雷和巴利尔开发(下载网址:https://sites.psu.edu/xxll3/lzsca/)(Lu,2010),把下载的数据包 L2SCA_R.zip 解压,并将解压文件放在 D:\Rpackage\L2SCA_R\文件夹中(Gaillat et al.,2019)。要正确使用本软件,需在操作系统中安装 Java 软件,下载网址为:https://www.java.com/en/download/。另外,用记事本打开句法分析脚本 L2SCA.R(路径为:D:\Rpackage\L2SCA_R\codes_R\L2SCA.R),把项目文件夹路径改为：project_directory = " D:\\Rpackage\\L2SCA_R\\"。

我们以上面使用的简短英文为例说明如何调用 L2SCA_R 包开展英语句法复杂度分析。把文字"Once there was a little boy who was hungry for success in sports. Winning was everything and success was measured by results."保存在文本文档中，取文件名 example. txt，然后把保存的文件放在 L2SCA_R 包自带的文件夹 corpusALE 中，文件存放的路径为：D:\Rpackage\L2SCA_R\corpusALE\。如果统计分析包括多个文件，可以把它们一起放在文件夹 corpusALE 中。注意，要使 L2SCA_R 包能够读入文件，必须先安装 R 数据包 readtext。

在放置好待处理的文本文档之后，打开 R 操作界面，输入 L2SCA_R 包的脚本路径：setwd("D:/Rpackage/L2SCA _ R/codes _ R/")。然后，在 R 操作界面输入：source("L2SCA. R")，按回车键即可执行脚本命令，执行命令后得到的统计分析结果保存在 L2SCA_R 包自带的文件夹 metrics_SCA 中，查看路径为：D:\Rpackage\L2SCA_R\。统计分析结果包括 23 个指标，其中前 9 个指标是基础测量，其他指标为派生测量。表 11.8 给出了文本句法复杂度分析的各个指标、对应的中文翻译和指标值。关于这些指标的定义，读者可参看"Automatic analysis of syntactic complexity in second language writing"（Lu,2010）。

表 11.8 例句的句法复杂度分析

编号	指标名	译名	指标值	编号	指标名	译名	指标值
1	W	词数	22	13	C/S	子句与句子比率	2
2	S	句子数	2	14	VP/T	动词短语与 T 单位比率	2
3	VP	动词短语数	4	15	C/T	子句与 T 单位比率	2
4	C	子句数	4	16	DC/C	从句与子句比率	0.5
5	T	T 单位数	2	17	DC/T	从句与 T 单位比率	1
6	DC	从句数	2	18	T/S	T 单位与句子比率	1
7	CT	复杂 T 单位数	2	19	CT/T	复杂 T 单位与 T 单位比率	1
8	CP	并列短语数	1	20	CP/T	并列短语与 T 单位比率	0.5
9	CN	复杂名词短语数	2	21	CP/C	并列短语与子句比率	0.25
10	MLS	平均句长	11	22	CN/T	复杂名词短语与 T 单位比率	1
11	MLT	平均 T 单位长度	11	23	CN/C	复杂名词短语与子句比率	0.5
12	MLC	平均子句长度	5.5				

在表 11.8 显示的结果中，W 代表词语数，即文本长度；S 代表句子数，匹配模式为 ROOT，即句子数等于根节点数。动词短语数用 VP 代表，匹配模式为 VP>S|SQ|SINV，其意为：匹配一个动词短语，该短语由一个简单陈述句或倒装陈述句、一般疑问句或特殊疑问句的主句支配。在本例中，该模式匹配 4 个动词短语：was a little boy, was hungry for

success,was everything 和 was measured by results。子句数用 C 代表,L2SCA_R 包的统计结果与利用软件 stanford-tregex 得到的匹配结果相同。一个 T 单位由一个主句外加一个从句或非从句组成。本例有两个句子,第一个句子只有一个 T 单位,第二个句子有两个 T 单位。L2SCA_R 包匹配 T 单位的模式为：S|SBARQ|SINV|SQ>ROOT|[$ --S|SBARQ|SINV|SQ!>>SBAR|VP],其意是：匹配一个子句(简单陈述句、wh-引导的直接问句、倒装陈述句、一般疑问句或 wh-问句的主句),它由一个根节点支配,或者它是另一个子句(简单陈述句、wh-引导的直接问句、倒装陈述句、一般疑问句或 wh-问句的主句)右边的姊妹,但是它不受从属连词引导的子句或动词短语的支配(Lu,2010)[482]。不过,这个表达式没有把第二个由"and"连接的两个并列句识别出来。一个解决方法是在原匹配模式后增加 |：(SBAR<<,VP),即新模式为：S|SBARQ|SINV|SQ>ROOT|[$ --S|SBARQ|SINV|SQ!>>SBAR|VP]|:(SBAR<<,VP)。增加的表达式旨在匹配并列句。

从句(DC)的匹配模式为"SBAR<(S|SINV|SQ[>ROOT<,(VP<#VB)|<#MD|VBZ|VBP|VBD|<(VP[<#MD|VBP|VBZ|VBD|<CC<(VP<#MD|VBP|VBZ|VBD)])])"。该模式匹配文本中的句子得到两个从句,而实际上只有一个从句(即"who was hungry for success in sports.")。因此,L2SCA_R 包中 T 单位和从句匹配存在的问题是如何区分由连词"and"引导的并列句和从属句。一个解决方法是在原匹配模式后增加选项：&! <<,VP,其意为：从属连词引导的子句最左边的子嗣不是动词短语。利用匹配模式 SBAR<(S|SINV|SQ[>ROOT<,(VP<#VB)|<#MD|VBZ|VBP|VBD|<(VP[<#MD|VBP|VBZ|VBD|<CC<(VP<#MD|VBP|VBZ|VBD)])])&! <<,VP 得到一个从句。

陆小飞提出复杂 T 单位指标(Complex T-unit,CT)是为了补充测量包括从句的 T 单位。这里的复杂 T 单位指包含一个从句的 T 单位(Lu,2010)。在前面对 T 单位匹配表达式的修改中,我们已经区分了并列句和从句,并把从句包含在了 T 单位测量中。陆小飞给出的复杂 T 单位匹配模式是：S|SBARQ|SINV|SQ[>ROOT|[$ --S|SBARQ|SINV|SQ!>>SBAR|VP]]<<(SBAR<(S|SQ|SINV<(VP<#MD|VBP|VBZ|VBD))),其意为：匹配一个子句(简单陈述句、倒装陈述句、wh-引导的直接问句、一般疑问句或 wh-问句的主句),它受一个根节点支配,或者它是另一个子句(简单陈述句、倒装陈述句、wh-引导的直接问句、一般疑问句或 wh-问句的主句)右边的姊妹,但是它不受从属连词引导的子句或动词短语的支配,或者它支配一个从属连词引导的子句,这个子句支配另一个子句(简单陈述句、倒装陈述句、一般疑问句或 wh-问句的主句),这个被支配的子句支配一个动词短语,而动词短语的直接中心词是情态动词、动词第三人称或非第三人称单数现在时或过去时。由这个表达式从本例文本中得到两个复杂 T 单位："Once there was a little boy who was hungry for success in sports."和"Winning was everything and success was measured by results."。如果我们只考虑包含从句的 T 单位数,可以使用以下匹配模式：S|SBARQ|SINV|SQ[>ROOT|[$ --S|SBARQ|SINV|SQ!>>SBAR|VP]]<<(SBAR<<,WHADVP|WHNP)。针对本例,使用修改的匹配模式得到一个复杂 T 单位。这个匹配模式是否适合更广泛的情形,尚需验证。

并列短语(CP)的匹配模式是 ADJP|ADVP|NP|VP<CC,其意为：匹配一个形容词短语、副词短语、名词短语或动词短语，这个短语直接支配并列连词。本例包含的一个并列短语是"everything and success"。

陆小飞利用三个模式匹配复杂名词短语(CN)：(1)NP!>NP[<<JJ|POS|PP|S|VBG|<<(NP$++NP!$+CC)];(2)SBAR[$+VP|>VP]&[<#WHNP|<#(IN<That|that|For|for)|<,S];(3)S<(VP<#VBG|TO)$+VP(Lu,2010)。第一个模式旨在匹配名词+形容词、名词+所有格、名词+介词短语、名词+关系从句、名词+分词或名词+同位语。该模式的含义是：匹配一个名词短语，该短语不受另一个名词短语的支配，但是支配一个形容词、所有格、介词短语、关系从句、动名词或现在分词，或者支配另一个名词短语，受支配的短语是另一个名词短语左边的姊妹，但不是并列连词左边直接的姊妹。针对本例，有两个匹配，每个句子各有一个匹配。第二个模式查找一个名词性从句，匹配一个从属连词引导的在主语或宾语位置的子句，该从句是动词短语左边直接的姊妹或由一个动词短语直接支配；从句的直接中心词是wh-名词短语或是补足语（如that引导的从句），或者它的第一个子女是简单陈述句。本例没有发现可匹配的模式。第三个模式的目的是匹配在主语位置的动名词或动词不定式，模式的含义是：匹配一个简单陈述句，陈述句直接支配一个动词短语，动词短语的中心词是动名词或不定式"to"，且陈述句是动词短语左边直接的姊妹。针对本例，使用这个模式没有发现匹配。我们使用一个不同的匹配模式：S<(VP,(NP<VBG))|<(VP,(S<(VP<TO))),其意为：匹配一个简单陈述句，陈述句直接支配一个动词短语，动词短语紧跟名词短语之后，名词短语直接支配一个动名词或现在分词，或者陈述句直接支配一个动词短语，动词短语紧跟另一个简单陈述句之后，这个陈述句直接支配另一个动词短语，动词短语又直接支配一个带"to"的动词不定式。使用调整的匹配模式,本例匹配词是"Winning"。如果使用调整的树形正则表达式匹配"Winning is everything. To succeed becomes a goal."这两个句子，那么"Winning"和"To succeed"均被匹配。若使用原表达式，则只匹配到"To succeed"。表11.8中的其他测量指标是比率测量，如在本例中，子句数是4,句子数是2,因而子句与句子比率(C/S)为2。

11.4 词语共现

第七章讨论了搭配,即前后相连的两个词之间的共现关系(co-occurrence)。我们在第七章调用了R数据包quanteda.textstats中的函数textstat_collocations()计算两个词之间搭配的频次和强度。当然,词语共现不限于前后出现的词配对,可以拓展到一个更大的语境,如句子、短文或同一个文档。数据包udpipe提供的函数cooccurrence()能够在句子、段落甚至文档等不同长度的语境中考察词语共现,也能够按照某种句法模式(如"形容词"+"名词")探究词语共现的特点,而且还能够设置词语之间的最大距离考察共现关系。该函数的主要结构是cooccurrence(x,order=TRUE,relevant,skipgram=0,order=TRUE,group,term),其中x是包括组别(group)和词项(term)的数据框或字符向量,

order=TRUE 指函数默认共现词频由高到低排序。变元 relevant 仅用于字符向量中的词语共现计算,可以设置共现计算不予考虑的停用词或选择特定句法结构模式(如形容词和名词的共现模式)。当 x 是字符向量时,变元 skipgram=0 指函数默认匹配前后连续的搭配(即跳字数为 0),字符距离是 1。当 x 是数据框时,变元 group 用于设置文档编号或句子编号等,是计算词语共现涉及的列向量;变元 term 用于设置数据框中的字符串列,每排包括一个词项。我们下面以第二章 2.6 节使用的复述故事文档 story.txt 为例,介绍函数 cooccurrence() 的三种基本用法。

第一种用法是利用词条(lemma)和词性(pos)检索在跳字数为 4 时文本词汇的共现情况。共现词语的词性包括形容词和名词,在"upos"中的标注符号分别为"ADJ"和"NOUN"。在这个例子中,我们利用数据框中的字符串向量"词条",通过变元 relevant 设置提取模式:upos %in% c("ADJ","NOUN")。由于共现搭配较多,我们在此只考虑共现频数至少为 2 的词配对。在本例中,计算词语共现频次和绘制频次条形图的 R 命令和执行结果如下:

```
> require(readtext)
> story<-readtext("D:/Rpackage/story.txt")$text
> require(udpipe)
> ud_en<-udpipe_load_model(file =
  "D:/Rpackage/udpipe.models.ud.2.5/english-ewt-ud-2.5-191206.udpipe")
> # cooccurrence
> story_anno<-udpipe(story,ud_en)
> story_oc<-cooccurrence(story_anno$lemma,
  relevant = story_anno$upos %in% c("ADJ","NOUN"), skipgram = 4)
> word<-paste(subset(story_oc, cooc>= 2)$term1,
  subset(story_oc, cooc>= 2)$term2,sep="_")
> value<-subset(story_oc, cooc>= 2)$cooc
> freq<-data.frame(word,value)
> # barchart
> library(lattice)
> freq$word <- factor(freq$word, levels = rev(freq$word))
> barchart(word ~ value, data = freq, col = "cadetblue",
  main = "", xlab = "Frequency")
```

图 11.6 显示,共现的两个词既可以都是形容词或都是名词,又可以一个是形容词,另一个是名词。在词语之间的距离不超过 5 个词时,有 10 个词配对出现的频数至少为 2,其中出现频次最高的共现词语是"little_boy"(出现 9 次)、"wise_man"(出现 8 次)。"wise_old"以及"old_man"分别共现 5 次和 4 次,其他词配对均出现 2 次。

第二种用法是调用函数 cooccurrence() 中的变元 group 和 term 考察词语共现频次。在

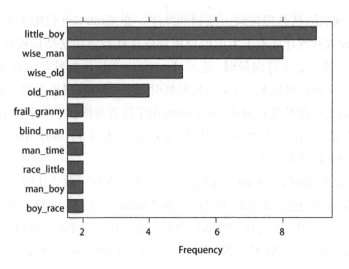

图 11.6 形容词与名词共现条形图

调用函数 cooccurrence( ) 之前，先调用基础函数 subset( ) 从数据框中筛选出要考察的句法模式，如"ADJ"和"NOUN"共现。我们把变元 term 设置为 lemma，把变元 group 设置为 sentence_id（句子编号）。针对文档 story.txt，接着上面的部分命令，计算词语共现频数和绘制词云图的 R 命令和执行结果如下：

```
> story_pattern <- subset( story_anno, upos %in% c( "ADJ" , "NOUN" ) )
> story_cooc<-cooccurrence( story_pattern, group =
    " sentence_id" , term = " lemma" )
> word<-paste( subset( story_cooc, cooc>= 2) $ term1,
    subset( story_cooc, cooc>= 2) $ term2, sep = " _ " )
> value<-subset( story_cooc, cooc>= 2) $ cooc
> freq<-data.frame( word, value)
> require( wordcloud2)
> wordcloud2( data = freq, size = 0.6, shape = " triangle" , fontFamily = " Segoe Script" )
```

这个例子没有像第一个例子那样绘制条形图，而是绘制了词云图。一个主要原因是，当词语共现的语境长度由 5 个词长扩大到句子时，共现词频不低于 2 次的词配对数由 10 个增加到 52 个。图 11.7 显示，共现频次最高的词配对是"boy_race"（12 次）和"boy_little"（10 次）。这两个词配对点明故事的主题是小男孩参加赛跑。其他 5 个频次高的词配对是"man_wise"（9 次）、"boy_man"（7 次）、"boy_other"（6 次）、"little_

图 11.7 形容词与名词共现词云图

race"(6次)和"other_race"(6次),表明故事的展开围绕小男孩和智慧老人之间的对话以及小男孩与其他选手之间的比赛。整体上看,共现词频词云图比单个词词频词云图(如第六章图6.1)不仅更能凸显故事主题,而且还能彰显词语之间的关系。

第三种用法探究某类词连续共现的特点。我们这里以名词为例,考察文档story.txt中名词连续出现的模式。在编写R代码时,调用基础函数subset()筛选udpipe数据框中的名词("NOUN"),再调用函数cooccurrence(),设置skipgram=0(这是默认设置,也可以省略)。

```
> story_N<- subset(story_anno, upos %in% c("NOUN"))
> story_N<-cooccurrence(story_N $ lemma,order=TRUE,skipgram = 0)
> word<-paste(subset(story_N, cooc>= 2) $ term1,
  subset(story_N, cooc>= 2) $ term2,sep="_")
> value<-subset(story_N, cooc>= 2) $ cooc
> freq<-data.frame(word,value)
> library(lattice)
> freq $ word <- factor(freq $ word, levels = rev(freq $ word))
> barchart(word ~ value, data = freq, col = "cadetblue", main = "", xlab = "Frequency")
```

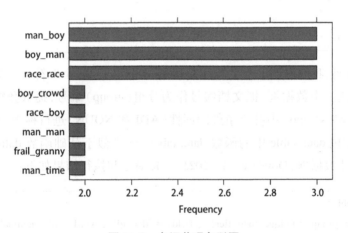

图11.8　名词共现条形图

图11.8显示,"frail_granny"是形容词与名词之间的搭配,udpipe包将它错误地标注为两个名词之间的共现。除此之外,其他7个名词配对被正确地识别,其中前三个配对"man_boy""boy_man"和"race_race"均出现3次,体现出故事的主要人物("boy"和"man")和事件("race"),其他出现2次的配对均围绕主要人物和事件。

11.5　快速自动关键词提取

11.5.1　快速自动关键词提取程序

快速自动关键词提取(Rapid Automatic Keyword Extraction,RAKE)技术由斯图亚特等

人提出(Stuart et al.,2010)。这里的关键词指连续出现的词语序列。RAKE 提取的关键词包括单个的词,但是更重视对短语的有效提取。这一算法的基本思路是,先从连续的相关词序列中提取备选关键词,再计算备选关键词中每个词出现的频率(或频次)以及与其他词连续共现的频率,然后计算每个相关词的分值,计算公式为不同词共现频率(或共现度)与词频的比率。最后,计算关键词中每个词的累计分值,由此得到关键词及其 RAKE 值。为了对 RAKE 算法有个直观的了解,我们利用第八章 8.2 节介绍潜在语义空间时使用的 3 个短句说明如何利用 RAKE 算法提取文本中的名词、形容词以及短语。

第一步,调用 udpipe 数据包中的函数和语言模型对文本开展形符化和词性标注,得到一个数据框 x,列名包括 doc_id(文档编号)、paragraph_id(段落编号)、sentence_id(句子编号)、token(形符)、lemma(词条)、upos(通用词性)和 xpos(语言特有词性)等。R 代码如下:

```
require(udpipe)
ud_en<-udpipe_load_model(file =
"D:/Rpackage/udpipe.models.ud.2.5/english-ewt-ud-2.5-191206.udpipe")
text<-"Shipment of gold damaged in a fire. Delivery of silver arrived in a silver truck. Shipment of gold arrived in a truck."
x<-udpipe(text,ud_en)
```

执行以上代码,我们得到 1 个文档、1 个段落、3 个句子、25 个形符和词类标注等。

第二步,创建一个数据框,把文档编号作为分组(group)编号,提取数据框 x 中的各个词条,列名为 word,对 upos 中符合条件的词性(ADJ 和 NOUN)进行标识,列名为 relevant。我们调用 R 数据包 data.table 中的函数 data.table()(类似于基础函数 data.frame(),但是计算更方便)创建数据框(Dowle et al.,2021)。R 命令和执行结果如下:

```
> require(data.table)
> x <- data.table(group = unique_identifier(x, fields = 'doc_id'), word = x[['lemma']], relevant = x
   $ upos% in% c("ADJ","NOUN"))
> x
      group     word    relevant
1:       1    shipment     TRUE
2:       1          of    FALSE
3:       1        gold     TRUE
4:       1      damage    FALSE
5:       1          in    FALSE
6:       1           a    FALSE
7:       1        fire     TRUE
8:       1           .    FALSE
9:       1    Delivery     TRUE
```

10：	1	of	FALSE
11：	1	silver	TRUE
12：	1	arrive	FALSE
13：	1	in	FALSE
14：	1	a	FALSE
15：	1	silver	TRUE
16：	1	truck	TRUE
17：	1	.	FALSE
18：	1	shipment	TRUE
19：	1	of	FALSE
20：	1	gold	TRUE
21：	1	arrive	FALSE
22：	1	in	FALSE
23：	1	a	FALSE
24：	1	truck	TRUE
25：	1	.	FALSE
	group	word	relevant

第三步，调用 data.table 包中的函数 rleid()，新建一个列变量 keyword_id（关键词编号），编号依据 relevant 列值。该函数生成游程或运行长度（run-length）型编号，即对连续出现的相同值赋值相同。这意味着，上面数据框中 relevant 所在列的前 7 个值"TRUE""FALSE""TRUE""FALSE""FALSE""FALSE""TRUE"在新增列 keyword_id 中的值为 1，2，3，4，4，4，5。这样做的目的是要在后面提取连续出现的关键短语。然后，调用 R 基础函数 subset()，利用"否定"逻辑符"！"，筛选"keyword_id"（关键词编号）和"word"（词语）列变量值，再根据 keyword_id 值在数据表格中增加列变量 keyword，并利用函数 paste() 把连续出现的编号相同的词合并在一起。R 命令和执行结果如下：

```
> x $ keyword_id <- rleid( x[ [ "relevant" ] ] )
> x <- subset( x, relevant ! = FALSE, select = c( "keyword_id" , "word" ) )
> x <- x[ , keyword： = paste( word, collapse = ' '), by = list( keyword_id ) ]
> x[ ]
```

	keyword_id	word	keyword
1：	1	shipment	shipment
2：	3	gold	gold
3：	5	fire	fire
4：	7	Delivery	Delivery
5：	9	silver	silver
6：	11	silver	silver truck
7：	11	truck	silver truck
8：	13	shipment	shipment
9：	15	gold	gold
10：	17	truck	truck

以上结果显示,词语"shipment"和"gold"的编号(keyword_id)不同,因而它们本身被视作 keyword(关键词)。第6排和第7排的两个词"silver"和"truck"编号相同,因而构成短语"silver truck"。

第四步,计算不同词(和短语)之间共现频率或共现度(co-occurrence frequency,即 degree),在数据框 x 中新增列 degree。在本例中,新增列依据分组编号 keyword_id 计算排数(.N),共现频率等同于 x[,degree:=.N-1L,by=list(keyword_id)](.N-1L 等于排数减去1,是共现度)。R 命令和执行结果如下:

```
> x <- x[ , degree: = .N-1L, by = list(keyword_id)]
> x[ ]
    keyword_id    word      keyword     degree
1:      1       shipment   shipment       0
2:      3        gold        gold         0
3:      5        fire        fire         0
4:      7      Delivery    Delivery       0
5:      9       silver      silver        0
6:     11       silver    silver truck    1
7:     11       truck    silver truck    1
8:     13      shipment   shipment       0
9:     15        gold        gold         0
10:    17       truck       truck        0
```

第五步,新建列表 word_score(词语值),列表变量包括每个词累计共现度(degree),并调用 udpipe 包中的函数 txt_freq 计算频率统计量,得到三列变量(键变量 key、频率 freq 和频率百分比 freq_pct)。R 命令和执行结果如下:

```
> word_score <- list( )
> word_score $ degree <- x[ , list(degree = sum(degree)), by = list(word)]
> word_score $ freq <- txt_freq( x $ word)
> word_score
$ degree
       word      degree
1:   shipment      0
2:    gold         0
3:    fire         0
4:  Delivery       0
5:   silver        1
6:   truck         1
$ freq
```

	key	freq	freq_pct
1	shipment	2	20
2	gold	2	20
3	silver	2	20
4	truck	2	20
5	fire	1	10
6	Delivery	1	10

第六步，调用 data.table 数据包中的函数 merge( )，合并 word_score 中两个数据表格包含的变量，得到列变量 word、degree、freq 和 freq_pct。函数 merge( ) 的基本结构是 merge(x, y, by.x=NULL, by.y=NULL, all.x=FALSE, all.y=TRUE, sort=TRUE)，其中 x 和 y 是数据表格，by.x 和 by.y 是合并两个数据表格的列向量。本例设置 by.x="word" 和 by.y="key"。变元 all.x=TRUE 表示 x 中与表格 y 不匹配的排也包括在新表格中；all.x=FALSE 指函数默认仅包括两个数据表格都包括的数据。变元 all.y 的设置方式与 all.x 的设置方式相同。变元 sort=TRUE 指函数默认按照键变量 key 值进行排序。此外，在新数据表格 word_score 中增加列名 rake_word_score（关键词 RAKE 值）。RAKE 值等于词共现度与词频的比率。R 命令和执行结果如下：

```
> word_score <- merge(word_score $ degree, word_score $ freq, by.x = "word", by.y = "key", all.x=FALSE, all.y=TRUE)
> word_score $ rake_word_score <- word_score $ degree/word_score $ freq
> word_score
```

	word	degree	freq	freq_pct	rake_word_score
1：	Delivery	0	1	10	0.0
2：	fire	0	1	10	0.0
3：	gold	0	2	20	0.0
4：	shipment	0	2	20	0.0
5：	silver	1	2	20	0.5
6：	truck	1	2	20	0.5

第七步，从数据表格 x 中提取 keyword 和 word 所在列，新增列变量 freq（频率，即词与关键词（含短语）共现的频率）。然后，调用函数 merge( )，得到 word、keyword、freq 和 rake_word_score 列，累计计算 RAKE 值（列名为 rake），并且增加 N 元组频数列。R 命令和执行结果如下：

```
> keywords <- x[, list(freq = length(unique(keyword_id))), by = list(keyword, word)]
> keywords<-merge(keywords,
    word_score[, c("word", "rake_word_score")], by="word", all.x=FALSE, all.y=TRUE)
> keywords <- keywords[, list(ngram = .N, rake = sum(rake_word_score)), by = list(keyword, freq)]
```

```
> keywords
       keyword    freq   ngram   rake
1:     Delivery    1       1      0.0
2:         fire    1       1      0.0
3:         gold    2       1      0.0
4:     shipment    2       1      0.0
5:       silver    1       1      0.5
6: silver truck    1       2      1
7:        truck    1       1      0.5
```

第八步,根据 N 元组和最小频数从数据表格 keywords 中筛选符合条件的关键词,调用 data.table 包中的函数 setorder()根据 RAKE 值对关键词按降序排列。调用 data.table 包中的函数 setcolorder()将列变量顺序重排为:"keyword""ngram""freq"和"rake",再调用 data.table 包中的函数 setDF()把 keywords 由数据表格转化为数据框。本例设置关键词最大 N 元组为三元组,最小词频为 1。R 命令和执行结果如下:

```
> keywords <- subset(keywords, ngram <= 3 & freq >=1)
> setorder(keywords, -rake)
> keywords <- setcolorder(keywords, neworder = c("keyword", "ngram", "freq", "rake"))
> keywords <- setDF(keywords)
> keywords
       keyword    ngram   freq   rake
1  silver truck    2       1     1.0
2        silver    1       1     0.5
3         truck    1       1     0.5
4      Delivery    1       1     0.0
5          fire    1       1     0.0
6          gold    1       2     0.0
7      shipment    1       2     0.0
```

以上结果显示,以形容词和名词为词类,由 3 个短句构成的短文的关键词是"silver truck""silver"和"truck"。

11.5.2 快速自动关键词提取案例分析

上一节利用很短的文本演示了快速自动关键词提取的程序。实际研究中文本会更长。本节从 R 数据包 quanteda 自带数据库 data_corpus_inaugural 中调取 2021 年美国总统拜登的就职演说,从中提取关键词,筛选 RAKE 值大于 1 的所有关键词,提取模式为"形容词"+"名词",关键词最大 N 元组为三元组,最小词频为 2。开展本例研究调用的 R 函数是

udpipe 包中的函数 keywords_rake(x,term,group,relevant,ngram_max=2,n_min=2,sep=" ")。在这个函数中,x 是 udpipe()返还的数据框,term 是数据框 x 中的列名,通常是 lemma(词条),group 指定计算关键词的范围,如指定文档编号(doc_id)或句子编号(sentence_id)计算关键词。变元 relevant 可用于把停用词排除在关键词的计算之外,也可用于选择特定的词性提取关键词。变元 ngram_max=2 指函数默认关键词中的最大 N 元组是二元组;n_min=2 是函数默认的关键词最小频数;sep=" "指函数默认使用一个空格把关键词中的词项粘贴在一起。

利用 RAKE 从拜登的就职演说中提取关键词的 R 命令和执行结果如下:

```
> require(quanteda)
> biden<-corpus_subset(data_corpus_inaugural,
   President% in% c("Biden"))
> require(udpipe)
> ud_en<-udpipe_load_model(file =
   "D:/Rpackage/udpipe.models.ud.2.5/english-ewt-ud-2.5-191206.udpipe")
> biden_ud<-udpipe(biden,ud_en)
> keywords <- keywords_rake(x = biden_ud, term = "lemma", group = "sentence_id", relevant =
   biden_ud $ upos % in% c("ADJ","NOUN"), ngram_max = 3, n_min = 2, sep = " ")
> subset(keywords, rake > 1)
          keyword  ngram  freq     rake
1   common object      2     2  2.000000
2     sacred oath      2     2  1.666667
3      whole soul      2     3  1.600000
4  american story      2     4  1.555556
5  racial justice      2     2  1.500000
6   silent prayer      2     2  1.500000
```

根据以上 RAKE 值,最能概括拜登就职演说的 6 个关键词是"common object""sacred oath""whole soul""American story""racial justice"和"silent prayer"。这些关键词全都是"形容词+名词"结构。我们在第六章调用 R 数据包 quanteda.textstats 中的函数 textstat_keyness()计算了前 20 个关键词(见图 6.3)。除"story"一词外,短语类关键词与单词类关键词没有重复之处。我们下面利用 R 命令以"common object"为例检索一下原文语境:

```
> biden<-corpus_subset(data_corpus_inaugural,
   President% in% c("Biden"))
> biden_tks<-tokens(biden,what="word")
> kwic(biden_tks,phrase("common object *"),window=10)
   Keyword-in-context with 2 matches.
```

[2021-Biden,1556:1557]　　wrote that a people was a multitude defined by the | common objects |
[2021-Biden,1565:1566]　　the common objects of their love. What are the | common objects |
of their love. What are the common objects we
we love that define us as Americans? I think

　　拜登就职演说的一个主题是确定任期内的奋斗目标。奥古斯丁(Augustine)曾经写道："人民就是一个由共同热爱的目标所界定的大众"，拜登由此出发指出美国大众所共同热爱的目标。

第十二章　在 R 中调用 Python 开展自然语言处理

R 是专业型语言，为统计学而设计。前面各章详细介绍与讨论了 R 语言及其数据包在自然语言处理各个领域中的应用功能。当然，任何一种编程语言都有其自身优势和局限性。作为一名研究者，最好能够掌握两种或两种以上编程语言。1989 年，荷兰籍计算机程序员吉多·范·罗苏姆（Guido van Rossum）开始构建一门新的编程语言，两年后发布了 Python。Python 是通用型语言，为编程而设计。同 R 相比，Python 更具多功能性，在自然语言处理技术方面有自己独特的优势。关于 R 和 Python 两种编程语言的详细比较，读者可以参考 *Python and R for the Modern Data Scientist*（Scavetta et al., 2021）。在掌握了 R 语言之后，通过交互界面学习 Python 语言相对容易，也会使读者对两种语言有更深入的了解，在自身实践中逐步实现融会贯通，取长补短。本章介绍如何利用 R 数据包 reticulate 实现 R 与 Python 的无缝对接。主要内容包括 Python 和 Python 库的安装、R 数据包 reticulate 的安装和基础操作、利用 NLTK 库和 spaCy 库开展自然语言处理。本章使用的数据包包括：reticulate、Python-docx、PyPDF2、re、string、nltk、wordcloud、PIL、numpy、matplotlib 和 spaCy。

12.1　安装 Python 和 Python 库

本节先介绍如何安装 Python 软件，再介绍 NLTK（Natural Language Toolkit）和 spaCy 两种自然语言处理库的安装。

要在本地电脑安装 Python 语言，比较简单的方法是安装 Python 的 Anaconda 版本，因为 Anaconda 版的 Python 会自动安装一些库。登录下面的 Anaconda 公司网站：https://www.anaconda.com/，可以看到适合于 Windows 系统的下载界面（图 12.1）。界面会因软件更新而不同。

点击下载界面上的"Download"，把下载的软件放在本地硬盘的一个文件夹中，如 D:\Rpackage\。按照系统默认的路径安装 Anaconda3，下载的版本号是 Anaconda3-2022.10-Windows-x86_64，Python 当前版本号为 Python 3.9。安装好 Python 后可以根据研究需要添加 Python 库。NLTK 库是最受欢迎的自然语言处理和文本挖掘 Python 库。雅各·帕金斯还专门写了一本书介绍 NLTK 数据包（Perkins，2014）。

以 DELL 电脑上的 Windows 11 系统为例，在"开始"菜单的子菜单"所有应用"中查找文件夹"Anaconda3（64-bit）"，在其列表中找到"Anaconda Prompt（anaconda3）"选项。点击这个选项，在打开的界面中输入以下命令：pip install nltk，按回车键执行该命令。若成功

图 12.1　Python 的 Anaconda 版本下载界面

安装,则界面上会显示以下提示行:Successfully installed nltk-3.7,如图 12.2 所示。若退出安装,则点击界面右上方的"✕"号关闭窗口即可。另外,R 数据包 reticulate 提供的函数 py_install()可直接安装 Python 库。若安装 NLTK 库,则可以在 R 控制台或操作界面输入以下代码,再按回车键进行在线安装:require(reticulate);py_install("nltk")。

图 12.2　NLTK 库安装界面

spaCy 库也是广受欢迎的 Python 库,其安装方式与 NLTK 库的安装方式大致相同,只不过是利用子菜单"所有应用"中文件夹"Anaconda3(64-bit)"下的"Anaconda Powershell Prompt(anaconda3)"选项,而不是"Anaconda Prompt(anaconda3)"选项。点击这个选项,在打开的界面中输入以下命令:pip install spacy,按回车键执行该命令。若安装成功,则界面上会显示 Successfully installed blis-0.7.9 等提示。若安装失败,则建议再试一次。在安装完 spaCy 库之后,接下来安装英语统计语言模型。在"Anaconda Powershell Prompt(anaconda3)"界面输入以下命令:python -m spacy download en_core_web_sm。若安装成功,则界面上会出现以下文字:You can now load the package via spacy.load('en_core_web_sm'),提示我们可以正常调用 spaCy 的英语模型了。

12.2　R 数据包 reticulate 的安装和基础操作

在 R 操作界面或控制台调用 Python 库及其函数需要安装 R 数据包 reticulate。在 R 控制台输入 install. packages("reticulate")即可实现在线安装。要调用 Python 库及其函数,则要先调用 R 数据包 reticulate,利用数据包中的函数 use_python()指定在本地电脑安装的 Python 软件默认路径("C:/Users/DELL/anaconda3/python. exe"),再利用 R 代码 repl_python()从 R 界面过渡到 Python 界面。这里的 repl 是 Read-Evaluate-Print Loop 的缩称。界面转化的典型标志是 R 提示符">"转化为 Python 提示符">>>"。初始 R 代码如下:

```
library(reticulate)
use_python("C:/Users/DELL/anaconda3/python. exe")
repl_python()
```

下面简要介绍 Python 对字符串的基本操作。

12.2.1 文档导入

第二章介绍了在 R 控制台中读入纯文本文档、word 文档和 PDF 文档,本节则介绍在 Python 界面读入这些文档的方法。

我们在第二章第一节利用 R 函数 readLines()将纯文本文档 text1(存储路径为:D:\Rpackage\text1.txt)读入 R。Python 的文档读取方式不同。把纯文本文档读入 Python 先要调用函数 open()打开文本文档。函数的基本结构是 open(file, mode='r'),其中的 file 是文档名或文档路径,mode='r' 是默认选项,指读入方式(r 为 read 首字母)。然后,调用 Python 函数 read()把加载的文本文档 file 读出来。操作时需要在 read()前加入文档名,中间用点号(.)隔开。这是 Python 调用函数的方法。把文档 text1 读入 Python 的代码和操作结果如下:

```
> library(reticulate)
> use_python("C:/Users/DELL/anaconda3/Python. exe")
> repl_python()
Python 3.9.13 (C:/Users/DELL/anaconda3/Python. exe)
Reticulate 1.27 REPL -- A Python interpreter in R.
Enter 'exit' or 'quit' to exit the REPL and return to R.
>>> filename = "D:/Rpackage/text1. txt"
>>> file= open(filename, 'r')
>>> text1 = file. read()
>>> print(text1)
Once there was a little boy who was hungry for success in sports.
>>> text1
'Once there was a little boy who was hungry for success in sports. '
```

以上结果显示,Python 的分配符是等号(=)。R 的分配符可以用等号,但是更常用的分配符是左向箭头(<-)。要退出 R 和 Python 的交互界面,可以键入 exit 或者 quit,再按回

车键。一旦退出,则 Python 提示符"＞＞＞"转化为 R 提示符"＞"。Python 通常使用函数 print()输出结果,但是有时也可以省略。

如果要读入的文档存放在 Python 当前工作簿,那么可以在函数 open()中直接输入文件名(文件名用单引号或双引号)。查找 Python 当前工作簿可以调用函数 os.getcwd(),如工作簿 'C:\\Users\\DELL\\Documents'。把 text1 从当前工作簿读入的 Python 代码和操作结果如下:

```
>>> file = open('text1.txt', 'r')
>>> text1 = file.read()
>>> text1
'Once there was a little boy who was hungry for success in sports.'
```

把 word 文档读入 Python 界面,需要安装 Python 库 Python-docx。可以在 R 操作界面键入 library(reticulate)之后直接输入 R 命令 py_install("Python-docx")进行在线安装。调用 R 数据包的函数是 require()或 library(),而 Python 库的调用则使用 import,如 import docx。要读取 word 文档,则需从 docx 中调用 Document,即键入以下代码: from docx import Document。然后,利用函数 open()创建文档对象。要创建保存在 'D:/Rpackage/text1.docx' 文档中的 word 文档 text1(文档后缀名为.docx),则调用函数 open(),数据读入方式为 'rb'(读入二进制数据,b 是 binary 的首字母缩写),再使用函数 docx.Document(表示从 docx 库中调入函数 Document)。Python 代码如下:

```
library(reticulate)
use_python("C:/Users/DELL/anaconda3/Python.exe")
repl_python()
import docx
from docx import Document
file = 'D:/Rpackage/text1.docx'
doc = open(file,'rb')
document = docx.Document(doc)
```

要从文档 document 中提取文本,可以利用 for 循环。我们要设定一个存储字符串的空文档(text1),然后使用 for 循环遍历文档 document 中的每一段,再把段落添加到 text1 中。Python 代码和操作结果如下:

```
>>> import docx
>>> from docx import Document
>>> file = 'D:/Rpackage/text1.docx'
>>> doc = open(file,'rb')
>>> document = docx.Document(doc)
>>> text1 = ''
```

```
>>> for para in document.paragraphs:
...     text1 = para.text
...
>>> text1
'Once there was a little boy who was hungry for success in sports.'
```

以上代码成功地读入 word 文档。注意,在使用 for 循环时,在输入 document. paragraphs 后要加上英文冒号(:),按回车键进入下一行,在空 4 个空格后再输入代码: text1 =para. text。输入完 para. text 之后,按回车键,出现三个点号(...),再按回车键,处理结束,出现 Python 提示符>>>。在界面输入 text1 或 print(text1)后按回车键即可查看结果。

读取 PDF 文档的方法是调用 Python 数据包 PyPDF2。在 R 操作界面键入 library (reticulate)之后直接输入 R 命令 py_install("PyPDF2")即可实现在线安装。我们以保存在 D:\Rpackage 文件夹中的 PDF 文档 text1(文档后缀名为. pdf)为例,把该文档读入 Python 界面。首先,调用 Python 库 PyPDF2,用函数 open()打开文档,并从 PyPDF2 库中调用函数 PdfFileReader,创建一个 PDF 文档对象。Python 代码如下:

```
library(reticulate)
use_python("C:/Users/DELL/anaconda3/Python.exe")
repl_python()
import PyPDF2
from PyPDF2 import PdfFileReader
file = 'D:/Rpackage/text1.pdf'
pdf = open(file,"rb")
pdf_reader = PdfFileReader(pdf)
```

接下来,调用函数 getPage()创建一个页码(page)对象,并利用文本提取函数 extractText()从中提取文档字符串。Python 代码如下:

```
page = pdf_reader.getPage(0)
text1 = page.extractText()
```

执行以上代码,输入 print(text1)后按回车键就会看到文本字符串。读者须注意,调用的函数随数据包的更新可能会有变化。

12.2.2 字符串操作

上一节介绍如何调用 Python 库和函数把纯文本文档、word 文档和 PDF 文档以字符串的形式读入 Python 操作界面。本节介绍对字符串的基本操作。

同 R 一样,Python 也使用正则表达式处理字符串。关于正则表达式及其用法,详见第二章。调用 Python 库 re 和 string,可以开展一些基础性的字符串操作。例如,下面是调用 re

库和 string 库计算文本词数的 Python 代码：

```
> library(reticulate)
> use_python("C:/Users/DELL/anaconda3/Python.exe")
> repl_python()
Python 3.9.13 (C:/Users/DELL/anaconda3/Python.exe)
Reticulate 1.27 REPL -- A Python interpreter in R.
Enter 'exit' or 'quit' to exit the REPL and return to R.
>>> text='Whenever Mr. Smith goes to Westgate, he stays at the Grand Hotel. In spite of its name, it is really not very "grand," but it is cheap, clean, and comfortable.'
>>> import re
>>> import string
>>> words = re.split('\W',text)
>>> words = [word for word in words if word not in string.punctuation]
>>> len(words)
30
```

在以上代码中，字符串被放在单引号中。单引号和双引号的使用必须前后一致，否则会报错。例如：

```
>>> print("To be or not to be.")
To be or not to be.
>>> print("To be or not to be.')
SyntaxError: EOL while scanning string literal (〈string〉, line 1)
```

如果在字符串中有所有格符号 's 时，使用双引号比较方便。例如：

```
>>> print("Sammy's balloon is green.")
Sammy's balloon is green.
```

若使用单引号，则需要使用退出符或转义符"\"。例如：

```
>>> print('Sammy\'s balloon is green.')
Sammy's balloon is green.
```

字符串形符化使用的函数是 re.split()，正则表达式是 '\W'（任意非词）。形符化后的形符包括词符和去除标点后留下的占位符("")。要去除占位符，则调用 string 包中的函数标点(punctuation)。最后，利用函数 len() 计算文本包含的词数。如果使用正则表达式 '\s'（任意空格符）开展形符化，标点和词合在了一起，但是文本长度不变。计算文本长度使用的 Python 函数是 len()，而 R 使用的函数是 length()。

要把文本 text 中的 'Smith' 改为 'Black'，可以调用 re 库中的替换函数 sub()：

```
>>> text = 'Whenever Mr. Smith goes to Westgate, he stays at the Grand Hotel. In spite of its name, it is
really not very " grand," but it is cheap, clean, and comfortable.'
>>> import re
>>> text = re.sub('Smith','Black',text)
>>> text
'Whenever Mr. Black goes to Westgate, he stays at the Grand Hotel. In spite of its name, it is really not
very " grand," but it is cheap, clean, and comfortable.'
```

若要查找字符串中某个字符出现的次数,则先调用 re 库中的函数 re.findall( ),再利用基础函数 len( )即可得到结果。例如:

```
>>> is_freq = re.findall('is',text)
>>> len(is_freq)
2
```

若我们想把上面文本中的词全部改为小写或大写字母,则可以直接使用基础函数 lower( )或 upper( )。例如:

```
>>> text.lower()
'whenever mr. black goes to westgate, he stays at the grand hotel. in spite of its name, it is really not
very" grand," but it is cheap, clean, and comfortable.'
```

如果我们要在上面得到的列表 words 中检索某个或某些形符,可以在 words 后面添加中括号([ ]),并在其中添加索引号。注意,与 R 的索引从 1 开始到 N(形符总数)结束不同的是,Python 索引从 0 开始到 N-1 结束。起始索引和终止索引之间用冒号(:)隔开。索引号为正值表示顺向提取,负值表示逆向提取。例如,[1:5]表示索引提取从第 2 个形符开始直至第 6 个形符结束,但是不包括第 6 个形符。试比较:

```
>>> words[0]
'Whenever'
>>> words[6:]
['he', 'stays', 'at', 'the', 'Grand', 'Hotel', 'In', 'spite', 'of', 'its', 'name', 'it', 'is', 'really', 'not', 'very',
'grand', 'but', 'it', 'is', 'cheap', 'clean', 'and', 'comfortable']
>>> words[:6]
['Whenever', 'Mr', 'Smith', 'goes', 'to', 'Westgate']
>>> words[-1]
'comfortable'
>>> words[:-7]
['Whenever', 'Mr', 'Smith', 'goes', 'to', 'Westgate', 'he', 'stays', 'at', 'the', 'Grand', 'Hotel', 'In',
'spite', 'of', 'its', 'name', 'it', 'is', 'really', 'not', 'very', 'grand']
```

```
>>> words[1:6]
['Mr', 'Smith', 'goes', 'to', 'Westgate']
```

我们可以将一个字符串或列表乘一个整数,整数表示一个字符串或列表重复的次数。例如:

```
>>> words[1:6] * 2
['Mr', 'Smith', 'goes', 'to', 'Westgate', 'Mr', 'Smith', 'goes', 'to', 'Westgate']
```

如果要把列表 words 中的各个形符转化成字符串,那么可以使用函数 join()。例如:

```
>>> " ".join(words)
'Whenever Mr Smith goes to Westgate he stays at the Grand Hotel In spite of its name it is really not very grand but it is cheap clean and comfortable'
```

注意,在以上代码中,引号之间留有一个空格(" "),表示前后相接的形符以一个空格隔开。

合并两个字符串可以使用加号(+)。Python 不会自动在字符串之间留有空格。如果需要空格或标点符号,可以使用空格符(如 ' ' 空一格)或添加标点(如 ',' 添加逗号)。例如:

```
>>> s1 = 'Whenever Mr. Smith goes to Westgate'
>>> s2 = 'he stays at the Grand Hotel.'
>>> s1+' '+s2
'Whenever Mr. Smith goes to Westgate he stays at the Grand Hotel.'
>>> s1+', '+s2
'Whenever Mr. Smith goes to Westgate, he stays at the Grand Hotel.'
>>> 'Yes, '+s2
'Yes, he stays at the Grand Hotel.'
```

12.2.3 Python 函数

函数是为完成具体工作而设计的有名称的代码块(blocks of code),如 re 库中的函数 split() 和 sub()。Python 和 R 一样有两种函数。一种函数是用户自定义函数,另一种函数是内置函数,如 print() 和 len()。本节简要介绍自定义函数。

在 Python 中,我们使用 def 先定义(define)一个函数,接下来给出函数名,如 hello,再在函数名后加上括号,在括号中写上函数使用的参数(也可以没有参数),并以冒号(:)结束。例如:

```
def hello():
```

在函数初始设置完成之后,我们另起一行,缩进 4 个空格后写入函数指令,如 'hello,Python':

```
def hello():
    print('hello,Python')
```

至此,函数 hello() 已经编写完成。要调用函数,则在控制台输入 hello(),按回车键执行操作:

```
> library(reticulate)
> use_python("C:/Users/DELL/anaconda3/Python.exe")
> repl_python()
Python 3.9.13 (C:/Users/DELL/anaconda3/Python.exe)
Reticulate 1.27 REPL -- A Python interpreter in R.
Enter 'exit' or 'quit' to exit the REPL and return to R.
>>> def hello():
...     print('hello, Python')
...
>>> hello()
hello,Python
```

上面的函数比较简单,实际研究中编写的函数要复杂一些。我们在第一章举例说明如何编写 R 函数 Min 计算向量的最小值。这里我们编写如下 Python 函数 Min 计算数值列表最小值:

```
def Min(x):
    min_val = x[0]
    for i in x:
        if i < min_val:
            min_val = i
    return min_val
```

在以上代码中,我们先用 def 定义一个函数 Min(),函数的参数是数值列表 x。在 def Min(x) 后添加一个冒号,按回车键进入下一行;空 4 格设定最小值 min_val 为 x 的第一个值(x[0]),然后按回车键到下一行;空 4 格后设定 for 循环 for i in x,遍历 x 的每个值,添加冒号后再按回车键到下一行;空 4 格后用 if 陈述条件表达式 if i<min_val,添加冒号后再按回车键到下一行;空 4 格设定条件满足时的最小值 min_val=i,输入完成后再按回车键到下一行;空 4 格后用 return 返还最小值 min_val。使用 return,后面添加一个待打印的变元,函数编写就结束了。我们以第一章 1.4.3 节的 y 变量值为例,利用新编写的 Min 函数计算最小值:

```
> library(reticulate)
> use_python("C:/Users/DELL/anaconda3/Python.exe")
> repl_python()
Python 3.9.13 (C:/Users/DELL/anaconda3/Python.exe)
Reticulate 1.27 REPL -- A Python interpreter in R.
Enter 'exit' or 'quit' to exit the REPL and return to R.
>>> def Min(x):
...     min_val = x[0]
...     for i in x:
...         if i < min_val:
...             min_val = i
...     return min_val
...
>>> y = [21.2, 19.2, 21.2, 20.2, 18.6, 21.2, 21.3, 19.5, 17.9, 19.0, 21.1, 21.6, 20.1, 20.2,
19.8, 14.4, 21.9, 20.0, 21.2, 23.0, 20.0, 22.8, 21.4, 20.5, 22.5, 22.5, 19.0, 18.9, 19.7, 21.3]
>>> Min(y)
14.4
```

以上结果显示，y 数值列表的最小值为 14.4。利用 Python 内置函数 min()，在 Python 交互界面输入 min(y) 会得到同样的结果。

针对字符串处理，我们下面来编写一个函数 ltr(text)，计算字符串中字母数小于 5 的所有词语数。

```
def ltr(text):
    import re
    toks = re.sub(r'\W',' ',text)
    s_split = toks.split()
    val = 0
    for token in s_split:
        if len(token) < 5:
            val = val+1
    return val
```

在以上代码中，ltr(text) 作为新建函数，计算字符串中字母数小于 5 的词语数，其中 text 是字符串。要计算文本词符或词语数，可以调用 Python 内置函数 split()。但是，函数 split() 的形符化操作没有把词符和连续出现的标点符号分隔开，因而需要调用 re 库中的 sub()，使用正则表达式去除字符串中的标点符号。设定计算字符串中字母数小于 5 的词语数的初始值为 0(val=0)，利用 for 循环和 if 条件表达式开展计数。最后，使用代码 return val 把计算的结果打印出来。例如：

```
> library(reticulate)
> use_python("C:/Users/DELL/anaconda3/Python.exe")
> repl_python()
Python 3.9.13 (C:/Users/DELL/anaconda3/Python.exe)
Reticulate 1.27 REPL -- A Python interpreter in R.
Enter 'exit' or 'quit' to exit the REPL and return to R.
>>> def ltr(text):
...     import re
...     toks = re.sub(r'\W',' ',text)
...     s_split = toks.split()
...     val = 0
...     for token in s_split:
...         if len(token) < 5:
...             val = val+1
...     return val
...
>>> string = 'Once there was a little boy who was hungry for success in sports. To him, winning was everything and success was measured by results.'
>>> ltr(string)
14
```

以上结果显示字符串 string 中字母数小于 5 的词语数是 14 个。因为函数 ltr(text) 只有一个参数 text，所以代码 ltr(text=string) 缩写成了 ltr(string)。

12.3 利用 NLTK 库的自然语言处理

本节以 2019 年全国英语专业四级口语测试中的听力文本为例（详见第二章），调用 NLTK 库中的函数绘制包括 20 个高频词词频数变化的线图，调用 wordcloud 库中的函数绘制词云图（最大词数为 200）。

12.3.1 NLTK 数据包下载和 wordcloud 库安装

利用 NLTK 库开展自然语言处理需要下载相关数据包。如果利用 R 控制台安装所有的数据包，R 代码如下：

```
library(reticulate)
use_python("C:/Users/DELL/anaconda3/Python.exe")
repl_python()
import nltk
```

```
nltk.download("all")
```

我们也可以利用 Spyder 等解释器安装 NLTK 数据包。以 Windows 11 系统为例,在"开始"菜单的子菜单"所有应用"中查找文件夹"Anaconda3(64-bit)",在其列表中找到"Spyder(anaconda3)"选项。点击该选项,自动安装 Spyder 解释器。在 Spyder 的控制台中输入以下代码安装 NLTK 库:

```
import nltk
nltk.download('all')
```

按回车键后,系统自动下载 NLTK 数据包。利用 R 安装 NLTK 数据包的进程如图 12.3 所示。

图 12.3　NLTK 数据包安装进程界面

对于 DELL 电脑的 Windows 11 系统,以上命令把数据包自动下载和解压到 C:\Users\DELL\AppData\Roaming\nltk_data 中。NLTK 下载的数据包较多,包括以下 10 个子文件夹:chunkers、corpora、grammars、help、misc、models、sentiment、stemmers、taggers 和 tokenizers。若成功下载,则界面会显示"Done downloading collection all"。在较长的安装过程中,网络连接可能会出现问题或报错。如果出现类似于"由于连接方在一段时间后没有正确答复或连接的主机没有反应,连接尝试失败"这样的问题,建议再试。如果不是下载所有的数据包,而是下载某个数据包,如 wordnet,那么可以使用以下代码减少下载时间:nltk.download('wordnet')。

Python 库 wordcloud 的安装方式与 NLTK 库的安装方式大致相同。在"开始"菜单的子菜单"所有应用"中查找文件夹"Anaconda3(64-bit)",在其列表中找到"Anaconda Prompt (anaconda3)"选项。点击这个选项,在打开的界面中输入以下命令:pip install wordcloud,按回车键执行该命令即可。

12.3.2 利用 NLTK 探索词义

NLTK 库是一个用于构建 Python 程序以处理人类语言数据的领先平台。根据官网介绍,NLTK 被称为"使用 Python 开展计算语言学教学和研究的绝好工具",是一个"处理自然语言的神奇库"。NLTK 现在已经为 50 多个语料库和词汇资源(如 WordNet)提供了易于使用的接口,以及一套用于分类、形符化和词性标注等的多个处理库。本节简要介绍如何利用 WordNet 探索词义。WordNet(词网)是面向语义的英语词典,也是大型的英语词汇库。NLTK 提供了一个简单的界面,利用该界面可以查询 WordNet 词典中词的词义。利用同义子集函数 synsets()可以查看一个词的同义词(或词条)集合。例如,'bank' 的词义检索如以下代码所示:

```
> library(reticulate)
> use_python("C:/Users/DELL/anaconda3/Python.exe")
> repl_python()
Python 3.9.13 (C:/Users/DELL/anaconda3/Python.exe)
Reticulate 1.27 REPL -- A Python interpreter in R.
Enter 'exit' or 'quit' to exit the REPL and return to R.
>>> from nltk.corpus import wordnet
>>> wordnet.synsets('bank')
[Synset('bank.n.01'), Synset('depository_financial_institution.n.01'), Synset('bank.n.03'), Synset('bank.n.04'), Synset('bank.n.05'), Synset('bank.n.06'), Synset('bank.n.07'), Synset('savings_bank.n.02'), Synset('bank.n.09'), Synset('bank.n.10'), Synset('bank.v.01'), Synset('bank.v.02'), Synset('bank.v.03'), Synset('bank.v.04'), Synset('bank.v.05'), Synset('deposit.v.02'), Synset('bank.v.07'), Synset('trust.v.01')]
```

在以上结果提供的 18 个 'bank' 语义中,有 10 个语义是名词类(计算代码为 len(wordnet.synsets('bank',pos = 'n'))),有 8 个语义是动词类(计算代码为 len(wordnet.synsets('bank',pos = 'v')))。利用函数 definition()可以查看某个意义上的词定义,利用函数 examples()可以查看语义例句。如果要查阅 'bank' 的第一个定义及其例句,既可以利用前面得到的列表名称 'bank.n.01'(词条名称为 'bank',可以通过代码 wordnet.synset('bank.n.01').lemma_names()查看),又可以在 wordnet.synsets('bank')之后给出索引[0]。Python 代码和执行结果如下:

```
>>> wordnet.synset('bank.n.01').definition()
'sloping land (especially the slope beside a body of water)'
>>> wordnet.synset('bank.n.01').examples()
```

```
['they pulled the canoe up on the bank', 'he sat on the bank of the river and watched the currents']
>>> syn = wordnet.synsets('bank')[0]
>>> syn.definition()
'sloping land (especially the slope beside a body of water)'
>>> syn.examples()
['they pulled the canoe up on the bank', 'he sat on the bank of the river and watched the currents']
```

以上结果显示,'bank' 的第一个词义是"堤"或"岸",用两个句子对这个词义进行解释。

同义子集中各个词语之间是按照上义词(hypernyms,较抽象的词)和下义词(hyponyms,较具体的词)进行排序的。例如:

```
>>> syn.hypernyms()
[Synset('slope.n.01')]
>>> syn.hypernyms()[0].hyponyms()
[Synset('ascent.n.01'), Synset('bank.n.01'), Synset('bank.n.07'), Synset('canyonside.n.01'), Synset('coast.n.02'), Synset('descent.n.05'), Synset('escarpment.n.01'), Synset('hillside.n.01'), Synset('mountainside.n.01'), Synset('piedmont.n.02'), Synset('ski_slope.n.01')]
```

以上结果显示,'bank' 的第一个词义("堤"或"岸")的上义词是作为名词(n,也可以通过代码 syn.pos() 查看)的第一个词义('slope.n.01')。执行代码 wordnet.synset('slope.n.01').definition() 得到的定义是 'an elevated geological formation'("斜坡")。'bank' 的第一个词义的下义词有 11 个,如作为名词的 ascent 的第一个词义('ascent.n.01',"上坡")。在 Python 交互界面中输入 syn.root_hypernyms(),按回车键后得到根词 'entity'('entity.n.01','实体')。利用函数 hypernym_paths() 可以跟踪从 'entity.n.01' 到 'bank.n.01' 的路径:

```
>>> syn.hypernym_paths()
[[Synset('entity.n.01'), Synset('physical_entity.n.01'), Synset('object.n.01'), Synset('geological_formation.n.01'), Synset('slope.n.01'), Synset('bank.n.01')]]
```

除了可以查阅词定义与范例、上义词和下义词之外,函数 synset() 还可以用来检索一个词的同义词和反义词集合。检索一个词的同义词或反义词可以利用 lemmas()。以 'evil' 为例,检索其同义词和反义词的 Python 代码和操作结果如下:

```
>>> synonyms = []
>>> antonyms = []
>>> for synset in wordnet.synsets('evil'):
...     for i in synset.lemmas():
...         synonyms.append(i.name())
...         if i.antonyms():
```

```
...            antonyms.append(i.antonyms()[0].name())
...
>>> print(set(synonyms))
{'malefic', 'evilness', 'malevolent', 'wickedness', 'evil', 'iniquity', 'malign', 'vicious', 'immorality'}
>>> print(set(antonyms))
{'goodness', 'good'}
```

在以上代码中,synonyms = [ ] 和 antonyms = [ ] 分别用于创建同义词和反义词列表,使用 synset.lemmas() 通过 for 循环和 synonyms.append() 把同义词存放在 synonyms 列表中。对反义词提取的方法大致相同:如果有某个反义词,那么通过 for 循环和 antonyms.append() 将之添加到 antonyms 列表中。最后,利用集合函数 set() 把不重复的同义词和反义词打印出来(print(set()))。统计结果显示有 9 个同义词和 2 个反义词,如 'evil' 的同义词包括名词(如 evilness)和形容词(如 malign)。WordNet 词典标注的词性包括名词('n')、动词('v')、形容词('a')和副词('r')。在本例中,如果我们只检索作为形容词 'evil' 的同义词和反义词,那么在函数 synsets() 中增加 pos = 'a' 即可:

```
>>> synonyms = [ ]
>>> antonyms = [ ]
>>> for synset in wordnet.synsets('evil', pos = 'a'):
...        for i in synset.lemmas():
...            synonyms.append(i.name())
...            if i.antonyms():
...                antonyms.append(i.antonyms()[0].name())
...
>>> print(set(synonyms))
{'malefic', 'malevolent', 'evil', 'malign', 'vicious'}
>>> print(set(antonyms))
{'good'}
```

以上结果显示,作为形容词,'evil' 有 5 个同义词和 1 个反义词。

12.3.3 短语提取

上一节介绍了如何使用 WordNet 词典开展词义分析,本节主要介绍如何利用 NLTK 库提供的词性标注提取文档中的简单名词短语。第十一章介绍了如何调用 R 数据包 udpipe 依据宾夕法尼亚大学树库使用的词性标注提取名词短语和动词短语。调用 NLTK 库提取名词短语则使用不同的代码,包括以下四个主要步骤。

第一步,确定名词短语的语法(grammar)。NLTK 库同 R 数据包 udpipe 一样使用宾夕法尼亚大学树库提供的英语词性标注方法。表 11.2 列出了 36 种主要标注方法,但是这个树库共有 44 种标注方法。未包括在表 11.2 中的其他 8 种方法涉及标点符号的标注，如美元符 '$' 表示 'A$' 'C$' 'HK$' 和 'M$' 等货币符号。关于这些标注方法的详情，可在 Python 交互界面输入代码 nltk.help.upenn_tagset() 查阅。若查找某个标注符号，如 'JJ'（形容词），则输入代码 nltk.help.upenn_tagset('JJ') 即可。短语语法体现词性标注，可以用正则表达式表示。提取简单名词短语的语法代码如下：

```
grammar = """
NP:{<DT>*<JJ>*<NN|NNP|NNPS|NNS>+}
"""
NPchunked = RegexpParser(grammar)
```

在以上代码中，名词短语（Noun Phrase，NP）及其结构置于三个引号（"""）内，且结构置于短语名称（NP）后的冒号（:）之后。每个结构置于大括号（{}）中。若有多个结构，结构之间分行显示，每个结构中元符号的意义等同于 R 软件中元符号的意义（详见第二章），如"*"表示匹配零次或多次。把以上结构命名为 grammar 之后，再从 NLTK 库中调用函数 RegexpParser() 生成短语规则，把结果命名为 NPchunked。名词的标注方法包括四种。第一种是 NN，表示单数物质名词或普通名词，如 cabbage、investment 和 machinist。第二种是 NNP，表示单数专有名词，如 Motown。第三种是 NNPS，表示复数专有名词，如 Americans 和 Animals。第四种是 NNS，表示复数普通名词，如 undergraduates 和 bodyguards。需要提醒的是，以上几种结构并未囊括所有的名词短语形式。例如，如果我们利用以上语法标注单句 'Beginning programmers will quickly learn to develop robust, reliable, and reusable Python applications.' 的词性，那么词性标注结果是：[[('Beginning', 'VBG'), ('programmers', 'NNS'), ('will', 'MD'), ('quickly', 'RB'), ('learn', 'VB'), ('to', 'TO'), ('develop', 'VB'), ('robust', 'JJ'), (',', ','), ('reliable', 'JJ'), (',', ','), ('and', 'CC'), ('reusable', 'JJ'), ('Python', 'NNP'), ('applications', 'NNS'), ('.', '.')]]。由于 'Beginning' 被标注为动词现在分词（'VBG'），因而本例语法不能检索出名词短语 'Beginning programmers'，只能提取出 'programmers'。同样，包含形容词并列的名词短语 'robust, reliable, and reusable Python applications' 也不能被提取，只能提取 'reusable Python applications'。

第二步，对字符串开展句法分析。对字符串开展词性标注，则先从 NLTK 库中调用函数 sent_tokenize，把字符串转化为句子，再调用 word_tokenize 对句子开展形符化，然后调用函数 pos_tag 对形符进行标注。Python 代码如下：

```
def chunked_text(text):
    sentences = sent_tokenize(text)
    sentences = [word_tokenize(sent) for sent in sentences]
    sentences = [pos_tag(sent) for sent in sentences]
    return sentences
```

第三步,提取名词短语。编写函数 NP_chunked(),利用 NPchunked.parse() 把名词短语分块,得到 Tree(树,句法分析树)对象。利用 for 循环从 Tree 对象中提取子树(subtree)中包括的名词短语(树叶是词符)。Python 代码如下:

```
def NP_chunked(sentences):
    nps = []
    for sent in sentences:
        tree = NPchunked.parse(sent)
        for subtree in tree.subtrees():
            if subtree.label() == 'NP':
                t = subtree
                t = ' '.join(word for word, tag in t.leaves())
                nps.append(t)
    return nps
```

第四步,利用以上编写的函数 chunked_text() 和 NP_chunked() 对字符串进行处理。下面以文本 'Natural Language Processing is the field of artificial intelligence and computational linguistics which works on the interactions between computers and human languages. It becomes important for computers to comprehend natural languages.' 为例,使用以上 Python 代码提取名词短语。完整的代码和名词短语提取结果如下:

```
> library(reticulate)
> use_python("C:/Users/DELL/anaconda3/Python.exe")
> repl_python()
Python 3.9.13 (C:/Users/DELL/anaconda3/Python.exe)
Reticulate 1.27 REPL -- A Python interpreter in R.
Enter 'exit' or 'quit' to exit the REPL and return to R.
>>> from nltk.tokenize import sent_tokenize
>>> from nltk.tokenize import word_tokenize
>>> from nltk.tag import pos_tag
>>> from nltk import RegexpParser
>>> grammar = """
... NP:{<DT>*<JJ>*<NN|NNP|NNPS|NNS>+}
... """
>>> NPchunked = RegexpParser(grammar)
>>> def chunked_text(text):
...     sentences = sent_tokenize(text)
...     sentences = [word_tokenize(sent) for sent in sentences]
...     sentences = [pos_tag(sent) for sent in sentences]
```

```
...         return sentences
...
>>> def NP_chunked(sentences):
...     nps = []
...     for sent in sentences:
...         tree = NPchunked.parse(sent)
...         for subtree in tree.subtrees():
...             if subtree.label() == 'NP':
...                 t = subtree
...                 t = ' '.join(word for word, tag in t.leaves())
...                 nps.append(t)
...     return nps
...
>>> text = 'Natural Language Processing is the field of artificial intelligence and computational linguistics which works on the interactions between computers and human languages. It becomes important for computers to comprehend natural languages.'
>>> sentences = chunked_text(text)
>>> np = NP_chunked(sentences)
>>> print(np)
['Natural Language Processing', 'the field', 'artificial intelligence', 'computational linguistics', 'the interactions', 'computers', 'human languages', 'computers', 'natural languages']
```

以上结果显示9个名词短语,包括一个<JJ><NNP><NNP>结构('Natural Language Processing')、一个<DT><NN>结构('the field')、一个 <DT><NNS>结构('the interactions')、一个<JJ><NN>结构('artificial intelligence')、三个<JJ><NNS>结构('computational linguistics'、'human languages' 和 'natural languages')、两个同样的<NNS>结构('computers')。

在前面的代码中,我们使用函数 RegexpParser 对标注的形符开展句法分析,得到了树形结构。以句子 'The quick brown fox jumped over the lazy dog.' 为例,执行以下代码能够绘制句法树:

```
>>> from nltk.tokenize import word_tokenize
>>> from nltk.tag import pos_tag
>>> from nltk import RegexpParser
>>> grammar = """
...     NP:{<DT>*<JJ>*<NN|NNP|NNPS|NNS>+}
...     """
>>> rp = RegexpParser(grammar)
>>> text = 'The quick brown fox jumped over the lazy dog.'
```

```
>>> sentence = word_tokenize(text)
>>> sentence = pos_tag(sentence)
>>> result = rp.parse(sentence)
>>> result.draw()
```

图 12.4　句法树

图 12.4 显示,这个句子包括两个名词短语,即 'The quick brown fox' 和 'the lazy dog'。

12.3.4　利用 NLTK 库和 wordcloud 库绘图

本节先以 2019 年全国英语专业四级口语测试中的听力文本为例,调用 NLTK 库中的函数绘制包括 20 个高频词词频数变化的线图,文本存放的路径为:D:\Rpackage\story.txt。绘制高频词变化线图包括以下三个步骤。

第一步,调用 R 数据包 reticulate,通过函数 use_python() 和 repl_python() 把 R 界面转向 Python 交互界面,然后把文档 story.txt 读入 Python。首先设置文件路径,利用 Python 基础函数 open(filename, 'r') 打开文档,然后调用基础函数 read() 读入该文档。与 R 操作不同的是,Python 的操作在 read() 前添加对象(file),并用点号把两个部分隔开。最后调用基础函数 lower() 把字符串中的字母统一转化为小写字母。Python 代码如下:

```
>>> library(reticulate)
>>> use_python("C:/Users/DELL/anaconda3/Python.exe")
>>> repl_python()
>>> filename = "D:/Rpackage/story.txt"
>>> file = open(filename, 'r')
>>> story = file.read()
>>> story = story.lower()
```

第二步,调用 NLTK 库中的函数 RegexpTokenizer(),利用正则表达式"[\w']+"对 story 中的字符串开展形符化。Python 代码如下:

```
>>> from nltk.tokenize import RegexpTokenizer
>>> tokenizer = RegexpTokenizer("\w+")
>>> story_toks = tokenizer.tokenize(story)
```

第三步,先调用 NLTK 库中的函数 stopwords.words(),排除 story_toks 中的停用词,再调用函数 FreqDist() 计算词频分布,最后调用函数 plot() 绘制 20 个高频词线图。NLTK 库中的英语停用词为179个(代码为stopwords.words('english'),计算停用词数调用Python基础函数len()。wordcloud库中的英语停用词为192个(调用函数STOPWORDS)。两者共有停用词为144个,Python代码如下:

```
>>> from nltk.corpus import stopwords
>>> stoplist1=stopwords.words('english')
>>> from wordcloud import WordCloud,STOPWORDS
>>> stoplist2 = STOPWORDS
>>> len([token for token in stoplist1 if token in stoplist2])
```

顺便提一下,stopwords是语料库,包括多种语言的停用词表。NLTK当前版本(3.7版)中停用词语料库由29个词表组成,其中包括中文停用词表。我们可以通过函数fileids()查看不同语言:

```
>>> stopwords.fileids()
['arabic', 'azerbaijani', 'basque', 'bengali', 'catalan', 'chinese', 'danish', 'dutch', 'english', 'finnish', 'french', 'german', 'greek', 'hebrew', 'hinglish', 'hungarian', 'indonesian', 'italian', 'kazakh', 'nepali', 'norwegian', 'portuguese', 'romanian', 'russian', 'slovene', 'spanish', 'swedish', 'tajik', 'turkish']
```

最后一步的Python代码和生成的线图(图12.5)如下:

```
>>> from nltk.corpus import stopwords
>>> stoplist1=stopwords.words('english')
>>> story_toks=[token for token in story_toks if token not in stoplist1]
>>> from nltk.probability import FreqDist
>>> word_dist= FreqDist(story_toks)
>>> word_dist.plot(20,cumulative=False)
```

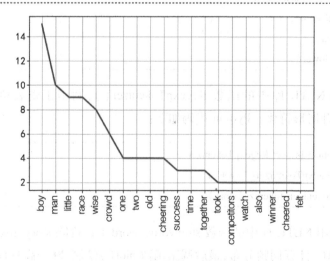

图12.5　20个高频词线图

如果我们执行 Python 代码 word_dist.most_common(20)，可以得到以下 20 个高频词：[('boy',15),('man',10),('little',9),('race',9),('wise',8),('crowd',6),('one',4),('two',4),('old',4),('cheering',4),('success',3),('time',3),('together',3),('took',2),('competitors',2),('watch',2),('also',2),('winner',2),('cheered',2),('felt',2)]。图 12.5 显示，'boy'、'man'、'little'、'race' 和 'wise' 出现的频次较高，'crowd' 出现的频次次之，'took' 和 'competitors' 等词只出现 2 次。这些词能够体现故事的主题。

下面调用 wordcloud 库绘制词云图（关于词云图，见第三章）。词云图的绘制可以分为四步。第一步与上面绘制线图的操作相同：

```
>>> library(reticulate)
>>> use_python("C:/Users/DELL/anaconda3/Python.exe")
>>> repl_python()
>>> filename = "D:/Rpackage/story.txt"
>>> file = open(filename, 'r')
>>> story = file.read()
>>> story = story.lower()
```

第二步，从 PIL 库中调用函数 Image，把自定义的小鸟图形读入 R（图形为 png 格式，存放路径为：C:\Users\DELL\Documents\bird.png。读者可以根据需要使用类似的图形）。然后，调用 Python 库 numpy 中的函数 array() 把图形转化为数组。Python 代码如下：

```
>>> from PIL import Image
>>> import numpy as np
>>> bird=np.array(Image.open("C:/Users/DELL/Documents/bird.png"))
```

第三步，从 wordcloud 库中调入函数 WordCloud()，设置绘图参数。要了解更多有关图形参数的信息，可以在 R 控制台中输入代码 help(WordCloud) 后按回车键查看。本例设置图形轮廓为自定义的 mask=bird，最大词数设为函数默认的 max_words=200，最小字体设为 min_font_size=8（函数默认值为 4），background_color='white' 设置白色背景（函数默认背景色为 'black'），把轮廓的宽度和颜色分别设置为 contour_width=3（函数默认值为 0）和 contour_color='skyblue'（函数默认轮廓颜色为 'black'）。图形设置完成后再调用函数 generate() 把字符串 story 转化为词云。函数默认去除停用词。Python 代码如下：

```
>>> from wordcloud import WordCloud
>>> wc=WordCloud(mask = bird, max_words = 200, min_font_size = 8, background_color = 'white', contour_width = 3, contour_color = 'skyblue')
>>> wc.generate(story)
```

第四步，从 matplotlib 库中调用 pyplot，利用函数 imshow，把其中参数 interpolation 设置为双线性插值（interpolation='bilinear'）。再设定 plt.axis("off") 以取消坐标显示。利用代

码 plt.show( )显示词云图。Python 代码和生成的词云图如下:

>>> import matplotlib.pyplot as plt
>>> plt.imshow(wc, interpolation = 'bilinear')
>>> plt.axis("off")
>>> plt.show( )

图 12.6　文本词语分布词云图

在词云图中,词的字体越大,在文中出现的频次就越高。图 12.6 显示,"boy""man" "race""little""wise"和"crowd"等词出现的频次较高。其他词,如"took""another"和 "granny",出现的频次较低。

12.4　利用 spaCy 库的自然语言处理

12.4.1　spaCy 库的初步使用

本章第一节介绍了 Python 库 spaCy 的安装,安装的版本号是 3.5.0,还介绍了英语统计语言模型的安装。安装的统计模型名称是 'en_core_web_sm',涉及语言代码、类型、体裁和容量。这里的语言代码是 'en'(英语), 'core' 指一般用途模型, 'web' 指模型识别的文本体裁是网络, 'sm' 指使用的统计语言模型容量小(sm 为 small 的缩写)。小模型的容量为 12 MB,中等模型的容量为 40 MB,而大模型的容量为 560 MB。读者可以根据需要使用中等容量('md')或大容量('lg')的统计语言模型。通常情况下,我们选择中等容量的统计语言模型。中等容量统计语言模型的安装见本章第一节。若安装不成功,建议登录网站 https:// github.com/explosion/spacy-models/releases?q=en_core&expanded=true,下载中等容量的模型 'en_core_web_md-3.5.0.tar.gz'。安装时,把下载的数据包放在系统盘中,如 'C:/Users/ Dell/Python/en_core_web_md-3.5.0.tar.gz',在"Anaconda Powershell Prompt (anaconda3)"的操作界面输入以下代码段: pip install C:/Users/Dell/Python/en_core_web_ md-3.5.0.tar.gz,按回车键即可执行自动安装。若安装中文统计语言模型,在浏览器中输入以下网址: https://github.com/explosion/spacy-models/releases?page=12,下载中等容量的中文统计模型压缩文件 'zh_core_web_md-3.5.0.tar.gz',安装方法同前。由于模型不断更新,建议使用最新版。

spaCy 是广受欢迎的工业级自然语言处理库,有专门的著作介绍 spaCy 的应用,如 *Natural Language Processing with Python and SpaCy: A Practical Introduction* 和 *Mastering SpaCy: An End-to-End Practical Guide to Emplementing NLP Applications Using the Python Ecosystem*(Vasiliev,2020;Altinok,2021)。本节介绍有关 spaCy 的基本操作。下面利用两个句子 'Once there was a little boy who was hungry for success in sports. To him, winning was everything and success was measured by results.'('曾经有一个小男孩,渴望在体育方面取得成功。对他来说,获胜就是一切,成功由结果来衡量。')简要介绍如何利用 spaCy 对中英文字符串开展形符化等操作。对英文句子的形符化操作使用的 Python 代码和处理结果如下:

```
> library(reticulate)
> use_python("C:/Users/DELL/anaconda3/Python.exe")
> repl_python()
Python 3.9.13 (C:/Users/DELL/anaconda3/Python.exe)
Reticulate 1.27 REPL -- A Python interpreter in R.
Enter 'exit' or 'quit' to exit the REPL and return to R.
>>> import spacy
>>> nlp = spacy.load('en_core_web_md')
>>> text = 'Once there was a little boy who was hungry for success in sports. To him, winning was everything and success was measured by results.'
>>> doc = nlp(text)
>>> sentences = list(doc.sents)
>>> print(sentences)
[Once there was a little boy who was hungry for success in sports. , To him, winning was everything and success was measured by results.]
>>> tokens = [token.text for token in doc]
>>> print(tokens)
['Once', 'there', 'was', 'a', 'little', 'boy', 'who', 'was', 'hungry', 'for', 'success', 'in', 'sports', '.', 'To', 'him', ',', 'winning', 'was', 'everything', 'and', 'success', 'was', 'measured', 'by', 'results', '.']
```

在以上代码中,首先按照通常的方式从 R 操作界面转向 Python 交互界面。然后,调用 spaCy 库和英语统计语言模型(即 spaCy 引擎),创建自然语言处理管道 nlp。利用 nlp 处理文本字符串,得到 Doc(文档,document 的前三个字母缩写)对象 doc。最后,利用函数 list()提取 doc 中的句子,利用 for 循环提取 doc 中的形符,并把结果打印出来。要统计文本包括的形符数,不包括标点符号,在 for 循环中利用 if not token.is_punct,再使用函数 len()即可:

```
>>> tokens_no_punct = [token.text for token in doc if
    not token.is_punct]
>>> print(len(tokens_no_punct))
24
```

Doc 是容器对象,以统一码(unicode)表征文本,可以被用于提取文本的语言特征,如句子、形符、词性、标注、依存关系、命名实体(named entities,指现实世界的具体对象)和名词短语等。词性标注包括通用词性标注(token.pos_)和精细词性标注(token.tag_)。对于不同标注的解释,可以用 spacy.explain()进行查阅。例如,执行下面的 Python 代码得到文本形符的不同词性标注:

```
# get universal and fine-rained pos tags
>>> tags = [(token.text, token.pos_, token.tag_,
spacy.explain(token.pos_), spacy.explain(token.tag_)) for token in doc]
>>> tags
[('Once', 'SCONJ', 'IN', 'subordinating conjunction', 'conjunction, subordinating or preposition'),
('there', 'PRON', 'EX', 'pronoun', 'existential there'), ('was', 'VERB', 'VBD', 'verb', 'verb, past tense'),
('a', 'DET', 'DT', 'determiner', 'determiner'), ('little', 'ADJ', 'JJ', 'adjective', 'adjective (English), other noun-modifier (Chinese)'),
('boy', 'NOUN', 'NN', 'noun', 'noun, singular or mass'),
('who', 'PRON', 'WP', 'pronoun', 'wh-pronoun, personal'), ('was', 'AUX', 'VBD', 'auxiliary', 'verb, past tense'),
('hungry', 'ADJ', 'JJ', 'adjective', 'adjective (English), other noun-modifier (Chinese)'),
('for', 'ADP', 'IN', 'adposition', 'conjunction, subordinating or preposition'), ('success', 'NOUN', 'NN', 'noun', 'noun, singular or mass'),
('in', 'ADP', 'IN', 'adposition', 'conjunction, subordinating or preposition'), ('sports', 'NOUN', 'NNS', 'noun', 'noun, plural'), ('.', 'PUNCT', '.', 'punctuation', 'punctuation mark, sentence closer'),
('To', 'ADP', 'IN', 'adposition', 'conjunction, subordinating or preposition'), ('him', 'PRON', 'PRP', 'pronoun', 'pronoun, personal'), (',', 'PUNCT', ',', 'punctuation', 'punctuation mark, comma'),
('winning', 'VERB', 'VBG', 'verb', 'verb, gerund or present participle'), ('was', 'AUX', 'VBD', 'auxiliary', 'verb, past tense'), ('everything', 'PRON', 'NN', 'pronoun', 'noun, singular or mass'),
('and', 'CCONJ', 'CC', 'coordinating conjunction', 'conjunction, coordinating'), ('success', 'NOUN', 'NN', 'noun', 'noun, singular or mass'), ('was', 'AUX', 'VBD', 'auxiliary', 'verb, past tense'),
('measured', 'VERB', 'VBN', 'verb', 'verb, past participle'), ('by', 'ADP', 'IN', 'adposition', 'conjunction, subordinating or preposition'), ('results', 'NOUN', 'NNS', 'noun', 'noun, plural'),
('.', 'PUNCT', '.', 'punctuation', 'punctuation mark, sentence closer')]
```

在第十一章提到,文档 text 中的 'Once' 为副词。但是,在以上输出结果中,'Once' 被标注为通用词性 'SCONJ'('subordinating conjunction',从属连词),在精细词性标注体系中被标注为 'IN'('conjunction, subordinating or preposition',从属连词或介词)。

词性标注和依存分析都是句法分析的类别。词性标注提供句子中邻近词的句法信息,而依存分析提供非邻近词之间的句法关系。例如:

```
# get dependency relations
>>> dep = [token.dep_ for token in doc]
>>> print(dep)
['mark', 'expl', 'ROOT', 'det', 'amod', 'attr', 'nsubj', 'relcl', 'acomp', 'prep', 'pobj', 'prep', 'pobj',
'punct', 'prep', 'pobj', 'punct', 'csubj', 'ROOT', 'nsubjpass', 'cc', 'conj', 'auxpass', 'ccomp', 'agent',
'pobj', 'punct']
```

我们可以从 spaCy 库中调用内置的可视化工具 displaCy,利用 Doc 对象绘制句子的依存树。以前面文档中的第一个句子为例,绘制并保存依存树图的 Python 代码和执行结果如下:

```
>>> from spacy import displacy
>>> from pathlib import Path
>>> doc = nlp('Once there was a little boy who was hungry for success in sports.')
>>> svg = displacy.render(doc, style = 'dep', jupyter = False, options = {'distance': 120})
>>> filename = 'boy.svg'
>>> output_path = Path('D:/Rpackage/images/'+filename)
>>> output_path.open('w', encoding='utf-8').write(svg)
10521
```

在以上代码中,前三行用于调用 Python 库和英语统计语言模型,得到 Doc 对象 doc。调用 displacy.render() 把依存分析结果保存为 svg 变量。设定文档名(filename)后,利用最后两行把结果保存在相应的路径里。打开保存的路径 'D:\Rpackage\images\boy.svg',可以看到如图 12.7 所示的依存树。

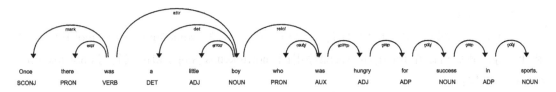

图 12.7 英文句依存树举例

关于图 12.7 中词性和依存关系的解释,可以调用函数 spacy.explain() 进行查看。例如,在 Python 交互界面执行代码 spacy.explain('SCONJ') 会得到 'subordinating conjunction'(从属连词)。

同样地,我们可以利用 doc.ents 提取 doc 中的命名实体(named entities,如人名、地点和机构名等),使用 doc.noun_chunks 提取 doc 中的名词短语。例如:

```
>>> doc = nlp('Once there was a little boy who was hungry for success in sports. To him, winning was everything and success was measured by results.')
>>> list(doc.noun_chunks)
[a little boy, who, success, sports, him, everything, success, results]
>>> string = 'Biden boarded a private jet for the short flight to Washington.'
>>> doc=nlp(string)
>>> doc.ents
(Biden, Washington)
```

使用第一行代码是因为前面在绘制树图时使用了一个句子而非两个句子。在第二行代码中，用 list() 显示前面 Doc 对象 doc 中包含的名词短语。注意，'who' 'him' 和 'everything' 都是代词(词性标注都是 'PRON')，但是在以上结果中均被视作名词短语。由于前面的文档没有命名实体，因此第三行代码使用了一个不同的字符串，第四行据此创建了 Doc 对象 doc，第五行利用 doc.ents 提取了两个名词实体，一个是人名('Biden')，另一个是地名('Washington')。这些实体的类型可以通过 token.ent_type_ 查看。例如，在本例中，输入代码 doc[0].ent_type_ 会得到人名 'PERSON'。此外，我们还可以利用 token.lemma_ 很方便地从 Doc 对象 doc 中提取词条(含标点符号)。以前面的文档 text 为例，依据由此得到的 Doc 对象 doc 得到词条的 Python 代码和执行结果如下：

```
>>> doc = nlp('Once there was a little boy who was hungry for success in sports. To him, winning was everything and success was measured by results.')
>> lemmas = [token.lemma_ for token in doc]
>>> lemmas
['once', 'there', 'be', 'a', 'little', 'boy', 'who', 'be', 'hungry', 'for', 'success', 'in', 'sport', '.', 'to', 'he', ',', 'win', 'be', 'everything', 'and', 'success', 'be', 'measure', 'by', 'result', '.']
```

以上结果显示 27 个词条(含三个标点符号)，如词符 'was' 被替换为 'be'，'measured' 被替换为 'measure'。

spaCy 的英语统计语言模型目前自带 326 个停用词，如 'sometime' 'while' 'too' 'fifteen' 'of' 和 'be' 等。我们可以在 for 循环中排除停用词(not token.is_stop)和标点符号(not token.is_punct)。例如：

```
>>> tokens_no_stop = [token.text for token in doc if not token.is_stop and not token.is_punct]
>>> tokens_no_stop
['little', 'boy', 'hungry', 'success', 'sports', 'winning', 'success', 'measured', 'results']
>>> len(tokens_no_stop)
9
```

以上结果显示，在排除 'Once'，'there'，'was'，'a' 和 'who' 等 15 个停用词后只剩下 9 个实词。

对于中文字符串的处理，我们只需把上面的英语统计语言模型换成中文统计语言模型，即可得到句子和形符等。例如：

```
>>> import spacy
>>> nlp = spacy.load('zh_core_web_md')
>>> text = '曾经有一个小男孩,渴望在体育方面取得成功。对他来说,获胜就是一切,成功由结果来衡量。'
>>> doc = nlp(text)
>>> sentences = list(doc.sents)
```

```
>>> print(sentences)
[曾经有一个小男孩,渴望在体育方面取得成功。对他来说,获胜就是一切,成功由结果来衡量。]
>>> tokens = [token.text for token in doc]
>>> print(tokens)
['曾经','有','一个','小','男孩',',','渴望','在','体育','方面','取得','成功','。',
'对','他','来说',',','获胜','就是','一切',',','成功','由','结果','来','衡量','。']
```

如果计算文本词符数,执行下面的代码得到22个中文词:

```
>>> tokens_no_punct = [token.text for token in doc if not token.is_punct]
>>> print(len(tokens_no_punct))
22
```

spaCy 的中文统计语言模型目前自带 1 891 个停用词,包括标点符号和其他符号,如'不管''正巧''[*]''若是''>''$''你们''去''5'等。我们只需在 for 循环中排除停用词(not token.is_stop)即可得到非停用词。例如:

```
>>> tokens_no_stop = [token.text for token in doc if not token.is_stop]
>>> print(tokens_no_stop)
['男孩','渴望','体育','成功','获胜','成功','衡量']
```

以上结果提供 7 个中文词,比排除停用词后的英文词少了 2 个('小'和'结果',即'little'和'results')。

如果要得到词性标注和依存分析结果,Python 代码和执行结果如下:

```
>>> tags = [(token.text, token.pos_, token.tag_, token.dep_) for token in doc]
>>> tags
[('曾经', 'ADV', 'AD', 'advmod'), ('有', 'VERB', 'VE', 'ROOT'), ('一个', 'ADV', 'AD',
'dep'), ('小', 'ADJ', 'JJ', 'amod'), ('男孩', 'NOUN', 'NN', 'dobj'), (',', 'PUNCT', 'PU',
'punct'), ('渴望', 'VERB', 'VV', 'conj'), ('在', 'ADP', 'P', 'case'), ('体育', 'NOUN', 'NN',
'compound:nn'), ('方面', 'NOUN', 'NN', 'nmod:prep'), ('取得', 'VERB', 'VV', 'ccomp'),
('成功', 'NOUN', 'NN', 'dobj'), ('。', 'PUNCT', 'PU', 'punct'), ('对', 'ADP', 'P', 'case'),
('他', 'PRON', 'PN', 'nmod:prep'), ('来说', 'PART', 'LC', 'case'), (',', 'PUNCT', 'PU', 'punct'),
('获胜', 'VERB', 'VV', 'dep'), ('就是', 'ADV', 'AD', 'advmod'), ('一切', 'PRON', 'PN',
'ROOT'), (',', 'PUNCT', 'PU', 'punct'), ('成功', 'ADV', 'AD', 'advmod'), ('由', 'ADP', 'P',
'case'), ('结果', 'NOUN', 'NN', 'nmod:prep'), ('来', 'PART', 'MSP', 'aux:prtmod'), ('衡量',
'VERB', 'VV', 'conj'), ('。', 'PUNCT', 'PU', 'punct')]
```

在以上结果中,'曾经'的词性被正确地标注为副词('ADV'和'AD'),依存关系被标注为状语修饰语('advmod')。绘制文档 text 中第一句依存关系树的 Python 代码和执行结果如下：

```
>>> import spacy
>>> from spacy import displacy
>>> from pathlib import Path
>>> doc = nlp('曾经有一个小男孩,渴望在体育方面取得成功。')
>>> svg = displacy.render(doc, style = 'dep', jupyter = False, options = {'distance': 140})
>>> filename = '男孩.svg'
>>> output_path = Path('D:/Rpackage/images/'+filename)
>>> output_path.open('w', encoding='utf-8').write(svg)
8838
```

图 12.8　中文句依存树举例

图 12.8 显示，副词'曾经'作为动词'有'的修饰语,'有'的直接宾语是名词'男孩','有'与动词'渴望'并列,'渴望'的子句补语是动词'取得','取得'的直接宾语是名词'成功'。

12.4.2　调用 Matcher 类提取名词短语

12.3 节介绍利用 NLTK 库提取文本中的短语。调用 spaCy 库中的 Matcher 类也可以根据匹配模式提取文中的短语和其他模式。我们先以 12.3.3 节中的例句为例提取名词短语。

上一节介绍利用 doc.noun_chunks 能够提取 DOC 对象中的名词短语。针对本节例子，从文本中提取名词短语的 Python 代码和执行结果如下：

```
> require(reticulate)
Loading required package: reticulate
> use_python("C:/Users/Dell/Anaconda3/Python.exe")
> repl_python()
Python 3.9.13 (C:/Users/DELL/anaconda3/Python.exe)
Reticulate 1.27 REPL -- A Python interpreter in R.
Enter 'exit' or 'quit' to exit the REPL and return to R.
>>> import spacy
>>> nlp = spacy.load('en_core_web_md')
```

```
>>> text = 'Natural Language Processing is the field of artificial intelligence and computational linguistics
which works on the interactions between computers and human languages. It becomes important for
computers to comprehend natural languages.'
>>> doc = nlp(text)
>>> list(doc.noun_chunks)
[Natural Language Processing, the field, artificial intelligence, computational linguistics, which, the
interactions, computers, human languages, It, computers, natural languages]
```

在以上提供的 11 个名词短语(含单个名词,如 'computers')中,'which' 和 'It' 被视作名词。但是,'which' 的词性实际标注为 'wh-determiner',即 wh-限定词,'It' 的词性实际标注为 'pronoun, personal',即人称代词。在名词短语研究中,这两个词通常不应视作名词短语。下面分六步调用 Matcher 类重新从以上文本中提取名词短语。

第一步,需要从 spacy.matcher 匹配器中调入 Matcher 类,用于对 Doc 对象 doc 的处理,用自然语言处理模型中的词汇对象对 matcher 对象初始化。

第二步,定义匹配模式 pattern。模式 pattern 是一个列表,列表项置于大括号内。这里要匹配的简单名词短语是限定词 'Determiner'(可选项)+形容词 'Adjective'(可选项)|(或者)名词 'NOUN'(可选项)+名词 'NOUN'(必选项)。如果我们采用精细词性标注,限定词的标注为 'DT',形容词的标注为 'JJ',可选项中的名词标注包括 'NN'(单数名词或物质名词)和 'NNP'(单数专有名词),中心名词包括 'NN' 和 'NNS'(复数名词)、'NNP' 和 'NNPS'(复数专有名词)。

第三步,调用函数 matcher.add() 把模式 pattern 增加到 matcher 对象,将模式名称定为 'NP',把 pattern 置于方括号内。

第四步,对 doc 对象调用 matcher 开展匹配操作。

第五步,为了避免分开提取同一个名词短语,如从 'artificial intelligence' 中提取 'artificial intelligence' 和 ' intelligence',设定跨度对象 spans,以便利用过滤函数 spacy.util.filter_spans() 排除对短语的重复提取。

第六步,利用 for 循环从跨度对象 spans 中提取名词短语。

接着前面的 doc 对象,本例提取名词短语的 Python 代码和操作结果如下:

```
>>> from spacy.matcher import Matcher
>>> matcher = Matcher(nlp.vocab)
>>> pattern = [{'TAG': 'DT', 'OP': '?'},
{'TAG': {'IN': ['JJ','NNP','NN']}, 'OP': '*'}, {'TAG': {'IN': ['NNP','NNPS','NN','NNS']},
'OP':'+'}]
>>> matcher.add('NP', [pattern])
>>> matches = matcher(doc)
>>> spans = [doc[start:end] for _id, start, end in matches]
```

```
>>> np =[ span for span in spacy.util.filter_spans( spans)]
>>> np
[Natural Language Processing, the field, artificial intelligence, computational linguistics, the interactions,
computers, human languages, computers, natural languages]
```

以上结果显示 9 个名词短语,排除了利用 doc.noun_chunks 提取的多余的 'which' 和 'It'。有关以上操作符 '?' '*' '+' 的意义,请参看第二章。

使用通用词性标注同样可以得到名词短语。只是在定义 pattern 时,用 'POS' 代替前面的 'TAG',把前面的形容词标注 'JJ' 改为通用标注 'ADJ',把前面的名词标注 'NNP''NNPS''NN' 和 'NNS' 改为通用标注 'PROPN' 和 'NOUN'。Python 代码和操作结果如下:

```
>>> from spacy.matcher import Matcher
>>> matcher = Matcher( nlp.vocab)
>>> pattern = [{'POS': 'DET', 'OP': '?'}, {'POS': {'IN': ['ADJ', 'PROPN','NOUN']},'OP': '*'},
{'POS': {'IN': ['PROPN','NOUN']}, 'OP': '+'}]
>>> matcher.add('NP', [pattern])
>>> matches = matcher(doc)
>>> spans = [doc[start:end] for _id, start, end in matches]
>>> np =[ span for span in spacy.util.filter_spans( spans)]
>>> np
[Natural Language Processing, the field, artificial intelligence, computational linguistics, the interactions,
computers, human languages, computers, natural languages]
```

12.4.3 调用 Matcher 类提取动词短语

我们可以使用与上一节相同的方法从文本中提取动词短语。下面以第二章使用的保存在 D:\Rpackage 文件夹中的故事复述文本 story.text 为例,从中提取动词短语。动词短语的构成比较复杂,本例提取的动词短语不包括前面考虑的名词短语,包括动词、动词+介词、动词+副词、副词+动词、系动词+形容词、系动词+形容词+介词、系动词+形容词+并列连词+形容词构成的并列短语、助动词+动词+介词、情态动词+副词+动词+副词和助动词+动词+副词+形容词等。本例 Python 代码和操作结果如下:

```
> require(reticulate)
Loading required package：reticulate
> use_python('C:/Users/Dell/Anaconda3/Python.exe')
> repl_python()
Python 3.9.13 (C:/Users/DELL/anaconda3/Python.exe)
Reticulate 1.27 REPL -- A Python interpreter in R.
Enter 'exit' or 'quit' to exit the REPL and return to R.
```

```
>>> import spacy
>>> nlp = spacy.load('en_core_web_md')
>>> from spacy.matcher import Matcher
>>> matcher = Matcher(nlp.vocab)
>>> filename = "D:/Rpackage/story.txt"
>>> file = open(filename, 'r')
>>> text = file.read()
>>> text = text.lower()
>>> doc = nlp(text)
>>> pattern1 = [{'POS': 'AUX','OP': '*'}, {'POS': 'VERB','OP': '+'}, {'POS': {'IN': ['ADJ', 'ADV']}},'OP': '*'}, {'POS': 'CCONJ','OP': '*'}, {'POS': {'IN': ['ADJ','VERB']},'OP': '+'}, {'POS': 'ADV','OP': '*'}]
>>> pattern2 = [{'POS': 'AUX','OP': '+'},{'POS': {'IN': ['ADJ','ADV']},'OP': '*'}]
>>> pattern3 = [{'POS': 'AUX','OP':'*'},{'POS': 'ADV','OP': '*'}, {'POS': 'VERB','OP': '+'}, {'POS': 'ADP','OP': '*'}, {'POS': 'AUX','OP': '*'}, {'POS': 'ADV','OP': '*'}]
>>> pattern4 = [{'POS': 'AUX','OP':'+'}, {'POS': 'ADJ','OP': '+'}, {'POS': 'ADP','OP': '+'}]
>>> matcher.add('VP', [pattern1])
>>> matcher.add('VP', [pattern2])
>>> matcher.add('VP', [pattern3])
>>> matcher.add('VP', [pattern4])
>>> matches = matcher(doc)
>>> spans = [doc[start:end] for _id, start, end in matches]
>>> vp =[span for span in spacy.util.filter_spans(spans)]
>>> print(vp)
[was, was hungry for, winning was, was measured by, took, were, gathered, watch, hearing about, also came from afar, watch, was, cheered and waved, felt proud and important, remained still and calm, showing, pleaded, hoping, impress, stepped forward and presented, was, finished, being left standing at, raised, was silent, happened, is, cheering, asked, replied, finish together, finish together, stood between, took, began, walked slowly, could also move together, finishing, crossed, cheered like, smiled, gently nodding, felt puzzled, do, understand, asked, is, cheering for, won, looked into, replied softly, remember, have won much more, ran before, is cheering]
```

在以上代码中，前三行代码从 R 界面转向 Python 交互界面。第四至第六行的代码调入 spacy 库、语言统计模型和 Matcher 类。第七行的代码利用语言统计模型中的词汇对 matcher 对象初始化。第八至第十一行的代码从本地硬盘文件夹中读取文档 story.txt, 并把字母统一改为小写字母。第十二行代码对文档进行处理，保存 Doc 对象 doc。第十三至第十六行的代码为编写的动词短语不同模式。在模式 pattern1 中使用了置于大括号中的操作符 'IN'，表示从列表中选择词性(如{'IN': ['ADJ','ADV']})。第十七至第二十行的代码把本例的四种动词短语模式添加到 matcher 对象中。剩下四行代码的解释同前面提取名词短语中对代码的说明。

以上结果显示,文档 story.txt 包含 55 个动词短语,如 'was hungry for'(系动词+形容词+介词)、'was measured by'(助动词+动词+介词)和 'cheered and waved'(动词+连词+动词)等结构。以上代码中对动词短语的设定限于文档 story.txt,没有穷尽动词短语模式。在实际研究中,建议读者根据研究需要设置特定的匹配模式。

12.4.4 调用 Matcher 类提取情态动词结构

Matcher 类的功能很强大。我们不仅可以像前两节那样利用通用词性标注和精细词性标注定义匹配模式从文档中提取名词短语和动词短语,还可以利用依存关系标注从文档中提取其他句法结构。下面引用 *Natural Language Processing with Python and SpaCy*: *A Practical Introduction* 书中的一小段话作为文档,提取其中使用 'can' 'could' 'may' 'might' 或 'must' 的情态动词结构(Vasiliev, 2020)。Python 代码和执行结果如下:

```
> require(reticulate)
Loading required package: reticulate
> use_python("C:/Users/Dell/Anaconda3/Python.exe")
> repl_python()
Python 3.9.13 (C:/Users/DELL/anaconda3/Python.exe)
Reticulate 1.27 REPL -- A Python interpreter in R.
Enter 'exit' or 'quit' to exit the REPL and return to R.
>>> import spacy
>>> from spacy.matcher import Matcher
>>> nlp = spacy.load("en_core_web_md")
>>> matcher = Matcher(nlp.vocab)
>>> pattern = [{'DEP': 'nsubj'}, {'TEXT':{'IN': ['can', 'could', 'may', 'might', 'must']}, 'OP':'+'},
{'DEP':'advmod', 'OP': '?'}, {'DEP': 'aux', 'OP': '?'}, {'DEP': 'aux', 'OP': '?'}, {'DEP': {'IN':
['ROOT', 'ccomp', 'conj']}, 'OP': '+'}, {'DEP': 'dative', 'OP': '?'}, {}, {'DEP': 'dobj', 'OP': '?'},
{'DEP':'pobj','OP': '?'}]
>>> matcher.add("modal", [pattern])
>>> text = "Increasingly, when you call the bank or your internet provider, you might hear something like the following on the other end of the line: 'Hello, I am your digital assistant. Please ask your question.' Today, robots can talk to humans using natural language, and they're getting smarter. Even so, very few people understand how these robots work or how they might use these technologies in their own projects."
>>> doc = nlp(text)
>>> matches = matcher(doc)
>>> spans = [doc[start:end] for _id, start, end in matches]
>>> modals = [span for span in spacy.util.filter_spans(spans)]
>>> modals
[you might hear something, robots can talk to humans, they might use these technologies]
```

在以上代码中,情态动词模式的定义利用依存关系。在一个包括情态动词'can''could''may''might' 或 'must' 的句子中,句子主句谓语动词是根词(依存关系标签为'ROOT'),做主语的名词或代词依存关系标签为 'nsubj'。各个情态动词在操作符 'IN' 之后以列表形式显示,属性为 'TEXT'(实际的形符)。如果情态动词出现在句子包含的从句中,如 'People understand that they could do the job well.',从句中的 'do' 的句法标签是 'ccomp'('clausal complement',子句补语)。但是,如果句子是 'People understand that they work hard and they could do the job well.',那么 'do' 的句法标签是 'conj'('conjunct',连接成分)。鉴于本例子句的特点,在代码定义中包括了结构{'DEP':{'IN':['ROOT', 'ccomp', 'conj']}, 'OP': '+'}。在该结构之后定义了三种宾语形式:'dative'(与格)、'dobj'(直接宾语)和 'pobj'(介词宾语),通配符 '{}' 表示匹配任一形符。

以上结果显示,本例包括三个含情态动词的子句。本例使用的代码主要依据本例使用的小段落,对其他情形未必合适。例如,利用以上代码不能从句子 'He saw a boy who could be hungry for success.' 中提取包括情态动词 'could' 的从句。如要提取此类句子中的情态动词结构,可以使用以下代码:

```
>>> pattern=[{},{'DEP': 'nsubj'},{'TEXT': {'IN': ['can', 'could', 'may', 'might', 'must']}, 'OP': '+'},{'DEP': 'relcl', 'OP': '*'},{'DEP': 'acomp','OP': '*'},{'DEP': 'prep', 'OP': '*'},{'DEP': 'pobj', 'OP': '*'}]
>>> matcher.add("modal", [pattern])
>>> doc = nlp('He saw a boy who could be hungry for success.')
>>> matches = matcher(doc)
>>> spans = [doc[start:end] for _id, start, end in matches]
>>> modals = [span for span in spacy.util.filter_spans(spans)]
>>> modals
[boy who could be hungry for success]
```

参考文献

ALTINOK D, 2021. Mastering spaCy: An end-to-end practical guide to implementing NLP applications using the Python ecosystem[M]. Birmingham: Packt Publishing Ltd.

ANDERSON J, 1983. Lix and Rix: Variations on a little known readability index[J]. Journal of Reading, 26(6): 490-496.

BATYRSHIN I Z, 2019. Data science: Similarity, dissimilarity and correlation functions[M]//OSIPOV G, PANOV A, YAKOVLEV K. Artificial intelligence. Cham: Springer: 13-28.

BENOIT K, OBENG A, 2021. Readtext: Import and handling for plain and formatted text files[Z/OL]. R package version 0.81. https://cran.r-project.org/package=readtext.

BENOIT K, WATANABE K, WANG H, et al, 2018. Quanteda: An R package for the quantitative analysis of textual data[J]. Journal of Open Source Software, 3(30): 774.

BLAHETA D, JOHNSON M, 2001. Unsupervised learning of multi word verbs[C]// Proceedings of the ACL 2001 workshop on collocation. Toulouse: Association for Computational Linguistics: 54-60.

CHALL J S, DALE E, 1995. Manual for use of the new Dale-Chall readability formula[M]. Cambridge: Brookline Books.

CHAMBERS J M, CLEVELAND W S, KLEINER B, et al, 1983. Graphical methods for data analysis[M]. New York: Chapman and Hall.

CHOI S S, CHA S H, TAPPERT C C, 2010. A survey of binary similarity and distance measures[J]. Systemics, Cybernetics and Informatics, 8(1): 43-48.

CICHOSZ P, 2015. Data mining algorithms: Explained using R[M]. West Sussex: John Wiley & Sons.

COHEN J, 1988. Statistical power analysis for the behavioral sciences[M]. 2nd ed. Hillsdale: Lawrence Erlbaum Associates.

COLEMAN M, LIAU T L, 1975. A computer readability formula designed for machine scoring[J]. Journal of Applied Psychology, 60(2): 283-284.

COVINGTON M A, MCFALL J D, 2010. Cutting the Gordian knot: The moving-average type-token ratio (MATTR)[J]. Journal of Quantitative Linguistics, 17(2): 94-100.

DESAGULIER G, 2017. Corpus linguistics and statistics with R[M]. Berlin: Springer.

DOWLE M, SRINIVASAN A, 2021. data.table: Extension of 'data.frame'[Z/OL]. R package version 1.14.2. https://cran.r-project.org/package=data.table

DUBAY W H, 2004. The principles of readability[R/OL]. Costa Mesa: Impact Information. http://www.impactinformation.com/impactinfo/readability02.pdf.12.12.2021.

FANG I E, 1966. The "easy listening formula"[J]. Journal of Broadcasting, 11(1): 63-68.

FARR J N, JENKINS J J, PATERSON D G, 1951. Simplification of Flesch reading ease formula[J]. Journal of Applied Psychology, 35(5): 333-337.

FELLOWS I, 2018. Wordcloud: Word clouds[Z/OL]. R package version 2.6. https://cran.r-project.org/package=wordcloud.

FIRTH J R, 1957. A synopsis of linguistic theory 1930—1955[M]//PALMER F. Selected papers of J. R. Firth 1952—1959. London: Longman: 168-205.

FLESCH R, 1948. A new readability yardstick[J]. Journal of Applied Psychology, 32(3): 221-233.

FLESCH R, 1949. The art of readable writing[M]. New York: Harper & Row.

GAILLAT T, BALLIER N, 2019. Expérimentation de feedback visuel des productions? Crites d'apprenants francophones de l'anglais sous moodle [C]// Actes de La conférence EIAH2019. Paris: Association des Technologies de l'Information pour l'Education et la Formation.

GEFEN D, ENDICOTT J E, FRESNEDA J E, et al, 2017. A guide to text analysis with latent semantic analysis in R with annotated code: Studying online reviews and the stack exchange community[J]. Communications of the Association for Information Systems, 41(11): 450-496.

GRIES S T, 2007. Coll. analysis 3.2a [Z/OL]. A program for R for Windows 2.x. https://stgries.info/teaching/groningen/index.html.

GRIES S T, 2013. 50-something years of work on collocations[J]. International Journal of Corpus Linguistics, 18(1): 137-165.

GRIES S T, ELLIS N C, 2015. Statistical measures for usage-based linguistics[J]. Language Learning, 65(S1): 228-255.

GRIES S T, STEFANOWITSCH A, 2004a. Extending collostructional analysis: A corpus based perspective on 'alternations' [J]. International Journal of Corpus Linguistics, 9(1): 97-129.

GRIES S T, STEFANOWITSCH A, 2004b. Covarying collexemes in the into causative[M]// ACHARD M, KEMMER S. Language, culture, and mind. Stanford: CSLI: 225-236.

GROSSMAN D A, FRIEDER O, 2004. Information retrieval: Algorithms and heuristics [M]. 2nd ed. Berlin: Springer.

GUNNING R. 1952. The technique of clear witing[M]. New York: McGraw-Hill.

GÜNTHER F, DUDSCHIG C, KAUP B, 2015. LSAfun-An R package for computations based on latent semantic analysis[J]. Behavior Research Methods, 47(4): 930-944.

HART-DAVIS G, HART-DAVIs T, 2022. Teach yourself visually™ python[M]. Hoboken: John Wiley & Sons.

HERDAN G, 1960. Type-token mathematics: A textbook of mathematical linguistics[M]. Hague: Mouton & Co.

HU M Q, LIU B, 2004. Mining and summarizing customer reviews[C]// Proceedings of the ACM SIGKDD international conference on knowledge discovery and data mining (KDD-2004). Seattle: The ACM SIGKDD.

HUNSTON S, 2002. Corpora in applied linguistics[M]. Cambridge: Cambridge University Press.

IHAKA R, GENTLEMAN R, 1996. R: A language for data analysis and graphics[J]. Journal of Computational and Graphical Statistics, 5(3): 299-314.

JOCKERS M L, 2014. Text analysis with R for students of literature[M]. New York: Springer.

JOCKERS M L, 2015. Syuzhet: Extract sentiment and plot arcs from text [Z/OL]. https://github.com/mjockers/syuzhet.

JOHNSTON M, Robinson D, 2022. Gutenbergr: Download and process public domain works from Project Gutenberg[Z/OL]. R package version 0.2.3. https://cran.r-project.org/package=gutenbergr.

KINCAID J P,FISHBURNE R P Jr,ROGERS R L,et al,1975. Derivation of new readability formulas (Automated Readability Index,Fog Count and Flesch Reading Ease Formula) for Navy enlisted personnel[R]. Millington:Naval Technical Training,U. S. Naval Air Station.

LANG D W,2023. Wordcloud 2:Create word cloud by htmlwidget[Z/OL]. R package version 0. 2. 2. https://github. com/lchiffon/wordcloud2.

LU X,2010. Automatic analysis of syntactic complexity in second language writing[J]. International Journal of Corpus Linguistics,15(4):474-496.

MAILUND T,2019. R Data science quick reference:A pocket guide to APIs,libraries,and packages[M]. New York:Apress.

MCCARTHY P M,JARVIS S,2007. Vocd:A theoretical and empirical evaluation[J]. Language Testing,24(4):459-488.

MCCARTHY P M,JARVIS S,2010. MTLD,vocd-D,and HD:A validation study of sophisticated approaches to lexical diversity assessment[J]. Behavior Research Methods,42(2):381-392.

MCLAUGHLIN G H,1969. SMOG grading:A new readability formula[J]. Journal of Reading,2(8):639-646.

MICHALKE M,2021. koRpus:Text analysis with emphasis on POS tagging,readability,and lexical diversity[Z/OL]. R package version 0. 13-8. https://reaktanz. de/?c=hacking&s=koRpus.

MOHAMMAD S,TURNEY P,2010. Emotions evoked by common words and phrases:Using Mechanical Turk to create an emotion lexicon[C]// Proceedings of the NAACL HLT 2010 Workshop on Computational Approaches to Analysis and Generation of Emotion in Text. LA:The NAACL HLT.

NEUWIRTH E,2022. R ColorBrewer:ColorBrewer palettes[Z/OL]. R package version 1. 1-3. https://cran. r-project. org/package=RColorBrewer.

NIELSEN F,2011. A new ANEW:Evaluation of a word list for sentiment analysis in microblogs[C]// Proceedings of the ESWC2011 Workshop on 'Making Sense of Microposts':Big things come in small packages. Heraklion:1st Workshop on Making Sense of Microposts:93-98.

OAKES M P,1998. Statistics for corpus linguistics[M]. Edinburgh:Edinburgh University Press.

PERKINS J,2014. Python 3 text processing with NLTK 3 cookbook[M]. Birmingham:Packt Publishing Ltd.

POWERS R D,SUMNER W A,KEARL B E,1958. A recalculation of four adult readability formulas[J]. Journal of Educational Psychology,49(2):99-105.

QIN W,WU Y,2019. jiebaR:Chinese text segmentation[Z/OL]. R package version 0. 11. https://cran. r-project. org/package=jiebaR.

RHYS H I,2020. Machine learning with R,the tidyverse,and mlr[M]. Shelter Island:Manning Publications Co.

SCAVETTA R J,ANGELOV B,2021. Python and R for the modern data scientist[M]. Sebastopol:O'Reilly Media.

SIGNORELL A,et al,2022. DescTools:Tools for descriptive statistics[Z/OL]. R package version 0. 99. 47. https://cran. r-project. org/package=DescTools.

SILGE J,2022. Janeaustenr:Jane Austen's complete novels[Z/OL]. R package version 1. 0. 0. https://cran. r-project. org/package=janeaustenr.

SMITH E A,SENTER R J,1967. Automated readability index[Z/OL]. AMRL-TR-66-22. Ohio:Wright-Paterson AFB. https://apps. dtic. mil/sti/pdfs/AD0667273. pdf.

STEFANOWITSCH A, GRIES S T, 2003. Collostructions: Investigating the interaction between words and constructions[J]. International Journal of Corpus Linguistics, 8(2): 209-243.

STUART R, DAVE E, CRAMER N, et al, 2010. Automatic keyword extraction from individual documents [M]// BERRY M W, KOGAN J. Text mining: Applications and theory. Chichester: John Wiley & Sons: 1-20.

TUKEY J W, 1977. Exploratory data analysis[M]. California: Addison Wesley Publishing Company, Inc.

TWEEDIE F J, BAAYEN R H, 1998. How variable may a constant be? Measures of lexical richness in perspective[J]. Computers and the Humanities, 32(5): 323-352.

VASILIEV Y, 2020. Natural language processing with Python and spaCy: A practical introduction[M]. San Francisco: No Starch Press, Inc.

WELBERS K, VAN ATTEVELDT W, 2022. Rsyntax: Extract semantic relations from text by querying and reshaping syntax[Z/OL]. R package version 0.1.4. https://cran.r-project.org/package=rsyntax.

WICKHAM H, 2016. ggplot2: Elegant graphics for data analysis[M]. 2nd Ed. New York: Springer-Verlag.

WICKHAM H, 2022. stringr: Simple, consistent wrappers for common string operations[Z/OL]. http://stringr.tidyverse.org.

WICKHAM H, GROLEMUND G, 2017. R for data science[M]. Sebastopol: O'Reilly Media, Inc.

WIJFFELS J, 2022. Textplot: Text plots[Z/OL]. R package version 0.2.2. https://cran.r-project.org/package=textplot.

WIJFFELS J, 2023. udpipe: Tokenization, parts of speech tagging, lemmatization and dependency parsing with the 'UDPipe' 'NLP' toolkit[Z/OL]. R package version 0.8.11. https://cran.r-project.org/package=udpipe.

WILKINSON L, 2005. The grammar of graphics[M]. 2nd Ed. New York: Springer-Verlag.

鲍贵, 魏新俊, 2020a. 二语习得研究中的常用统计方法[M]. 2版. 北京: 北京大学出版社.

鲍贵, 张蕾, 2020b. 语言学研究统计分析方法[M]. 南京: 南京大学出版社.

徐琳宏, 林鸿飞, 潘宇, 等, 2008. 感词汇本体的构造[J]. 情报学报, 27(2): 180-185.